CONTINUED FRACTIONS

CONTINUED FRACTIONS

Doug Hensley
Texas A&M University, USA

World Scientific

NEW JERSEY · LONDON · SINGAPORE · BEIJING · SHANGHAI · HONG KONG · TAIPEI · CHENNAI

Published by

World Scientific Publishing Co. Pte. Ltd.
5 Toh Tuck Link, Singapore 596224
USA office: 27 Warren Street, Suite 401-402, Hackensack, NJ 07601
UK office: 57 Shelton Street, Covent Garden, London WC2H 9HE

British Library Cataloguing-in-Publication Data
A catalogue record for this book is available from the British Library.

CONTINUED FRACTIONS

Copyright © 2006 by World Scientific Publishing Co. Pte. Ltd.

All rights reserved. This book, or parts thereof, may not be reproduced in any form or by any means, electronic or mechanical, including photocopying, recording or any information storage and retrieval system now known or to be invented, without written permission from the Publisher.

For photocopying of material in this volume, please pay a copying fee through the Copyright Clearance Center, Inc., 222 Rosewood Drive, Danvers, MA 01923, USA. In this case permission to photocopy is not required from the publisher.

ISBN 981-256-477-2

Preface

The main justification for this book is that there have been significant advances in continued fractions over the past decade, but these remain for the most part scattered across the literature, and under the heading of topics from algebraic number theory to theoretical plasma physics.

We now have a better understanding of the rate at which assorted continued fraction or greatest common denominator (gcd) algorithms complete their tasks. The number of steps required to complete a gcd calculation, for instance, has a Gaussian normal distribution.

We know a lot more about *badly approximable* numbers. There are several related threads here. A badly approximable number is a number x such that $\{q|p - qx| \colon p, q \in \mathbb{Z} \text{ and } q \neq 0\}$ is bounded below by a positive constant; badly approximable numbers have continued fraction expansions with bounded partial quotients, and so we are led to consider a kind of Cantor set E_M consisting of all $x \in [0, 1]$ such that the partial quotients of x are bounded above by M. The notion of a badly approximable *rational* number has the ring of crank mathematics, but it is quite natural to study the set of rationals r with partial quotients bounded by M. The number of such rationals with denominators up to n, say, turns out to be closely related to the Hausdorff dimension of E_M, (comparable to $n^{2\dim E_M}$) which is in turn related to the spectral radius of linear operators $L_{M,s}$, acting on some suitably chosen space of functions f, and given by $L_{M,s} f(t) = \sum_{k=1}^{m} (k + t)^{-s} f(1/(k + t))$. Similar operators have been studied by, among others, David Ruelle, in connection with theoretical one-dimensional plasmas, and they are related to entropy.

Alongside these developments there has been a dramatic increase in the computational power available to investigators. This has been helpful on the theoretical side, as one is more likely to seek a proof for a result when, following computations and graphical rendering of the output, that result leaps off the screen.

Consider, for instance, the venerable Hurwitz complex continued fraction algorithm. This algorithm takes as input a complex number ξ (say, inside the unit square centered on 0), and returns a sequence $\langle a_n \rangle$ of Gaussian integers $a_1, a_2, \ldots$, all outside the unit disk, such that

$$\xi = \cfrac{1}{a_1 + \cfrac{1}{a_2 + \ddots}}.$$

The algorithm uses an auxiliary sequence $\langle \xi_n \rangle$, with $\xi_1 = \xi$ and $\xi_{n+1} = 1/\xi_n - a_n$. Is there any particular pattern to the distribution of the ξ_k's? What sorts of numbers have atypical expansions?

These questions are analogs of questions for which, in the case of the real numbers and the classical continued fraction expansion, answers are known or suspected. Almost always, the expansion of a randomly chosen real input $\xi \in (0, 1)$ will have the property that if $\xi = 1/(a_1 + 1/(a_2 + \cdots))$, then the ξ_n given by the same recurrence relation as mentioned above are distributed according to the *Gauss density* $1/((1+x) \log 2)$. Quadratic irrationals have ultimately periodic continued fraction expansions, and therefore, their ξ_n are not so distributed, but in the case of real inputs these seem to be the only algebraic exceptions. Back in the complex case, to assemble some tens of thousands of data points (a bare minimum considering that a 1-megapixel image is hardly high resolution) can require extensive computations. But once this is done, it turns out there are some surprises-there are algebraic numbers with expansions atypical of randomly chosen inputs, yet not of degree 2. This is discussed in Chapter 5.

Passing from the complex numbers, at once one-dimensional and two-dimensional, we turn our attention to simultaneous diophantine approximation of real n-tuples $\xi = (\xi_1, \ldots, \xi_n)$. Here, we are looking for a positive integer q, and further integers $(p_1, \ldots, p_n)$, such that $e(q, \xi) := \max\{|p_j - q\xi_j|, 1 \leq j \leq n\}$ will be 'small'. The Dirichlet principle guarantees that there are infinitely many choices of q such that, in combination with the unique sensible choice of the p_j's, gives $e(q, \xi) \ll q^{-1/n}$. (If ξ contains only rational entries, these q are eventually just multiples of a common-denominator representation of ξ, and the errors are zero.)

Computing good choices of q by head-on search is computationally prohibitively expensive, as the sequence of good q tends to grow exponentially. We discuss two algorithms for this task. Both rely upon the insight that approximation of ξ is related to the task of finding *reduced* bases of $(n+1) \times (n+1)$ lattices of the form

$$\begin{pmatrix} 1 & 0 & \ldots & 0 \\ 0 & 1 & \ldots & 0 \\ \vdots & 0 & \ddots & \vdots \\ -\xi_1 & -\xi_2 & \ldots & \epsilon \end{pmatrix}.$$

There are different ways to give exact meaning to the notion of a reduced lattice basis. The general idea is that the vectors should form an integral basis of the lattice, and they should be short.

The Gauss lattice reduction algorithm treats the two-dimensional case, and has recently been analyzed by Daudé, Flajolet, and Vallée. It is discussed in Chapter 2.

The Lagarias geodesic multidimensional continued fraction algorithm uses a form of Minkowski reduction, which is computationally feasible for modest dimensions and gives in a sense best-possible answers, while the Lenstra-Lenstra-Lovasz algorithm gives much quicker answers when the dimension is large, but at the risk that the results obtained may not be quite so good. These are discussed in Chapter 6.

Some numbers, for instance, e, have continued fraction expansions featuring fairly frequent, ever-larger partial quotients. (Liouville numbers take this to an extreme!) Others, for instance, $\sqrt{2}$, have continued fraction expansions with bounded partial quotients. Chapter 3 is dedicated to this latter type of number, further broken down as the union $\cup_{M=2}^{\infty} E_M$ of continued fraction Cantor sets. Of course, even E_2 is uncountable, and quadratic irrationals are but the tip of this iceberg. The size of the E_M is best understood in the context of Hausdorff dimension, and we discuss this. The *discrepancy* of the sequence $\langle n\alpha \rangle$, as well as the behavior of related sums, is discussed as well. Chapters 4 and 9 also treat the topic of E_M.

In Chapter 4, we look at the ergodic theory of continued fractions. (There is a recent book by Dajani and Kraaikamp which treats the topic more extensively.) Portions of this chapter first appeared in *New York J. of Math.* **4**, pp. 249-258.

Chapter 5 is devoted to the complex continued fraction algorithms of Asmus Schmidt and of Adolf Hurwitz. Interest in the former has perhaps suffered from the lack of a convenient algorithms for computer implementation, while it seems not to have been recognized that the latter enjoys many good properties beyond those initially established by Hurwitz. In particular, there is an analog to the Gauss density for the Hurwitz algorithm; it even makes a pretty picture, and is featured on the cover.

Chapter 6 is devoted to multidimensional Diophantine approximation, and in particular, to the so-called *Hermite approximations* to numbers and vectors, and the Lagarias geodesic multidimensional continued fraction algorithm.

Chapter 7 discusses an interesting generalization of the approximation properties of quadratic irrationals. The field $\mathbb{Q}(\sqrt{2})$, seen as a vector space over $\mathbb{Q}$, has the canonical basis $\{1, \sqrt{2}\}$. The field $\mathbb{Q}(2^{1/3})$, seen also as a vector space over $\mathbb{Q}$, has canonical basis $\{1, 2^{1/3}, 2^{2/3}\}$. Thus from a certain point of view, we should expect theorems about quadratic irrationals to have analogues not in the context of rational approximations to a single number such as $2^{1/3}$, but rather in the context of simultaneous diophantine approximation to $(2^{1/3}, 2^{2/3})$. And so it is.

Chapter 8 discusses Marshall Hall's theorem concerning sums of continued fraction Cantor sets. This theorem has undergone various iterations and the current strongest version seems to be due to Astels. We give a taste of his approach.

Chapter 9 discusses the functional-analytic techniques arising out of work by K. I. Babenko, E. Wirsing, D. Ruelle, D. Mayer, and others, or, if one goes back all the way, out of the conjecture by Gauss concerning the frequency of the partial quotients in continued fraction expansions of typical numbers. Combined with modern computing power, it becomes possible to evaluate, say, the Wirsing constant, to many digits. Portions of this chapter first appeared in Number Theory for the Millennium, Vol II, pp. 175-194.

Chapter 10 discusses a dynamical-systems perspective, related to Chapter 9 but bringing new tools to the analysis. This approach has scored a real triumph recently, with the result by V. Baladi and B. Vallée that all the standard variants of the Euclidean algorithm have Gaussian normal distribution statistics for a wide variety of measures of the work they must do on typical inputs.

Chapter 11 discusses so-called *conformal iterated function systems*. Much of the material of continued fractions can be seen as an instance of such systems. In this topic, the names Mauldin and Urbański are prominent.

Finally, Chapter 12 discusses convergence of continued fractions, in the spirit of Perron's classic book, and the later classic by Jones and Thron.

Little in this chapter is new, but it would be a pity to omit all mention of these wonderful results. One tidbit that emerges from an extensive theory going back to Jacobi and Laplace is a (standard) continued fraction expansion for $e^{2/k}$. Also discussed are analytical continued fraction expansions, such as those for $\log(1+z)$, $\tanh(z)$, and e^z. This also gives us the classical Euler expansion of e itself as a continued fraction, so that at the end of the book we have come full circle to the beginnings of the subject matter.

What's New Quite a bit, actually. Theorems 3.2, Theorems 4.2 and 4.3, Theorems 5.1-5.5, Theorem 6.2, Theorems 7.1-7.3, the estimate for the Wirsing constant in Chapter 9, Theorem 9.5, the proof of Theorem 12.1, and Theorem 12.5, are, so far as the author is aware, new.

What's Wrong Nothing that I know of. But there is scant chance that the text is free of errors. Any errors remaining after proofreading are the author's. Where the works of others have been restated, if there is a mistake, it is in my restatement of their work. As to algorithms, they are presented here solely for purposes of illustration. No warranty, express or implied, is made that these algorithms are free of all defect. (And with algorithms, even a typographical error invalidates the algorithm.)

Acknowledgments My thanks first to colleagues who gave advice or encouragement, among them O. Bandtlow, I. Borosh, P. Cohen, H. Diamond, O. Jenkinson, J. C. Lagarias, D. Mauldin, and B. Vallée.

In a sense, it should go without saying that no book is written from scratch. Everything has roots, and the content presented here is a distillation and compilation of the work of hundreds of researchers, over and above the author's own contribution. Many of their names have been left out, but only because citation chains must be pruned, or through an oversight.

It should also go without saying that authors do not work in a vacuum. For creating and explaining LATEX, (particularly as it applies to books,) my thanks to the Knuths and to George Grätzer, respectively. For editorial assitance and help with LATEX, my thanks to Mary Chapman and Robin Campbell, respectively.

It should further go without saying that authors are far from disembodied scholarly entities. The kilo-hours that are devoted to preparing a manuscript are a gift from their families. As leaf litter forms the floor of a forest, on occasion, yellow legal pad litter formed a floor of sorts across one room and another. My thanks to Pam for the gifts of time to work, and patience with the worker.

My thanks finally to Lucille, who always held that scholars should write books, but would have had to wait until her 114th birthday or so to see the result. These things should go without saying, but they should not go unsaid.

Contents

Preface   v

1. Introduction   1
   1.1 The Additive Subgroup of the Integers Generated by a and b 4
   1.2 Continuants   6
   1.3 The Continued Fraction Expansion of a Real Number . . .   7
   1.4 Quadratic Irrationals   8
       1.4.1 Pell's equation   10
       1.4.2 Linear recurrence relations   11
   1.5 The Tree of Continued Fraction Expansions   12
   1.6 Diophantine Approximation   13
   1.7 Other Known Continued Fraction Expansions   19
       1.7.1 Notes   21

2. Generalizations of the gcd and the Euclidean Algorithm   23
   2.1 Other gcd's   23
   2.2 Continued Fraction Expansions for Complex Numbers . . .   26
       2.2.1 Computing the gcd of a pair of algebraic integers . .   28
   2.3 The Lattice Reduction Algorithm of Gauss   30

3. Continued Fractions with Small Partial Quotients   33
   3.1 The Sequence $(\{n\alpha\})$ of Multiples of a Number   39
       3.1.1 The sequence of increments of $\lfloor n\alpha \rfloor$ 40
   3.2 Discrepancy   45
   3.3 The Sum of $\{n\alpha\}$ from 1 to N   46

4. Ergodic Theory — 49

- 4.1 Ergodic Maps — 49
- 4.2 Terminology — 50
- 4.3 Nair's Proof — 52
- 4.4 Generalization to E_M — 53
- 4.5 A Natural Extension of the Dynamic System (E_M, μ, T) — 58
  - 4.5.1 Variants on the theme — 61

5. Complex Continued Fractions — 67

- 5.1 The Schmidt Regular Chains Algorithm — 67
- 5.2 The Hurwitz Complex Continued Fraction — 71
- 5.3 Notation — 73
- 5.4 Growth of $|q_n|$ and the Quality of the Hurwitz Approximations — 75
- 5.5 Distribution of the Remainders — 79
- 5.6 A Class of Algebraic Approximants with Atypical Hurwitz Continued Fraction Expansions — 82
- 5.7 The Gauss-Kuz'min Density for the Hurwitz Algorithm — 86

6. Multidimensional Diophantine Approximation — 99

- 6.1 The Hermite Approximations to a Real Number — 107
- 6.2 The Lagarias Algorithm in Higher Dimensions — 111
- 6.3 Convexity of Expansion Domains in the Lagarias Algorithm — 117

7. Powers of an Algebraic Integer — 127

- 7.1 Introduction — 127
- 7.2 Outline and Plan of Proof — 130
- 7.3 Proof of the Existence of a Unit $\mu \in \mathbb{Q}(\alpha)$ oF Degree n — 132
- 7.4 The Sequence $\nu[k]$ of Units with Comparable Conjugates — 133
- 7.5 Good Units and Good Denominators — 135
- 7.6 Ratios of Consecutive Good q — 137
- 7.7 The Surfaces Associated With the Scaled Errors — 138
- 7.8 The General Case of Algebraic Numbers in $\mathbb{Q}(\alpha)$ — 141

8. Marshall Hall's Theorem — 145

- 8.1 The Binary Trees of E_N — 145
- 8.2 Sums of Bridges Covering $[\alpha_N, \omega_N]$ — 150
- 8.3 The Lagrange and Markoff Spectra — 153

9.	Functional-Analytic Techniques	155
	9.1 Continued Fraction Cantor Sets	161
	9.1.1 Hausdorff dimension of continued fraction Cantor sets	163
	9.1.2 Minkowski dimension	166
	9.2 Spaces and Operators	167
	9.2.1 A simple linear space and the fruits of considering it	167
	9.2.2 A Hilbert space of power series	168
	9.3 Positive Operators	176
	9.4 An Integral Representation of $g_{M,\sigma}$	177
	9.5 A Hilbert Space Structure for G when $s = \sigma$ is Real	181
	9.6 The Uniform Spectral Gap	190
	9.7 Log Convexity of λ_M	195
	9.7.1 Applications of strict log convexity	202
10.	The Generating Function Method	205
	10.1 Entropy	206
	10.2 Notation	207
	10.3 A Sampling of Results	208
11.	Conformal Iterated Function Systems	213
12.	Convergence of Continued Fractions	217
	12.1 Some General Results and Techniques	218
	12.2 Special Analytic Continued Fractions	224

Bibliography   233

Index   241

Chapter 1

Introduction

The Euclidean algorithm is the epitome of elegance in computational mathematics: it is clean and simple, much faster than factoring the inputs one by one and then casting out common factors, and requires little memory. As mathematics it shines too: there are numerous fruitful generalizations, it can be used to prove things outside its own orbit, and tools from the wide reaches of mathematics are needed to ferret out its deeper aspects. It even lends itself to striking graphics.

The algorithm goes back to antiquity and was known to Euclid. It takes as input a pair of integers, not both zero, and returns their *greatest common divisor* $\gcd[a,b]$. Many books express algorithms in pseudocode. Here we sometimes use *Mathematica* in place of pseudocode. This has the advantage of being executable on a machine.

```
EuclideanAlgorithmGCD[p_Integer,q_Integer]:= Module[{a,b},
    If[p==0,
        If[q==0,Return[Infinity]];
        Return[Abs[q]]
        ];
    a=Abs[p];b=Abs[q];
    While[a>0,{a,b}={Mod[b,a],a}];
    b]
```

This book is about that algorithm, and about the deep and intricate theory that has grown up in association with it. For all its origin in antiquity and its accessibility to schoolchildren, there is still an open research frontier. For instance, it is not known whether, for every $b \geq 2$, there exists $a < b$ so that every approximation c/d to a/b with $0 < d < b$ satisfies $|c/d - a/b| > 1/(10d^2)$, and it has only recently been established that the number of steps needed to complete the algorithm, averaged over pairs (a, b) with $a, b \leq x$, has asymptotically a Gaussian distribution.

The greatest common divisor of two integers, not both zero, has been of interest to mathematicians for millennia. For $a, b \in \mathbb{Z}$, not both zero, $\gcd[a, b]$ is the largest integer d so that $d \mid a$ and $d \mid b$. An equivalent definition is that it is the least positive integer d in $\Lambda(a, b) := \{ja + kb : j, k \in \mathbb{Z}\}$, for if $d \mid a$ and $d \mid b$, then $d \mid ja + kb$. In the other direction, if u is the least positive element of $\Lambda(a, b)$, then u divides all the rest because otherwise, there would be an element $v \in \Lambda(a, b)$ not divisible by u. But then $v \mod u \in \Lambda(a, b)$ and $0 < v < u$, a contradiction.

One can also speak of the gcd of a set of integers, and again, the two definitions, as the greatest integer that divides them all, and as the least positive element of the set of finite integer linear combinations of them, are equivalent.

The gcd algorithm given above (in Mathematica code) discards potentially valuable ancillary information. There are variants of the algorithm which keep and return this information. Our first variant starts with a pair $(p_0, q_0) := (p, q)$ of positive integers, sorted so that $p \leq q$, and generates a list of working pairs (p_j, q_j) of non-negative integers. It halts when $p_n = 0$. The loop generates successive pairs using the rule

$$(p_{n+1}, q_{n+1}) := (q_n \mod p_n, p_n).$$

Equivalently,

$$a_{n+1} := \lfloor q_n / p_n \rfloor$$
$$(p_{n+1}, q_{n+1}) := (q_n - a_{n+1} p_n, p_n).$$

It also keeps a working matrix M_j defined by $M_0 = \begin{pmatrix} 1 & 0 \\ 0 & 1 \end{pmatrix}$ and $M_{j+1} = M_j \begin{pmatrix} 0 & 1 \\ 1 & a_{j+1} \end{pmatrix}$. The algorithm terminates when $p_n = 0$ and returns a list with five integer entries: (d, u, v, p', q'), where $d = \gcd(p, q)$, (u, v) is a pair of integers so that $qu - pv = (-1)^n d$, $0 \leq u < q$, $0 \leq v < p$, and $p' = p/d, q' = q/d$. Here, $d = p_n$ while $M_n = \begin{pmatrix} u & p' \\ v & q' \end{pmatrix}$.

The reason the algorithm returns the gcd is that with each step, $d \mid p_j$ and $d \mid q_j$ if and only if $d \mid q_j - kp_j$ and $d \mid p_j$ so that in particular $d \mid p_{j+1}$ and $d \mid q_{j+1}$ if and only if $d \mid p_j$ and $d \mid q_j$. Thus,

$$\gcd(p_{j+1}, q_{j+1}) = \gcd(p_j, q_j).$$

At the end, the working pair is $(0, q_n) = (0, d)$ which clearly has greatest common divisor d. The reason the algorithm is fast is that $p_{j+1}q_{j+1} < (1/2)p_j q_j$. (The reader will see the proof as readily by thinking about it a bit, as by reading about it.)

There are various modified versions of this algorithm which achieve slightly improved efficiency, at the price of slightly greater complexity and a certain loss of elegance. But inasmuch as the algorithm is a workhorse of computational number theory, every bit of efficiency counts. Instead of taking the remainder $a \bmod b$ to be the non-negative integer $a' \equiv a \pmod{b}$, one can take the integer, positive or negative, nearest zero and in the required congruence class. This sometimes gets us two steps for the price of one.

The *centered algorithm*, which is another way to present this same idea, proceeds by replacing a working pair (u, v) of integers satisfying $0 \leq u < v$ by the pair $(|v - nu|, u)$, where $n = \lfloor v/u \rfloor + 1/2$, until $u = 0$. It is on average faster than the classical algorithm by a factor of $\log(2)/\log(\phi)$, where $\phi = (1 + \sqrt{5})/2$.

Division of one large integer by another, to obtain the quotient and remainder, is the most time-consuming part of executing the classic Euclidean algorithm on large integer input pairs. This can be circumvented with the *binary shift* Euclidean algorithm.

Suppose we are representing our integers in binary notation, as strings of 0's and 1's. Extracting common factors of two is then trivial. The binary shift Euclidean algorithm takes up where stripping out the 0's leaves off, accepting inputs of the form (u, v) where u and v are odd positive integers with $0 < u \leq v$. The algorithm has a double loop structure. Let $\mathrm{Val}_2(n)$ denote the largest b so that $2^b \mid n$.

The algorithm takes as input a pair (u_0, v_0) of odd positive integers with $u_0 \leq v_0$. It terminates when $u_n = v_n$, returning u_n, and if desired, performance statistics such as n. If $u_n < v_n$, the outer loop passes the current pair (u_n, v_n) to an inner loop.

The inner loop generates a list $w_{n,j}$ beginning with $w_{n,0} := v_n$ which proceeds while $w_{n,j} > u_n$ by

$$x := (w_{n,j} - u_n)$$
$$b_{n,j} := \mathrm{Val}_2(x)$$
$$w_{n,j+1} := 2^{-b_{n,j}} x$$

and exits when $w_{n,j} \leq u_n$, returning $(u_{n+1}, v_{n+1}) := (w_{n,j}, u_n)$ to the outer loop.

Performance bounds are easy. Since each step of the inner loop reduces the number of bits in w by at least one, the sum of the bit-string lengths of the pair being processed decreases with each step, so that the total number of subtractions effected by the inner loop is on the order of $\log v_0$ or less. Clearly there are no more exchanges than subtractions, so the total number of steps is also $O(\log v_0)$.

A more exact analysis has recently been made by Vallée [Va2], who showed that the number of subtraction steps, averaged over all odd input pairs (u, v) with $v \leq x$, is asymptotically $A \log x$ where A is a positive constant related to a certain linear operator. She has a similar result about the number of exchanges.

There are other variations on this theme. To determine whether or not n is a quadratic residue mod p (Legendre symbol), one uses the Kronecker-Jacobi symbol $\left(\frac{a}{b}\right)$. The Kronecker algorithm (see p 29 of [Co]) has at its heart a cycle of steps of the form $(a, b) \to (b \bmod a, a)$, but with additional steps which remove powers of 2 from a and b between divisions, so that all divisions involve a pair of odd numbers. Vallée discusses averages related to this algorithm in [Va2].

1.1 The Additive Subgroup of the Integers Generated by a and b

Another way to look at what is achieved by computing the gcd of two integers is that the set of all integer linear combinations of a and b forms an additive subgroup of the integers, this subgroup has a single generator, and we find it by finding a series of pairs that generate the same subgroup, culminating in the pair $(0, d)$. At first sight, this amounts to nothing better than throwing around big words, but there is an important fact implicit in this perspective. Our output d belongs to the subgroup, so there must exist integers x and y so that $ax + by = d$. The basic Euclidean algorithm, unlike the matrix-based version, does not provide this ancillary information, but it can be readily adapted to do so. The *extended gcd* algorithm takes input (a, b) with $0 \leq a \leq b$ and returns d, x, and y.

Algorithm 1.1 Extended gcd algorithm. Inputs: Nonnegative integers a and b. Outputs: Integers x and y, and a positive integer d, so that $d = \gcd(a, b)$ and $ax + by = d$, with $|x| \leq b/d$, $|y| \leq a/d$. (If a and b are both zero, which they should not be, rather than crash the algorithm we specify that it return ∞ for d, and $(0,0)$ for x and y.)

Input a, b. Set $p' = 1, p = 0, q' = 0$, and $q = 1$. Set $u = a$, $v = b$, and $r = 0$.

While $u > 0$, set $c = \lfloor v/u \rfloor$, $(u,v) \leftarrow (v - cu, u)$, $(p', p) \leftarrow (p, cp + p')$, and $(q', q) \leftarrow (q, cq + q')$. Increment: $r \leftarrow r + 1$.

If r is even, return $(-q', p', v)$ as the values of (x, y, d). If r is odd, return $(q', -p', v)$ as values of (x, y, d).

Variant: Keep track of p, q, and c. Set $r = 1$, $p_{-1} := 1, p_0 := 0, q_{-1} := 0$ and $q_0 := 1$. Then while $u > 0$ set $a_r := \lfloor v/u \rfloor$, $p_r := a_r p_{r-1} + p_{r-2}$, and $q_r := a_r q_{r-1} + q_{r-2}$. Return the lists $(a_1, \ldots a_r)$, $(p_1 \ldots p_r)$, and $(q_1, \ldots q_r)$, as well as the final value of v, which is the gcd of a and b.

Remark 1.1 *This algorithm carries over without significant modification to the case of polynomials in one variable over a field. When working over, for instance, the field Q, there is a practical difficulty in that the coefficients can become so unwieldy as to obviate the formal speed and simplicity of the algorithm. In a finite field, where exact arithmetic is realizable in practice as well as in principle, the algorithm is generally simpler than with integers, mainly because all polynomials of the same degree are of the same size, in the sense that the number of congruence classes modulo such polynomials depends only on the degree and on the underlying field, see [FriHe].*

The list (p_j, q_j) of intermediate values of p and q, and the list (a_j) of values of c, give valuable information. Suppose $0 < a < b$. The fractions $p_j/q_j, 1 \leq j < r$ are especially good approximations p/q of a/b with $q \leq b$. The final p_r/q_r is the reduced value of a/b. The value of r is the number of steps needed to execute the algorithm. The identity $p_{r-1} q_r - p_r q_{r-1} = (-1)^r$, together with $p_r = a/\gcd[a, b]$ and $q_r = b/\gcd[a, b]$, give x and y so that $ax + by = d = \gcd[a, b]$.

An important special case is $d = 1$. Two randomly chosen positive integers will, more likely than not, be relatively prime. This goes back to Euler, who calculated the fraction of such pairs to be $\prod_{p \text{ prime}} (1 - p^{-2}) = 6/\pi^2$. Now there is scant point in computing the gcd of numbers a and b if they are known to be relatively prime. But when we get x and y, we will have found $a^{-1} \mod b$ (that is x), and vice-versa. This algorithm thus lies at the heart of computational number theory, for it allows us to find multiplicative inverses in finite fields of prime order. (For finite fields of order $q = p^n$, finding the multiplicative inverse of an element amounts to solving a system of n linear equations in n variables, in the ground field $\mathbb{Z}/p\mathbb{Z}$. Straight Gaussian reduction suffices.)

1.2 Continuants

The denominator q_r of the finite continued fraction

$$[\mathbf{u}] = [u_1, \ldots u_r] = [0; u_1, \ldots, u_r] = \cfrac{1}{u_1 + \cfrac{1}{u_2 + \cfrac{1}{\ddots + 1/u_r}}}$$

is a function of the integer list $\mathbf{u} = (u_1, u_2, \ldots u_r)$. We call this number the *continuant* of $\mathbf{u}$ and write $|(u_1, u_2, \ldots u_r)|$ or, if there is no risk of confusion, simply $|\mathbf{u}|$. The continuants satisfy a number of useful identities. By convention, the continuant of the empty list is 1. For $\mathbf{u} \in \mathbb{Z}^{+r}$, we write $\mathbf{u}^-$ for $(u_1, u_2, \ldots, u_{r-1})$ and $\mathbf{u}_-$ for $(u_2, \ldots, u_r)$. We write $\{\mathbf{u}\}$ for the continued fraction $[u_r, \ldots, u_1]$. (When dealing with a single real number x, we use $\{x\}$ as usual to denote the fractional part of x.) For the empty list $\mathbf{z} = ()$, we declare $|\mathbf{z}| = 1$, $[\mathbf{z}] = \{\mathbf{z}\} = 0$.

Proposition 1.1 *Suppose* $\mathbf{u} = (u_1, u_2, \ldots, u_r)$ *and* $\mathbf{v} = (v_1, \ldots, v_s)$ *are lists of positive integers. Let* $\mathbf{uv}$ *denote the concatenation* $(u_1, \ldots, u_r, v_1, \ldots, v_s)$ *of* $\mathbf{u}$ *and* $\mathbf{v}$*. Then*

(i) $|\mathbf{u}| = \begin{vmatrix} u_1 & 1 & 0 & 0 \ldots \\ -1 & u_2 & 1 & 0 \ldots \\ 0 & -1 & u_3 & 1 \ldots \\ & & \vdots & \\ 0 & 0 & \cdots -1 & u_r \end{vmatrix} = q_r,$

(ii) $|\mathbf{uv}| = |\mathbf{u}||\mathbf{v}| + |\mathbf{u}^-||\mathbf{v}_-| = |\mathbf{u}||\mathbf{v}|(1 + \{\mathbf{u}\}[\mathbf{v}]),$

(iii) $p_r = |\mathbf{u}_-| = [\mathbf{u}]|\mathbf{u}|,$

(iv) $q_{r-1} = |\mathbf{u}^-| = \{\mathbf{u}\}|\mathbf{u}|,$

(v) $|u_1, u_2, \ldots u_r| = |u_r, u_{r-1}, \ldots u_1|.$

The proof is by induction. The identity $q_{j+1} = u_{j+1}q_j + q_{j-1}$ determines the successive denominators, together with $q_0 := 1$ and $q_{-1} := 0$. Clearly, the determinant obeys this same recursion, and agrees with q_r for $r = 1$ and $r = 2$. Item (ii) is an immediate consequence of (i) and basic properties of determinants. Item (iii) is a consequence of the construction of p_r, which is governed by the same recursion as for q_r but beginning with u_2. Items (iv) and (v) are an immediate consequence of (i).

1.3 The Continued Fraction Expansion of a Real Number

The gcd algorithm is closely related to the continued fraction expansion of a real number. Given a rational number $\alpha = a/b$ with $0 < a < b$, if $c_1, c_2, \ldots, c_r$ are the partial quotients in the continued fraction expansion of a/b, then $a/b = [(c_1, c_2, \ldots, c_r)] = 1/(c_1 + 1/(c_2 + 1/(c_3 + \cdots + 1/c_r)))$ is exactly $p_r/q_r = a/b$. Truncating the expansion at some earlier value $r' < r$ gives $p_{r'}/q_{r'}$, a rational approximation of a/b by a simpler rational number.

For general real numbers α, a relentlessly constructivist approach would be to insist that since we know real numbers by their rational approximations, we should ask for the continued fraction expansion of a rational interval, and then take limits. This is actually not such a bad idea, for it forces us to consider something we should eventually have to think through in any case: what is the quality of the approximations $p_r/q_r = 1/(c_1 + 1/(c_2 + \ldots 1/c_r))$ as we go along? But this author is not a relentless constructivist. Here is the conventional continued fraction expansion algorithm for a real number:

Algorithm 1.2 (Continued fraction expansion of a real number α.) Input α, and a limit R to your patience. Set $p_{-1} = 1$, $p_0 = \lfloor \alpha \rfloor$, $q_{-1} = 0$, and $q_0 = 1$. Set $a_0 = \lfloor \alpha \rfloor$, and set $\alpha_0 = \alpha - a_0$ so that $0 \leq \alpha_0 < 1$. Set $r = 0$. While $\alpha_r > 0$ and $r \leq R$, set $\beta = 1/\alpha_r$, $a_{r+1} = \lfloor \beta \rfloor$, and $\alpha_{r+1} = \beta - a_{r+1}$. Set $p_{r+1} := a_{r+1} p_r + p_{r-1}$ and $q_{r+1} = a_{r+1} q_r + q_{r-1}$. Increment r. Return the lists $(a_0, a_1, a_2, \ldots a_R)$, $(p_0, p_1, p_2, \ldots p_R)$, and $(q_0, q_1, q_2, \ldots q_R)$.

Remark 1.2 *The list $(a_0, a_1, a_2, \ldots a_r)$ gives the continued fraction expansion of α to depth r: $\alpha = a_0 + 1/(a_1 + 1/(a_2 + 1/(a_3 + 1/(a_4 + \ldots))))$. The integers p_r and q_r are the numerators and denominators respectively of the rational numbers $a_0 + 1/(a_1 + 1/(a_2 + 1/(a_3 + \cdots + 1/a_r)))$. Both this algorithm, and most of the associated results below, carry over without significant modification to the case of power series with coefficients over Q or over a finite field, on taking the stance that higher powers of z are 'smaller'.*

Theorem 1.1 *The integers a_r, p_r, and q_r for the continued fraction expansion of a rational number $\alpha = a/b$ are the same as the corresponding numbers generated by the gcd algorithm given input (a, b). For both rational* and irrational α, and for all relevant r,

(i) $p_{2r}/q_{2r} < \alpha < p_{2r+1}/q_{2r+1}$,
(ii) $\alpha = \dfrac{p_r + \alpha_r p_{r-1}}{q_r + \alpha_r q_{r-1}}$,
(iii) $\alpha_r = [a_{r+1}, a_{r+2}, \ldots]$,
(iv) $\dfrac{1}{(2 + a_{r+1})q_r^2} < \left|\alpha - \dfrac{p_r}{q_r}\right| < \dfrac{1}{a_{r+1}q_r^2}$,
(v) $0 \le \alpha_r < 1$.

The proofs are straightforward induction. The estimate for $|\alpha - p/q|$ can be given a sharper constant. Hurwitz [Sc1] proved in 1891 that for every irrational number α there are infinitely many distinct rationals p/q with $|\alpha - p/q| < 1/(\sqrt{5}q^2)$. The constant $\sqrt{5}$ is best possible; any real number with arbitrarily long strings of consecutive $a_r = 1$ in its continued fraction expansion provides a counterexample to the form the result would take with a smaller constant in place of $1/\sqrt{5}$.

1.4 Quadratic Irrationals

There is a nice analogy: Terminating binary decimal expansions↔ dyadic rationals, periodic binary expansions↔ rational numbers: terminating continued fraction expansion ↔ rational number, periodic continued fraction expansion ↔ quadratic irrational.

Theorem 1.2 *Given an eventually periodic (infinite) continued fraction*

$$u_0 + [\mathbf{u}\bar{\mathbf{v}}] = u_0 + [\mathbf{u}\mathbf{v}\mathbf{v}\ldots] = [u_0; u_1, u_2, \ldots u_s, v_1, \ldots v_t, v_1, \ldots v_t, \ldots]$$

the corresponding real number $\alpha = \alpha(\mathbf{u}, \mathbf{v})$ *is a quadratic irrational* $\alpha = a + b\sqrt{d}$ *where a and b are rational and d is square-free, and conversely.*

Proof. Let $\alpha = u_0 + [\mathbf{u}\bar{\mathbf{v}}]$. Since every rational number has a terminating continued fraction expansion, α is irrational. Now $\alpha = u_0 + [\mathbf{u} + \beta]$ where $\beta = [\mathbf{v} + \beta]$. From theorem 1.1,

$$\beta = (p_t + \beta p_{t-1})/(q_t + \beta q_{t-1})$$

so that

$$q_{t-1}\beta^2 + (q_t - p_{t-1})\beta - p_t = 0.$$

Thus β is a quadratic irrational and hence α is as well. In the other direction, suppose α is a quadratic irrational, with $\alpha = \theta + \phi\sqrt{w}$, θ and ϕ rational and w a non-square positive integer, and $0 < \alpha < 1$. We leave as an exercise for the reader, to put α first in the form $(s \pm \sqrt{t})/r$ with integer s, t and $r \neq 0$, and then in the form $\alpha = (a \pm \sqrt{d})/b$ with integers a, b and d, $d > 0$, $b > 0$, and $b \mid (d - a^2)$.

Consider first the case in which $x = (\sqrt{d} + a)/b$ with $b > 0$, $b \mid (a^2 - d)$, $d > a^2$ and $0 < x < 1$. A single step
$$x \to \{1/x\} = 1/x - \lfloor 1/x \rfloor = 1/x - k$$
takes x to $x' = (\sqrt{d} + (-a - kb'))/b'$ where $b' = (d - a^2)/b > 0$. Let $a' = -a - kb'$. Then $0 < x' < 1$ so
$$\sqrt{d} + a' > 0, \quad a' > -\sqrt{d}.$$
On the other hand, $-a < \sqrt{d}$ so $-a - kb' < \sqrt{d}$ so $|a'| < \sqrt{d}$, and trivially $b' \mid (d - a'^2)$. Thus all subsequent iterates of $x \to \{1/x\}$ will again be of this form. Since there are only finitely many pairs (a, b) of integers with $a^2 < d$ and $b \mid (d - a^2)$, some value of $x = (\sqrt{d} - a)/b$ must eventually repeat, and thereafter, the continued fraction expansion of x will be periodic.

This proof has the advantage that it gives an upper bound to the length of the period, and to the size of the integers $a_j := \lfloor 1/x_j \rfloor$ that occur in the periodic part of the expansion, in terms of d alone.

Next, consider the case $(x = a \pm \sqrt{d})/b$, with $a^2 > d$ but as before with $0 < x < 1$, $b > 0$, and $b \mid (a^2 - d)$. This time, $b' = (a^2 - d)/b$ and $a' = a - kb'$ where $k = \lfloor 1/x \rfloor$, so that $x' = (a' \pm \sqrt{d})/b'$. Since $a' < a$, there can be but finitely many consecutive steps $x \to \{1/x\}$ of this type, and after that, we are in the first case, in which the number of steps needed to finish is bounded by $O(d)$ [this could be sharpened but we have other fish to fry]. Therefore, the expansion must go into a loop. □

Proposition 1.2 *The purely periodic expansions are those in which the loop begins immediately. That is, if r_0 is the quadratic irrational to be expanded, and $x_{j+1} := \{1/x_j\}, a_{j+1} := \lfloor 1/x_j \rfloor$, then $a_{j+p} = a_j$ and $x_{j+p} = x_j$ for all $j \geq 0$. The expansion of x is purely periodic if and only if the algebraic conjugate $\overline{x} = (a \mp \sqrt{d})/b$ of $x = (a \pm \sqrt{d})/b$ with $0 < x < 1$ satisfies $\overline{x} < -1$.*

Proof. If x satisfies the conditions, then $x_{j+1} = 1/x_j - a_{j+1}$ so that $\overline{x_{j+1}} = 1/\overline{x_j} - a_{j+1}$. Inductively, then, x_j satisfies the same conditions. Now because x_0 is a quadratic irrational, the sequence (x_j) is at any rate eventually periodic. But if $x_j = x_{j+p}$ for some p, then $\overline{x_j} = \overline{x_{j+p}}$. But $\overline{x_j} < -1$ and $\overline{x_j} = 1/\overline{x_{j-1}} - a_j$ and likewise $\overline{x_{j+p}} = 1/\overline{x_{j+p-1}} - a_{j+p}$. On the other hand, since $\overline{x_{j-1}} < -1$, $-1 < 1/\overline{x_{j-1}} < 0$ so that $a_j = \lfloor -\overline{x_j} \rfloor$ and $a_{j+p} = \lfloor -\overline{x_{j+p}} \rfloor$. Thus $a_j = a_{j+p}$ and so $\overline{x_{j-1}} = \overline{x_{j-1+p}}$ so $x_{j-1} = x_{j-1+p}$. From this it follows that $x_{j+p} = x_j$ for all $j \geq 0$ as claimed.

In the other direction, if x has a purely periodic expansion, then $x = [a_1, a_2, \ldots a_{r-1}, a_r + x]$ for some $r \geq 1$ and some list of r positive integers. Thus $x = (p_r + xp_{r-1})/(q_r + xq_{r-1})$. Consequently,

$$x = \frac{(p_{r-1} - q_r) \pm \sqrt{q_r^2 - 2q_r p_{r-1} + 4q_{r-1}p_r + p_{r-1}^2}}{2q_{r-1}}$$

and since $-2q_r p_{r-1} + 4q_{r-1}p_r = 2q_r p_{r-1} + 4(-1)^{r-1}$,

$$x = \frac{(p_{r-1} - q_r) \pm \sqrt{(q_r + p_{r-1})^2 + 4(-1)^{r-1}}}{2q_{r-1}}$$

The $+$ in $\pm$ gives x, which we already know to lie between 0 and 1, while the $-$ in $\pm$ gives the conjugate. But with this minus sign, for $r \geq 2$ at any rate, $(p_{r-1} + q_r)^2 - 4 > (p_{r-1} + q_r - 1)^2$ and so the conjugate satisfies $\overline{x} < -(q_r - 1)/q_{r-1} \leq -1$ as required. If the base period of the continued fraction is one, one nevertheless has $x = [a_1, a_1 + x]$ so the argument still works. □

1.4.1 Pell's equation

We can use this to solve Pell's equation, $X^2 - DY^2 = 1$. This is linked to the continued fraction expansion of $\sqrt{D}$. The numerators and denominators p_n and q_n in the continued fraction convergents p_n/q_n to $\sqrt{D}$ serve as reasonable candidates for integers x and y, and it turns out that one need only extract this expansion to the depth of a full-period.

The continued fraction expansion of $n + \sqrt{D}$, where $n = \lfloor \sqrt{D} \rfloor$, is purely periodic. That is,

$$n + \sqrt{D} = 2n + 1/(a_1 + 1/(a_2 + \cdots + 1/(a_r + 1/(2n + 1/(a_1 + 1/(a_2 + \ldots))))))$$

so that if $\xi := n + \sqrt{D}$ and (x_r/y_r) is the rth convergent to $\sqrt{D} - n$ then

$$\xi = 2n + \frac{x_r + \xi^{-1}x_{r-1}}{y_r + \xi^{-1}y_{r-1}}.$$

From this it follows that

$$\sqrt{D} - (n + x_r/y_r) = (\sqrt{D} - n) - x_r/y_r = \frac{(-1)^r}{\xi y_r + y_{r-1}}$$

Thus

$$|(x_r + ny_r)^2 - Dy_r^2| < \frac{(x_r + ny_r) + \sqrt{D}y_r}{(n + \sqrt{D})y_r} < 2,$$

this last because $x_r < y_r$. Thus taking $X := x_r + ny_r$ and $Y = y_r$ gives the fundamental solution to $X^2 - Dy^2 = \pm 1$.

For instance, to solve $x^2 - 76y^2 = \pm 1$ in integers, one calculates the continued fraction expansion of $\sqrt{76}$. In these calculations there is no need of decimal approximations, and we can proceed with exact calculations. Thus,

$$\sqrt{76} = 8 + (\sqrt{76} - 8) = 8 + 1/((\sqrt{76} + 8)/12)$$

Continuing in this vein, we obtain a series of identities of the form $\sqrt{76} = a_0 + 1/(a_1 + 1/(a_2 + \cdots + 1/(\sqrt{76} + u_r)/v_r)\ldots)$ where $q_r \mid (76 - u_r^2)$, $v_r > 0$, $0 < u_r < \sqrt{76}$, and where $p_r^2 - 76q_r^2 = (-1)^r v_r$. The continued fraction expansion is periodic, (and would be purely periodic if we started with $8 + \sqrt{76}$), and one eventually reaches $u_r = \lfloor\sqrt{76}\rfloor$, $v_r = 1$ which gives the fundamental unit. As it happens, the continued fraction expansion is that $\sqrt{76} = 8 + [1, 2, 1, 1, 5, 4, 5, 1, 1, 2, 1, 16, \ldots]$. The sequence is purely periodic, and the numerator and denominator 57799 and 6630 respectively of $[1, 2, 1, 1, 5, 4, 5, 1, 1, 2, 1]$ give the fundamental units $57799 \pm 6630\sqrt{76}$.

What we have done here is not the last word in computing solutions to the Pell equation, though. For large D, the approach taken here requires too many steps. There are better ways. For further information, see [Vardi].

1.4.2 Linear recurrence relations

The continued fraction algorithm uses the recurrence relations $p_n = a_n p_{n-1} + p_{n-2}$, $q_n = a_n q_{n-1} + q_{n-2}$ to generate the numerator and denominator sequences (p_n) and (q_n) for the convergents to a real number α. In the special case that the sequence (a_n) of partial quotients for α is constant, this provides, forthwith, a second order linear recurrence with constant coefficients for the corresponding (p_n) and (q_n). If the (a_n) vary, then there is no such second order linear recurrence. But this does not foreclose the possibility of a higher order linear recurrence with constant coefficients governing (p_n) and (q_n). A recent result of Lenstra and Shallit [LeSh] provides a nice counterpart to our other characterization of quadratic irrationals, as numbers with an ultimately periodic continued fraction expansion. A linear recurrence with constant complex coefficients, for a sequence (z_n), is a recurrence of the form $z_n = \sum_{k=1}^{N} \lambda_k z_{n-k}$, with fixed complex numbers $\lambda_k, 1 \le k \le N$, that holds for $n > N$.

Theorem 1.3 *Let α be an irrational number, with continued fraction expansion $\alpha = [a_0; a_1, a_2, \ldots]$. Let (p_n) and (q_n) be the numerator and denominator sequences of the convergents to α. If either (p_n) or (q_n) satisfy a linear recurrence with constant complex coefficients, then α is a quadratic irrational, and conversely, if α is a quadratic irrational, then both (p_n) and (q_n) satisfy a linear recurrence with constant complex coefficients.*

It may come as a bit of a surprise that this was not known since classical times, but the only known proof requires a difficult result of van der Poorten, known as the Hadamard Quotient Theorem. [vdP2; vdP3].

1.5 The Tree of Continued Fraction Expansions

The continued fraction expansion of a real number is a path along a tree. At each step, the tree branches countably many ways, with one branch for each positive integer. The vertices of this rooted tree are the finite lists of positive integers $[a_1, a_2, \ldots a_r]$, of whatever length, and the edges go from each vertex (list) to each list got by appending a single positive integer to the original list. The root of this tree is the empty list. The set $I_{\mathbf{a}} = I_{(a_1, a_2, \ldots a_r)}$ of all real numbers $\alpha, 0 \le \alpha < 1$ such that the continued fraction expansion begins with the list $\mathbf{a} = \langle a_1, a_2, \ldots a_r \rangle$ is an interval with endpoints $p_r/q_r = [\mathbf{a}]$ and $(p_r + p_{r-1})/(q_r + q_{r-1})) = [\mathbf{a}, 1]$, closed at the former end and open at the latter. If r is odd, the open end of the interval is its lower end and $I_{\mathbf{a}}$ has the form $(x, y]$, while if r is even, the interval $I_{\mathbf{a}}$ has the form $[x, y)$. The length of $I_{\mathbf{a}}$ is $1/(q_r(q_r + q_{r-1}) = 1/|\mathbf{a}|^2(1 + \{\mathbf{a}\})$.

Every interval $(p/q, p'/q')$ of length $1/qq'$ occurs as the interior of $I_{\mathbf{a}}$ for a unique list $\mathbf{a}$ of positive integers; $[0, 1)$ is the interval corresponding to the empty list.

The union of all the intervals corresponding to lists of length r is the whole interval $[0, 1)$, and for any r and any list $\mathbf{a} = \langle a_1, a_2, \ldots a_r \rangle$, the set of extensions of this list to arbitrary fixed depth generate a partition of the interval $I_\mathbf{a}$. If α is rational, the number α will appear as p_r/q_r for some r; as we have already observed, in this case generating the continued fraction expansion of α, and finding the gcd of the pair of integers whose ratio is α together with the auxiliary information, are essentially identical computations. The expansion terminates when this fraction is reached, which will also be when $\alpha_r = 0$.

For irrational α, the expansion is infinite, and there is then a one-to-one correspondence between paths to infinity in the tree (that is, infinite sequences of positive integers), and irrational numbers.

1.6 Diophantine Approximation

Diophantine approximation takes for its subject the approximation of real numbers or vectors by rational numbers or vectors. The particular case of real numbers (dimension 1) is particularly well understood because of its connection to continued fractions.

Given an irrational number α, and given a positive integer Q, the list

$$(x_1, x_2, \ldots x_Q) := (\alpha \mod 1, 2\alpha \mod 1, \ldots Q\alpha \mod 1)$$

has an element nearest zero mod 1. (That is, the distance $\|r\alpha\|$ from the fractional part of one of these numbers $r\alpha \mod 1$, to 0 or 1 whichever is closer, is minimal among the numbers in the list.) Consider the sequence of successive minima of $\|r\alpha\|$. This sequence, it turns out, is just another facet of the continued fraction expansion of α. [L]. The *discrepancy* of the list $(n\alpha \mod 1, 1 \leq n \leq N)$ is also essentially governed by the continued fraction expansion of α. Discrepancy is a measure of how unevenly the list is distributed in the unit interval, but we defer further discussion until the necessary background is in place.

Proposition 1.3 *For all positive integers Q, there exists integer q, $1 \leq q \leq Q$ so that $\|q\alpha\| < 1/Q$.*

Proof. The sequence $q\alpha, 0 \leq q \leq Q$ has $Q+1$ elements. Some two entries, say $q_1\alpha$ and $q_2\alpha$, must fall within the same interval $[k/Q, (k+1)/Q) \mod 1$, and these two will differ by strictly less than $1/Q$. Take $q = |q_1 - q_2|$. □

Remark 1.3 *If $\alpha = a/b$ is rational, and if $Q \geq b$, the result reduces to the trivial observation that $\|b(a/b)\| = 0$.*

From theorem 1.1(iv), it follows that the continued fraction convergents p_r/q_r furnish approximations to α that satisfy $\|q_r\alpha\| < 1/q_r$, and that the ratio by which q_r undercuts this bound, is effectively given by a_{r+1}. It can happen that fractions of the form $(ap_r + p_{r-1})/(aq_r + q_{r-1})$, with $1 \leq a < a_{r+1}$, also slip under the wire. That is, in certain cases, there do exist positive integers a and r, with $1 \leq a < a_{r+1}$ so that $\|(aq_r + q_{r-1})\| < 1/(aq_r + q_{r-1})$. In any event, for these a, as a consequence of our upcoming Theorem 1.4,

$$\|q_{r+1}\alpha\| < \|(aq_r + q_{r-1})\alpha\| < \|q_{r-1}\alpha\|.$$

Furthermore, we have at least this result to the effect that good approximations come only from convergents:

Proposition 1.4 *Let $\alpha \in (0, 1/2)$ be a real number, and p and q be positive integers. Let $(0/1, p_1/q_1, p_2/q_2, \ldots)$ be the convergents of α. If $q > q_1$ and $|\alpha - p/q| \leq 1/(2q^2)$, then p/q is a continued-fraction convergent of α.*

Proof. Assume p/q is not a convergent, yet $|\alpha - p/q| \leq 1/(2q^2)$. Choose $j \geq 2$ such that $q_{j-1} < q < q_j$. Since $q > q_1$, we can do this. Consider the open interval A with endpoints p_{j-1}/q_{j-1} and p_{j-2}/q_{j-2}. We must have $p/q \in A$, because otherwise either $|p/q - \alpha| > |p_{j-1}/q_{j-1} - p/q|$ or $|p/q - \alpha| > |p/q - p_{j-2}/q_{j-2}|$, and in either case, the latter difference is at least $1/qq_{j-1}$ and thus more than $1/2q^2$.

Thus, (p, q) can be written as

$$(p, q) = c(p_{j-1}, q_{j-1}) + d(p_{j-2}, q_{j-2})$$

with integers $c, d > 0$. Now there is a positive integer n such that

$$(p_j, q_j) = n(p_j - 1, q_{j-1}) + (p_{j-2}, q_{j-2}),$$

and $\theta \in (0, 1)$ such that

$$\alpha = \frac{(n+\theta)p_{j-1} + p_{j-2}}{(n+\theta)q_{j-1} + q_{j-2}}.$$

Since $q < q_j$, $c < n$ so $n \geq 2$. Now

$$\left| \frac{(n+\theta)p_{j-1} + p_{j-2}}{(n+\theta)q_{j-1} + q_{j-2}} - \frac{cp_{j-1} + dp_{j-2}}{cq_{j-1} + q_{j-2}} \right| = \frac{(n+\theta)d - c}{q((n+\theta)q_{j-1} + q_{j-2})}.$$

This exceeds $1/2q^2$ because on expanding and clearing fractions, the claim amounts to $2(cq_{j-1} + dq_{j-2})((n+\theta)d - c) > (n+\theta)q_{j-1} + q_{j-2}$. The worst case is $d = 1$, but even then, the coefficients for both q_{j-1} and q_{j-2} are greater on the left hand side than on the right, in view of $1 \leq c \leq n-1$ and $0 < \theta < 1$. □

Now let $\rho_j := q_j\alpha - p_j$, with $\rho_{-1} := -1$ and $\rho_0 := \alpha$, be the signed distance from $q_j\alpha$ to the nearest integer. Note that the distances to the nearest integer, though not the sign, are identical for α and $1 - \alpha$.

Theorem 1.4 *For $0 < \alpha < 1/2$ and $j \geq 1$,*

(i) $\|q_j\alpha\| = (-1)^j \rho_j$,
(ii) *If $1 \leq q < q_j$ then $\|q\alpha\| > |\rho_j|$,*
(iii) $\rho_j = a_j \rho_{j-1} + \rho_{j-2}$,
(iv) $\rho_j = -\dfrac{\alpha_j}{q_j + \alpha_j q_{j-1}}$,
(v) $\dfrac{1}{3(a_{j+1}q_j)} < |\rho_j| < \dfrac{1}{a_{j+1}q_j}$,
(vi) $a_j = \lfloor |\rho_{j-1}/\rho_{j-2}| \rfloor$,
(vii) *For $1 \leq k < a_j$, $|k\rho_{j-1} + \rho_{j-2}| > |\rho_{j-1}|$.*

Note: Taken together, (i) and (ii) say that the nearest approaches to 0 mod 1 in the sequence $\langle q\alpha \rangle$ are alternately from the right and the left, and that the sequence of integers q_j at which successive minima of the unsigned distance occur, is the same sequence as that generated by the continued fraction algorithm.

Proof. We begin with (iii). The recurrence follows from the fact that $\langle p_j \rangle$ and $\langle q_j \rangle$ obey that recurrence, and ρ_j is a linear combination of these. Now (vi) follows from (iii), and (vii) from (vi).

For (ii), we have a calculation beginning with

$$\alpha = \frac{p_j + \alpha_j p_{j-1}}{q_j + \alpha_j q_{j-1}}.$$

Thus $(q_j + \alpha_j q_{j-1})\alpha = (p_j + \alpha_j p_{j-1})$, so $(q_j\alpha - p_j) = -\alpha_j(q_{j-1}\alpha - p_{j-1})$ as required.

Part (iv) holds by induction. For $j = 1$, $q_1 = a_1$, $q_0 = 1$, $\rho_1 = a_1\alpha - 1$, and $\alpha_1 = 1/\alpha - a_1$ so that

$$\frac{(-1)\alpha_1}{q_1 + \alpha_1 q_0} = \frac{(-1)(1/\alpha - a_1)}{a_1 + (a/\alpha - a_0)} = a_1\alpha - 1$$

as required. Now suppose the inductive hypothesis holds for $j-1$. Then

$$\rho_j = -\alpha_j \rho_{j-1} = \frac{-\alpha_j(-1)^{j-1}\alpha_{j-1}}{q_{j-1} + \alpha_{j-1}q_{j-2}}$$

and we must show that this is equal to $(-1)^j \alpha_j/(q_j + \alpha_j q_{j-1})$. Equivalently, we need to show that $\alpha_j = (q_{j-1} + \alpha_{j-1} q_{j-2})/(q_j + \alpha_j q_{j-1})$. But

$$q_j + \alpha_j q_{j-1} = (a_j q_{j-1} + q_{j-2}) + \alpha_j q_{j-1} =$$

$$q_{j-1}/\alpha_{j-1} + q_{j-2} = \frac{q_{j-1} + \alpha_{j-1} q_{j-2}}{\alpha_{j-1}}$$

as needed.

The inequalities, parts (i) and (v) here, are as usual the more difficult parts. An heuristic argument for (i) both lends plausibility to the conclusion, and motivates the proof. The plausibility argument begins with the observation that since $\|q_j \alpha\|$ is small, $nq_j \alpha \mod 1 = n\rho_j$ for n small. Thus, the sequence $(nq_j \alpha \mod 1)$ marches across the unit interval in steps of directed magnitude ρ_j, heading for the opposite end. That value of n so that $-\rho_{j-1} \mod 1$ lies between $n\rho_j$ and $(n+1)\rho_j$ provides the first n so that $\|(nq_j + q_{j-1})\alpha\| \, \|n\rho_j + \rho_{j-1}\| < |\rho_j|$. A little calculation shows that $n = a_{j+1} = \lfloor 1/\alpha_j \rfloor$. But while it is clear that this affords *one* way to find values of q for which $\|q\alpha\| < |\rho_j|$, we have as yet no guarantee that no smaller choice of q works.

Suppose, then, for purposes of deriving a contradiction, that such a q exists. Let q' be the least integer between q_j and q_{j+1} so that $\|q'\alpha\| < |\rho_j|$. Let p' be the integer nearest $q'\alpha$. If $q'\alpha - p'$ and $q_j \alpha - p_j$ have the same sign, then $|(q_j - q')\alpha - (p_j - p')| < |\rho_j|$, a contradiction. If, on the other hand, $q'\alpha - p'$ and $q_j\alpha - p_j$ have opposite signs, then $q_{j-1}\alpha - p_{j-1}$ and $q'\alpha - p'$ have the same sign, with $|q'\alpha - p'| < |q_{j-1}\alpha - p_{j-1}|$. Thus (q', p') lies in the positive cone of the vectors (q_{j-1}, p_{j-1}) and $(1, \alpha)$ and so it is a positive integer combination of (q_{j-1}, p_{j-1}) and (q_j, p_j). But for $n < a_{j+1}$, $\|(nq_j + q_{j-1})\alpha\| > |\rho_j|$ which eliminates the most promising candidate linear combinations. What of the less promising ones?

If $1 \leq n < a_{j+1}$ and $m > 1$, then

$$(nq_j + mq_{j-1}, np_j + mp_{j-1}) = (nq_j + q_{j-1}, np_j + p_{j-1}) + (m-1)(q_{j-1}, p_{j-1}).$$

Since both vectors lie on the same side of the ray $t(1, \alpha), t \geq 0$, the vertical distance from $n(q_j, p_j) + m(q_{j-1}, p_{j-1})$ to the ray $t(1, \alpha)$ is $\|(nq_j + mq_{j-1})\alpha\|$, and it is greater than the vertical height of the parallelogram with vertices

$0, (q_j, p_j), (q_{j-1}, p_{j-1})$ and $(q_j, p_j) + (q_{j-1}, p_{j-1})$ which is $1/q_{j-1} > |\rho_{j-1}|$. Thus, $\|(nq_j + q_{j-1})\alpha\| > \|q_{j-1}\alpha\| = |\rho_{j-1}|$, a contradiction. This proves (i). Finally for the inequality (v), we first note that

$$|\rho_j| = \frac{\alpha_j}{q_j + \alpha_j q_{j-1}} = \frac{1}{q_j/\alpha_j + q_{j-1}}$$

Now $a_{j+1} \leq 1/\alpha_j$ so

$$|\rho_j| < \frac{1}{a_{j+1}q_j + q_{j-1}} < \frac{1}{a_{j+1}q_j}$$

In the other direction, we need $q_j/\alpha_j + q_{j-1} < 3a_{j+1}q_j$. But $q_{j-1} < q_j$, so $q_j/\alpha_j + q_{j-1} < (1 + 1/\alpha_j)q_j$. Thus it will suffice that $1/\alpha_j < 2a_{j+1}$. But $1/2 < \alpha_j a_{j+1}$ because $1/2 < \alpha_j(1/\alpha_j - \{1/\alpha_j\})$ where $\{u\}$ (in this context) denotes the fractional part of u. □

There is a remarkable recent elementary result due to F.E. Su[Su] concerning the distribution of $\{n\alpha : 1 \leq n \leq q_k\}$ mod 1, where (p_k/q_k) are the successive convergents to an irrational α. The result is simplest in the case where the fractional part of α is less than $1/2$, so we make that assumption.

As we have seen in Theorem 1.4, the successive minima of $\|q\alpha\|$, $q \geq 1$ occur at $q = q_j, j \geq 0$ and are given by $\|q_j\alpha\| = |\rho_j|$. Su's result begins with the notion of dividing the open unit interval $(0, 1)$ into countably many half-open intervals, or *bins* B_j, $j \geq 0$, with $B_j := [|\rho_j|, |\rho_{j-1}|)$. Next, Su asks: how do the numbers $q\alpha$ mod 1 distribute themselves into these bins for $1 \leq q \leq q_j$? Following his notation, we let $N_\alpha(m, n)$ denote the number of q with $1 \leq q \leq m$ so that $q\alpha$ mod $1 \in B_\alpha(n)$.

Consider the illustrative example $\alpha = \sqrt{2}$. All partial quotients a_j are 2. By convention, $q_{-1} = 0$ and $q_0 = 1$. The next few q_j are $2, 5, 12, 29$ and 70, while the first few ρ_j are a conventional $\rho_0 = -1$ followed by $\rho_1 = \sqrt{2} - 1$, $\rho_2 = 2\sqrt{2} - 3$, $\rho_3 = 5\sqrt{2} - 7$ and $\rho_4 = 12\sqrt{2} - 17$, so that the bins are

$$\ldots (17 - 12\sqrt{2}, 5\sqrt{2} - 7], (5\sqrt{2} - 7, 3 - 2\sqrt{2}], (3 - 2\sqrt{2}, \sqrt{2} - 1], (\sqrt{2} - 1, 1]$$

The numbers $k\sqrt{2}$ mod 1, $1 \leq k \leq 12$ fall into bins $0, 1, 1, 1, 2, 0, 2, 1, 1, 2, 0$, and 3 respectively. One number fell into bin 3, three into bin 2, five into bin 1, and three into bin 0. Thus for $q = 12 = q_3$ the answer to Su's question can be given as a list

$$(\ldots 0, \ldots 0, 1, 3, 5, 3)$$

We form a similar list for each q_j and arrive at a table of which the preceding calculation forms the basis for the entries in the fourth row.

$$\begin{array}{ccccccccccc|c}
 & & & & & & & & & & & q_j \\
0 & 0 & 0 & 0 & 0 & 0 & 0 & 0 & 0 & 0 & 1 & 1 \\
0 & 0 & 0 & 0 & 0 & 0 & 0 & 0 & 0 & 1 & 1 & 2 \\
0 & 0 & 0 & 0 & 0 & 0 & 0 & 0 & 1 & 3 & 1 & 5 \\
0 & 0 & 0 & 0 & 0 & 0 & 0 & 1 & 3 & 5 & 3 & 12 \\
0 & 0 & 0 & 0 & 0 & 0 & 1 & 3 & 5 & 15 & 5 & 29 \\
0 & 0 & 0 & 0 & 0 & 1 & 3 & 5 & 15 & 33 & 13 & 70 \\
0 & 0 & 0 & 0 & 1 & 3 & 5 & 15 & 33 & 83 & 29 & 169 \\
0 & 0 & 0 & 1 & 3 & 5 & 15 & 33 & 83 & 197 & 71 & 408 \\
0 & 0 & 1 & 3 & 5 & 15 & 33 & 83 & 197 & 479 & 169 & 985 \\
\end{array}$$

Inspection reveals an apparent recurrence relation and Su proves it. The details of the result are these: Put a conventional row for q_{-1} on top, with a single nonzero entry: $N_\alpha(q_{-1}, 0) := -1$. Observe that $N_\alpha(q_0, 0) = 1$ while $N_\alpha(q_j, k) = 0$ if $k > j$ and $N_\alpha(q_j, j) = 1$ for $j \geq 1$. The recurrence relation applies to the non-trivial entries in the table. Column zero (the rightmost column as we have displayed it) lists the number of $q \leq q_j$ for which $\|q\alpha\|$ falls in bin 0, the interval $[\alpha, 1)$. It is governed by the recurrence

$$N_\alpha(q_{j+1}, 0) = a_{j+1}(N_\alpha(q_j, 0) + ((-1) + (-1)^j)/2) + N_\alpha(q_{j-1}, 0)$$

while for $k \geq 1$, the recurrence relation is

$$N_\alpha(q_{j+1}, k) = a_{j+1}(N_\alpha(q_j, k) + (-1)^{j-k}) + N_\alpha(q_{j-1}, k).$$

Although this result is of interest in its own right, Su had an application in mind. A *random walk* on the circle starts somewhere and moves clockwise or counterclockwise, each with probability 1/2, by an angle $2\pi\alpha$. If α is rational, the walk is a discrete walk and the distribution of frequencies with which the walk is in any one state converges to $1/N$ where N is the denominator of α. But if α is irrational, the walk has a dense orbit with probability one, and the appropriate measure of the extent to which the distribution after r moves differs from uniform, is the *discrepancy* of the sequence. (See section 3.3) If α is badly approximable, and in particular if it is a quadratic irrational, then the discrepancy of the distribution is comparable to best possible. This result has been generalized to the case of a random walk in which the step size is chosen at random from a list of n possible steps; here one does best to take the list to be a *badly approximable vector*[HenSu]. These are discussed in Chapter 6.

1.7 Other Known Continued Fraction Expansions

Let ϕ be the golden mean constant $(1 + \sqrt{5})/2$. Then

$$\alpha := \sum_{n=1}^{\infty} 2^{-\lfloor n/\phi \rfloor} = [0; 2^0, 2^1, 2^1, 2^2, 2^3, 2^5, 2^8, 2^{13} \ldots]$$

where the exponents are the Fibonacci numbers. This result, and a generalization, goes back to Böhmer[B]. There is a nice proof of the more general result in [AB]. One key tool is an old observation of H. J. S. Smith, Note on continued fractions, Messenger Math. 6 (1876), 1-14. We quote this result from [AB]:

Let α be an irrational number with $0 < \alpha < 1$. Let $\alpha = [0, a_1, a_2, \ldots]$ and $p_n/q_n = [0, a_1, a_2, \ldots a_n]$, $n \geq 0$, where p_n, q_n are relatively prime non-negative integers. (As usual, we put $p_{-1} = 0, p_0 = 1, q_{-1} = 1, q_0 = 0$, so that $p_n = a_n p_{n-1} + p_{n-2}$, $q_n = a_n q_{n-1} + q_{n-2}$ for all $n \geq 1$.) For $n \geq 1$, define $f_\alpha(n) = \lfloor (n+1)\alpha \rfloor - \lfloor n\alpha \rfloor$, and consider the infinite binary sequence $f_\alpha(n)_{n \geq 1}$, which is sometimes called the *characteristic sequence* of α. Define binary words X_n, $n \geq 0$, by $X_0 = 0$, $X_1 = 0^{a_1-1}1$, $X_n = X_{n-1}^{a_n} X_{n-2}$ for $n \geq 2$, where X^a denotes the word X repeated a times, and $X_1 = 1$ if $a_1 = 1$. Then for each $n \geq 1$, X_n is a prefix of f_α. That is, $X_n = f_\alpha(1) f_\alpha(2) \ldots f_\alpha(s)$ where s is the length of X_n.

Remark 1.4 *The proof of this last statement is a matter of attention to detail once the key is at hand: the progression of $(n\alpha)$ mod 1 brings one back to near zero for each q_k, so that the further progress of the sequence will duplicate its initial segment, until such time as the difference between $q_k \alpha$ mod 1, and exactly zero, becomes large enough to affect the result.*

There is another more recently discovered class of particular numbers defined in some other way than by their continued fraction expansion, for which the continued fraction is known. The best known instance is

$$\beta := \sum_{n=0}^{\infty} 2^{-2^n} = [0; 1, 4, 2, 4, 4, 6, 4, 2, 4, 6, 2, 4, 6, 4, 4, 2, \ldots]$$

due originally to M. Kmošek [Km] and independently by J. Shallit [Sh1]. The most recent work along these lines of which the author is aware is H. Cohn's paper [Ch]. Some of the key ideas can be stated briefly. First, one may study the more general situation

$$\sum_{n=0}^{\infty} 1/f^n(x)$$

where $f^n(x)$ denotes the n-fold iteration of f on x, and f is a polynomial function which takes integer values at integers. Secondly, there is a *folding lemma* at the heart of these expansions.

Folding Lemma: If $p_n/q_n = [a_0, \vec{w}]$, $\overleftarrow{v} = (-w_n, -w_{n-1}, \cdots - w_1)$, and $0 < x < 1$ then

$$\frac{p_n}{q_n} + \frac{(-1)^n}{xa_n^2} = [a_0, \vec{w}, x, \overleftarrow{v}].$$

This lemma is also used in an analysis of the intriguing number, call it the van der Poorten-Shallit constant,

$$\sum_{n=0}^{\infty} 2^{-F_n} = [1, 10, 6, 1, 6, 2, 14, 4, 124, 2, 1, 2, 2039, 1, 9, 1, 1, 1, 262111, \ldots].$$

It is unusual to find numbers outside the familiar classes of rational and quadratic irrational, defined other than by their continued fraction expansions, for which that continued fraction expansion is atypical of randomly chosen real numbers. This one results from taking the special case $X = 2$ in a formal identity for the continued fraction expansion of the Laurent series $\sum X^{-F_j}$ [vdPS].

One application of this to matters outside the immediate topic of continued fractions is a proof that every prime p of the form $p = 4n + 1$ is the sum of two squares. See [CELV] for a very simple and clear proof. The basic idea is that corresponding to every integer $2 \le b < p/2$ there is a list $(a_1, a_2, \ldots a_r)$ so that $[a_1, a_2, \ldots a_r] = b/p$. Since there are $2n - 1$ such lists, (one for each value of b) and since $2n - 1$ is odd, if we pair each off with its reversal $(a_r, a_{r-1}, \ldots a_1)$ there must be one case of a list paired with itself. That list will be palindromic, reading the same forward and backward, and with a little more work, one sees that it must have $r = 2s$ even. Thus, $p = |a_1, a_2, \ldots a_s, a_s, a_{s-1}, \ldots a_1|$ and $p = |a_1, a_2, \ldots a_s|^2 + |a_1, a_2, \ldots a_{s-1}|^2$. While this proof does not give any efficient way to find the two squares, another continued-fraction idea does. Half the residues mod p are quadratic nonresidues. Take a random $1 < b < p$ and check whether it is a quadratic nonresidue by checking whether $b^{(p-1)/2} \equiv -1 \mod p$. Half the time, this will be the case, so with reasonable luck, such a b can be found quickly.

Retrieve $n = b^{(p-1)/4} \mod p$, and observe that $n^2 \equiv -1 \mod p$. The lattice generated by $(n,1)$ and $(p,0)$ is a set of integer pairs (u,v) such that $u^2 + v^2 \equiv 0 \mod p$. It has determinant p, so the shortest nonzero vector in the lattice lies within the square $-\sqrt{p} < x < \sqrt{p}, -\sqrt{p} < y < \sqrt{p}$. The Gauss lattice reduction algorithm is an efficient method for finding this shortest vector, and the resulting (u,v) will satisfy $u^2 + v^2 = p$.

1.7.1 Notes

Edward R. Burger [Bg] shows that there are real numbers for which all the denominators of convergents are squares, cubes, or values of other such polynomials, or even prime. Unsolved is whether these things can be achieved using real numbers for which the continued fraction expansion has bounded partial quotients.

Glyn Harman and Kam C. Wong [HW] show that almost all real α have infinitely many even numbered convergents with even numerator, and they give a natural generalization.

Chapter 2

Generalizations of the gcd and the Euclidean Algorithm

2.1 Other gcd's

There are several mathematical settings in which it can make sense to speak of greatest common divisors, and of algorithms for determining them. Indeed, the term "Euclidean domain" is defined precisely so that the Euclidean algorithm will work in any such setting.

A Euclidean domain is a ring R with identity, and equipped with a function λ from $R\backslash\{0\}$ to $\mathbb{N}$ so that if $a, b \in R$ with $b \neq 0$ then there exist $q, r \in R$ so that $a = qb + r$, and either $r = 0$ or $\lambda(r) < \lambda(b)$. Given any Euclidean domain, the gcd algorithm of Chapter 1 can be used (almost) verbatim. Suppose, then, that we have a Euclidean domain, and we have an algorithm 'r' that takes as inputs a pair (a,b) and returns the r such that $\lambda(r) < \lambda(b)$. Then our new version of the Euclidean algorithm is given below. It returns an element a of R which in a sense is the greatest common divisor of the inputs u and v. That is, $a \mid u$, $a \mid v$, and if $d \mid u$ and $d \mid v$, then $d \mid a$ and $\lambda(d) \leq \lambda(a)$.

```
EuclideanAlgorithmGCD[u_,v_]:=
Module[{a,b},
    If[u==0,
        If[v==0,Return[0]];
        Return[v]
        ];
    a=u;b=v;
    While[lambda[b]>0,{a,b}={b,r[a,b]}];
    a]
```

The Gaussian integers form a Euclidean domain, with $\lambda(z) = z\bar{z} = |z|^2$.

There are several other Euclidean domains of the same sort (quadratic extension of $\mathbb{Z}$); they are $\mathbb{Q}[\sqrt{D}]$ where $D = -11, -7, -3, -2, -1, 2, 3, 5, 6, 7, 11, 13, 17, 19, 21, 29, 33, 37, 41, 57$, or 73.

Another example is the ring of polynomials in a single formal variable X over a field K. A polynomial in X over K is a formal expression of the form $a_0 + \cdots + a_n X^n$, where n is a nonnegative integer and all the $a_i \in K$. We require also that $a_n \neq 0$, except that if $n = 0$ we allow the zero polynomial. The *degree* of a polynomial $p = \sum_0^n a_i X^i$ is n; but if $p = 0$, the degree is by convention set to $-\infty$. There is a division 'algorithm' for polynomials: given polynomials p, and $q \neq 0$, there exist unique polynomials b and r so that $p = bq + r = qB(p,q) + R(p,q)$ and the degree of r is less than that of q.

In practice, real or complex numbers given as the limit of a sequence of rational numbers do not lend themselves to the exact calculations called for in the division algorithm, so that the procedures involved are arguably not exactly algorithms. If we brush aside this point, we can go ahead and think about a continued fraction expansion

$$\frac{p}{q} = b_0 + \cfrac{1}{b_1 + \cfrac{1}{b_2 + \cfrac{1}{\ddots + 1/b_r}}}$$

where $b_0 = b$ and where the successive b_k are determined by the recursion

$$p_k = q_{k-1}, \quad q_k = R(p_{k-1}, q_{k-1}), \quad b_{k-1} = B(p_{k-1}, q_{k-1})$$

with initial conditions $p_0 = p$, $q_0 = q$, and terminating when $q_r = 0$. If the coefficients are algebraic numbers, and particularly if they are rational, it is possible to carry out the exact computations and implement the algorithm. There is, however, a tendency to computational explosion, as the coefficients become complicated.

Another way to determine the successive b_k involves matrices. We begin with $p_0 = p$ and $q_0 = q$ as before, and we set

$$\begin{pmatrix} p_j \\ q_j \end{pmatrix} = \begin{pmatrix} 0 & 1 \\ 1 & -b_{j-1} \end{pmatrix} \begin{pmatrix} p_{j-1} \\ q_{j-1} \end{pmatrix}.$$

As before, $b_j = B(p_j, q_j)$ is the unique polynomial b so that $\deg(p_j - bq_j) < \deg(q_j)$. The algorithm terminates when $q_r = 0$.

If we are working in a finite field, this all works beautifully. There is no combinatorial explosion. If $f/g = [b_1; b_2, \ldots, b_r] = b_1 + 1/(b_2 + 1/(\cdots + 1/b_r))$ then the list of convergents $b_1 + 1/(b_2 + 1/(b_3 + \ldots + 1/b_m)) =$

$[b_1; \ldots, b_m]$, $1 \leq m \leq r$ may be calculated recursively. Each convergent is a good approximation to f/g, in the sense that $[b_1; \ldots, b_m]$ is the best approximation to f/g by a rational function with numerator of degree less than or equal to $\sum_1^m \deg b_k$. There are applications. Of particular interest are rational functions with partial quotients $(b_1, \ldots, b_r)$ of small degree. [Ni2]. One way to generate pseudorandom numbers, due to Tausworthe [Tau], involves a recursively defined sequence of integers between 0 and $p-1$, where p is prime. Fix integers $a_0 \not\equiv 0 \pmod{p}$, $a_1, \ldots, a_{k-1}$, and let $y_0, y_1, \ldots$ be the sequence of integers in $[0, p-1]$ generated by the recurrence

$$y_{n+k} = \sum_{j=0}^{k-1} a_j y_{n+j} \pmod{p}$$

with initial conditions $y_n = 1$, $0 \leq n \leq k-1$. The characteristic polynomial of this recurrence relation is

$$f_a(x) = x^k - \sum_0^{k-1} a_j x^j$$

seen as a polynomial over the finite field with p elements. Successive pairs of entries in the (ultimately periodic) sequence (y_n) are more nearly statistically independent if the continued fraction expansion of $f(x)/x^k$ has all partial quotients of degree one. Niederreiter proves that in the case $p = 2$, if r is the largest integer so that $2^r \leq k$, then

$$f_k(x) = \sum_{j=0}^{r+1} x^{k-\lfloor k/2^j \rfloor}$$

is the unique polynomial mod 2 with these properties and with $f_k(0) = 1$. He gives other results in this vein. By relaxing the condition on the largest degree of a partial quotient, it is possible to impose the further condition on f that it be a *primitive* polynomial.

There are results about the statistics of the continued fraction expansion: What is a typical value for r, when we look at the pool of all f and g of degree n, for example?

Suppose we are working over the finite field $\mathcal{F}_q$ of order q. There is a $q-1$ to 1 correspondence between pairs (f, g) of polynomials over $\mathcal{F}_q$, with $\deg(f) = n$ and $\deg(g) < n$ with f and g relatively prime, and lists (of length $r \geq 1$) $(b_1, b_2, \ldots, b_r)$ of polynomials over $\mathcal{F}_q$ with positive degrees that sum to n. The correspondence pairs (f, g) with that list

$\mathbf{b} = (b_1, \ldots, b_r)$ so that

$$f/g = b_1 + 1/(b_2 + 1/(\ldots + 1/b_r)).$$

For $1 \le k \le n$, let $F(n, k, q)$ denote the number of such lists for which the maximum degree of any b_j is at most k. Then we have this result, due to C. Friesen and the author : [FriHe]

Theorem 2.1 *There are positive constant $\lambda(k, q)$ and $C(k, q)$ so that*

$$|F(n, k, q) - C(k, q)\lambda(k, q)^n q^n| \le (2/3) q^n.$$

The constant $\lambda(k, q)$ is an algebraic integer and indeed a **PV** *number: it is the principal root of a monic irreducible polynomial, and the remaining roots all lie strictly within the unit circle in* $\mathbb{C}$. *The polynomial in this case is $z^k - (q-1)\sum_0^{k-1} z^j$ and its principle root is approximately equal to $q - q^{1-k}$. The other constant $C(k, q)$ is given by $C(k, q) = (\lambda(k, q) - 1)/(\lambda(k, q) - k(q - \lambda(k, q)))$ and is positive and approximately equal to $1 - 1/q$.*

2.2 Continued Fraction Expansions for Complex Numbers

Algebraic number theory has its roots in the observation that the Gaussian integers, that is, the complex numbers of the form $a + bi$ where $a, b \in \mathbb{Z}$, have a theory very similar to that of the (rational) integers. We shall see that in some respects, continued fractions for Gaussian integers work just about the same as in the classical case, while in other respects, matters are different and apparently considerably more complicated. The whole of Chapter 5 is given over to this topic, but we make a start on it now.

Our first feat, as a warmup, will be to compute the gcd of two Gaussian integers.

Given Gaussian integers u, v, we arrange them so that $|u| \ge |v|$, and then compute successive pairs (u_k, v_k) with steps of the form

$$(u_{k+1}, v_{k+1}) = (v_k, u_k - gv_k)$$

where g is chosen to be the Gaussian integer nearest u_k/v_k, rounding real and imaginary parts down if there are ties. We halt when some $v_r = 0$. Equivalently, we could keep a parallel column of computations in which, after arranging that $|u| \ge |v|$, we set $z = v/u$, and proceed by means of steps $z \to 1/z - g$, where g is the Gaussian integer nearest $1/z$, halting when $z = 0$. Clearly, $|1/z - g| \le 1/\sqrt{2}$, so in the other column of execution, where pairs (u, v) of Gaussian integers are kept track of, we have $|u_{k+1}v_{k+1}| \leq (1/\sqrt{2})|u_k v_k|$. Thus, it is a fast algorithm. (In fact, the ratio is generally better than this, because after the first step, we have $|u| \geq \sqrt{2}|v|$, so that $|u_{k+1}v_{k+1}| \leq (1/2)|u_k v_k|$ for $k \geq 1$.)

Consider the example in which we start with $(u, v) = (u_0, v_0) = (77 + 190i, 20+204i)$. Our successive pairs are $(20+204i, 57-14i)$, $(57-14i, -22+33i)$, $(-22+33i, 2-3i)$, and finally, $(2-3i, 0)$.

This Gaussian-integer analog of the centered continued fraction algorithm serves nicely the purpose of extracting gcd's, and its diophantine approximation behavior has been studied already in the 19th century, by Hurwitz. We have much more to say about this topic in Chapter 5.

Work by Asmus Schmidt has thrown considerable further light on the topic of diophantine approximation by Gaussian rationals. This, too, is discussed further in Chapter 5. [Sch] [Sch2] For now, we just sketch the basic idea, which bears similarities to one way of looking at the familiar real case.

Even in the setting of real numbers, one can restate all questions of continued fractions as questions about invertible integer matrices. Instead of having approximations p_{n-1}/q_{n-1} and p_n/q_n, one has an integer matrix

$$M_n = \begin{pmatrix} p_{n-1} & q_{n-1} \\ p_n & q_n \end{pmatrix}$$

and the twin recurrence relations that generate the sequences (p_n) and (q_n) are replaced by a matrix-valued recurrence relation: $M_{n+1} = AM_n$, in which A is a matrix of the form

$$\begin{pmatrix} 0 & 1 \\ 1 & a \end{pmatrix}.$$

This setup can be reduced to a still simpler scheme in which A is always either

$$\begin{pmatrix} 0 & 1 \\ 1 & 0 \end{pmatrix} \quad \text{or} \quad \begin{pmatrix} 1 & 1 \\ 0 & 1 \end{pmatrix},$$

with no two consecutive instances of the first choice.

For the complex case, Schmidt uses instead a list of 8 matrices with Gaussian integer entries, each with a determinant equal to one of ± 1 or $\pm i$. These take the place of the two choices above.

Another way to compute good Gaussian-rational approximations to a general complex number z_0 involves lattice reduction. One may use either the lattice reduction algorithm of Lenstra, Lenstra, and Lovasz, [Co] or Minkowski reduction. One starts with the matrix

$$M_0(\theta) = \begin{pmatrix} 1 & 0 & 0 & 0 \\ 0 & 1 & 0 & 0 \\ -\theta_1 & -\theta_2 & t & 0 \\ \theta_2 & -\theta_1 & 0 & t \end{pmatrix}$$

and the resulting reduced matrix will have for its first entry an integer linear combination of the four initial row vectors, say, $n_1 v_1 + n_2 v_2 + n_3 v_3 + n_4 v_4$. (The dual vector $n_2 v_1 - n_1 v_2 + n_4 v_3 - n_4 v_3$ will be perpendicular and have the same Euclidean norm, so it too must go into the reduced basis.) The complex-number perspective on this computation in a real setting is that

$$|(n_1 + in_2) - (n_3 + in_4)(\theta_1 + i\theta_2)|^2 + t^2(n_3^2 + n_4^2)$$

is small, which means that

$$\theta_1 + i\theta_2 \approx \frac{n_1 + in_2}{n_3 + in_4}.$$

If we put aside the goal of finding 'diophantine approximations' along with finding greatest common divisors, we can with reasonable efficiency compute the gcd of two arbitrary algebraic numbers.

2.2.1 Computing the gcd of a pair of algebraic integers

The perspective we want is one that is available to us even in the simplest of settings, the rational integers. Associated with any integer n is the *ideal* (n) consisting of all integer multiples of that integer. An integer d is a proper divisor of an integer n if and only if (n) is a proper subset of (d). The gcd of two nonzero integers a and b is the (single) positive integer d that generates the ideal $(a) \cap (b)$. Computing the gcd, then, is a matter of passing from one generating set $\{a, b\}$ for an ideal, to another, and another, arriving eventually at a simplest generating set $\{d\}$.

An algebraic number field K is an algebraic extension of $\mathbb{Q}$ of finite degree. We bring in some standard facts. See, e.g. H. Cohen [Co].

An algebraic integer is a real or complex number α that is a root of a monic irreducible (over $\mathbb{Q}$, that is) polynomial p_α of positive degree, in a single variable, with integer coefficients. If the degree is one, then it is a rational integer. If the degree n of α is greater than one, then α has n conjugates, among them α itself; these are the roots of p_α, and they are distinct. All symmetric integer polynomials of the conjugates are rational integers; the elementary symmetric functions of the conjugates are, to within a factor of ± 1, the coefficients of p_α.

An algebraic number field K is an extension of $\mathbb{Q}$ of finite degree. Every algebraic number field K can be expressed as $\mathbb{Q}[\alpha]$ where α is an algebraic integer. The set $O[\alpha]$ of algebraic integers in $\mathbb{Q}[\alpha]$ is a ring which includes $\mathbb{Z}$ as a proper subset.

It need not be the case that $O[\alpha] = \mathbb{Z}[\alpha]$, that is, there may be integers of $O[\alpha]$ not of the form $\sum_0^{n-1} a_k \alpha^k$.

An *ideal* in $O[\alpha]$ is an additive subgroup I of $O[\alpha]$ that is closed under multiplication by $O[\alpha]$. That is, if $\beta \in I$ and $\gamma \in O[\alpha]$ then $\beta\gamma \in I$. The set $O[\alpha]$ is itself an ideal. It is the largest ideal of $O[\alpha]$, while $\{0\}$ is the smallest.

Every ideal of $O[\alpha]$ is finitely generated. That is, for every ideal I of $O[\alpha]$, there exists a finite list $(\beta_1, \ldots, \beta_m)$ so that every element of I has the form $\sum_1^m \gamma_k \beta_k$ for some integers γ_k of $O[\alpha]$. Every ideal also has an *integral basis*. That is, there exists a set $\iota_1, \ldots, \iota_n$ of n elements of I, so that every $\gamma \in I$ can be written as $\gamma = \sum_1^n a_k \iota_k$, with all a_k rational integers.

The *product* of two ideals $I_1 I_2$ is the set (it is again an ideal) consisting of all sums of products $\alpha\beta$ with $\alpha \in I_1$ and $\beta \in I_2$. This ideal, too, is finitely generated of course. The set of n^2 elements got by multiplying each element of an integral basis of I_1 by each element of an integral basis of I_2 is an integral generating set for $I_1 I_2$, but with considerable redundancy. To each such product we may associate the coefficients of its representation in terms of a fixed integral basis of $O[\alpha]$. This gives us a list of n^2 vectors in $\mathbb{Z}^n$. It is the province of *lattice reduction* theory to provide us with a lattice basis, of just n elements, for this lattice. There is a modified version of the LLL algorithm (algorithm 2.6.8 page 94 of [Co]) which does the job. When we have our lattice basis of n integer vectors, the n algebraic numbers which are formed by using as coefficients on the integral basis of $O[\alpha]$, the entries of each of the n integer vectors, constitute an integral basis of the product ideal.

One ideal I_1 *divides* another I_2 if and only if there exists an ideal D so that $I_1 D = I_2$; equivalently, though this is not obvious, I_1 divides I_2 if and only if $I_2 \subseteq I_1$. And this, finally, brings us to the gcd. The lcm of two ideals is simply their intersection. The gcd is the product divided by the lcm. But from the computational perspective, there is work to do. We require not just a description of the gcd, but an integral basis for it. We may assume that I_1 and I_2 come to us with integral bases

B_1 and B_2, respectively. We must purge redundant elements from the generating set $B_1 \cup B_2$ of $G = \gcd(I_1, I_2)$. This, too, is a lattice reduction question.

The Gaussian integers are a principal ideal domain ; every ideal D has a single generator d if we allow Gaussian integer multiples of a generator. If we want an integral basis, then (d, id) will serve. The obvious choice for an integral basis of the Gaussian integers is $(1, i)$, and with this choice, the coefficient vector for $d = d_1 + id_2$ is (d_1, d_2), and the integral basis for the ideal generated by d is $((d_1, d_2), (-d_2, d_1))$. Thus, to find the gcd of $(77 + 190i)$ and $(20 + 204i)$, say, using our general methods, we would reduce the two by four matrix

$$\begin{pmatrix} 77 & 190 \\ -190 & 77 \\ 20 & 204 \\ -204 & 20 \end{pmatrix} \to \begin{pmatrix} -2 & 3 \\ 3 & 2 \end{pmatrix}$$

and conclude that the ideal $(3 + 2i)$ is the required gcd. Our earlier computation with this same example can, in retrospect, be seen also to have done just this. Our successive pairs (u_k, v_k) had associated with them a set of four lattice vectors. If $u_k = x_k + iy_k$, with $x, y \in \mathbb{Z}$, then the integer vectors (x_k, y_k) and $(y_k, -x_k)$ belong to the lattice associated with the principle ideal generated by u_k. Thus, we were in effect generating a list of four-element generating sets for a two-dimensional lattice, and reduction was complete when we arrived at a set in which two of the vectors were zero and so the remaining two formed a basis, which, conveniently, consisted of two orthogonal vectors.

2.3 The Lattice Reduction Algorithm of Gauss

Given a pair of positive integers, the Euclidean algorithm finds a least common denominator. Another way to look at this is to consider the pair as generators of the *lattice* that includes the given pair and is closed under addition and multiplication by integers. The Euclidean algorithm then amounts to a sort of one-dimensional lattice reduction algorithm. In the context of $\mathbb{R}^2$, a lattice over the integers is a discrete subset of $\mathbb{R}^2$ which is closed under addition and integer scalar multiplication. The set of all Gaussian integer multiples of a given Gaussian integer affords a natural example of such a lattice if we identify $\mathbb{C}$ with $\mathbb{R}^2$.

For a related example, consider a prime $p \equiv 1 \mod 4$, and a quadratic nonresidue k. Take $j = k^{(p-1)/4} \mod p$. Then $j^2 \equiv -1 \mod p$. The closure, under addition and multiplication by rational integers, of the set $\{(j,1),(p,0)\}$ is a lattice Λ of determinant p and with the property that for $(a,b) \in \Lambda$, $a^2+b^2 \equiv 0 \mod p$. We can use this to find an integer solution to $a^2+b^2 = p$ because the shortest nonzero vector in the lattice must give this result. To find this shortest vector, we need to reduce the lattice. In two dimensions, that amounts to finding a basis for Λ so that the first vector is the shortest vector in the lattice, and the second vector is the shortest vector which is not a multiple of the first.

It should be noted that in higher dimensions, lattice reduction is much more difficult. The Lenstra-Lenstra-Lovasz reduction algorithm [Co] is computationally feasible and gives a useful, though not necessarily best-possible, reduction. Best-possible reduction is in general an NP-class problem. This to some extent overstates the difficulty of the task. For any fixed dimension, the computational complexity does not rise exponentially with the length of the input data. There is an algorithm, the Fincke-Pohst algorithm [Co], which finds all lattice elements with norm below a fixed upper bound, and it can be applied to the task of lattice reduction.

The Gauss algorithm proceeds along very much the same lines as the classical Euclidean algorithm: given a pair of vectors (a,b) and (c,d), one subtracts an appropriate multiple of the shorter from the longer, to arrive at a new pair. When no further progress can be made, we are essentially done. (There is a bit of an 'endgame', but that need not concern us here.) But what can this have to do with our g and G and so on? Daudé, Flajolet and Vallée [DFV] have found that there is a close connection. They first observe that the workings of this algorithm is isomorphic to the action of the following algorithm acting on the single complex input z, after scaling and rotating so that the second input becomes 1.

The reduction step takes as input a complex number $z \in D$ where D is the open disk in $\mathbb{C}$ with diameter $(0,1)$. It returns the complex number $1/z - \lfloor \Re(1/z) \rfloor$. The iteration repeats the reduction step until the output no longer lies in this disk. The set of points within the disk so that the algorithm takes at least $k+1$ steps is the union of the set of disks with diameters of the form (a,b) where (a,b) is a real pair of the form $([\mathbf{v}],[\mathbf{v}+1])$ or $([\mathbf{v}+1],[\mathbf{v}])$, depending on the parity of k, with $\mathbf{v} \in V_{\mathbb{Z}^+}(k)$. Color coding the regions by parity for graphic effect gives the following picture (which B. Vallée has made her personal research logo) of the region where the number of steps is odd (black) or even (white): They prove that the probability p_r

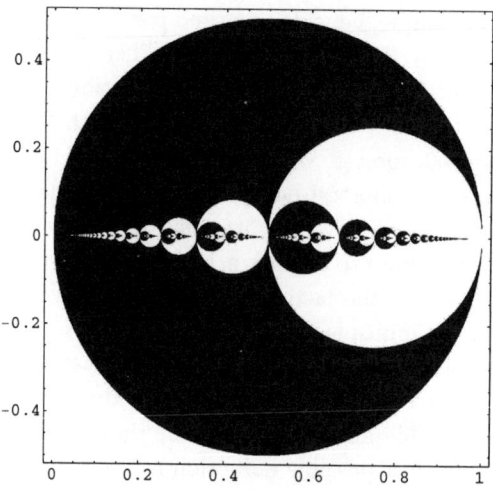

Fig. 2.1   The Gauss lattice reduction algorithm.

that more than r steps will be needed is $\sum_{\mathbf{v} \in V_{\mathbb{Z}^+(r)}} |\mathbf{v}|^{-2}|\mathbf{v}+1|^{-2}$. The rate at which p_r declines (and a look at the figure will persuade the reasonably credulous reader that indeed it does decline exponentially) is thus given by $p_r = O(\lambda^r)$ where λ is the spectral radius of a certain linear operator $G_{4,\mathbb{Z}^+}$. Where is z likely to be after r reduction steps $z \to 1/z - \lfloor \Re(1/z) \rfloor$, given that it has not yet exited the disk? The answer revolves around the dominant eigenfunction g for $G_{4,\mathbb{Z}^+}$. We refer the reader to the excellent survey article [DFV] for further information on this and related topics.

This lattice reduction algorithm ties into another one-dimensional 'continued fraction' algorithm, curiously enough. A rational number p/q is a *Hermite approximation* to α if $|p - q\alpha| + tq^2$ is, for all t in some interval $(a,b) \subseteq \mathbb{R}^+$, minimal. We then term q a *Hermite denominator*. The Hermite denominators of α are continued fraction denominators of α, but not vice-versa. Computing Hermite denominators is computationally more laborious than computing the continued fraction denominators, but not radically so. We discuss this in some detail in Chapter 6, (multidimensional continued fraction algorithms), because it turns out that the Hermite approach can be generalized to the task of simultaneous diophantine approximation, in which we want a single q so that fractions of the form p_j/q afford approximations to each of several targets α_j.

Chapter 3

Continued Fractions with Small Partial Quotients

Harald Niederreiter has written a major paper on this subject. [Ni3]. Both his paper, and this section, have mainly to do with what uses we can make of continued fractions with small partial quotients, and how common such fractions are. Just what 'small' means will vary from case to case. The continued fractions can be integer-based, or fractions with the numerator and denominator taken from polynomials over a finite field, or formal power series.

Given a pair (a, m) of integers with $m \geq 2$ and $\gcd(a, m) = 1$, one can write

$$\frac{a}{m} = a_0 + [a_1, a_2, \ldots a_r] = a_0 + \cfrac{1}{a_1 + \cfrac{1}{a_2 + \cfrac{1}{a_3 + \cfrac{1}{\ddots + 1/a_r}}}}$$

where $a_0 = \lfloor a/m \rfloor$ and where the a_j, $1 \leq j \leq r$ are positive integers. There are two such expansions. If $a/m = a_0 + [a_1, a_2, \ldots a_r]$, and $a_r = 1$, then also $a/m = a_0 + [a_1, a_2, \ldots 1 + a_{r-1}]$. Let $\mathbf{v}[a, m] = (a_1, a_2, \ldots a_r)$ with $a_r = 1$ be the first of these two expansions. We recall our notation that for $\mathbf{v} \in \mathbb{Z}^{+r}$, $\mathbf{v} = (a_1, \ldots, a_r)$ say, $|\mathbf{v}|$ is the denominator of the continued fraction $[a_1, a_2, \ldots a_r]$. By convention, assign to the empty list the 'denominator' 1. We continue to use $\mathbf{v}_- = (a_2, a_3, \ldots a_r)$, $\mathbf{v}^- = (a_1, a_2, \ldots a_{r-1})$, and $\mathbf{v}' = (a_r, a_{r-1}, \ldots a_1)$. By further convention, assign to either decrement of an empty list, the 'denominator' zero.

For $0 \leq \theta \leq 1$, let $\mathbf{v} + \theta = (a_1, a_2, \ldots a_{r-1}, a_r + \theta)$. Let $[\mathbf{v}] = [a_1, a_2, \ldots a_r]$, and let $\{\mathbf{v}\} = [\mathbf{v}'] = [a_r, a_{r-1}, \ldots a_1]$. If $\mathbf{w} = \mathbf{uv}$ is the concatenation of two lists of positive integers, then by prop. 1.1,

$$|\mathbf{w}| = |\mathbf{u}||\mathbf{v}| + |\mathbf{u}^-||\mathbf{v}_-| = (1 + \{\mathbf{u}\}[\mathbf{v}])|\mathbf{u}||\mathbf{v}|.$$

Furthermore, $|\mathbf{v}| = |\mathbf{v}'|$.

Let $K(a,m) = \max\{a_1, a_2, \ldots a_r\}$. Let $K_m = \min K(a,m)$, where the minimum is taken over all $1 \le a < m$ so that $\gcd(a,m) = 1$. Let $S(a,m) = \sum_{1 \le j \le r} a_j$, where $(a_1, a_2, \ldots a_r) = \mathbf{v}[a,m]$. Note that $S(a,m)$ does not depend upon which of the two continued fractions representing a/m are used.

Zaremba [Za] conjectured that for all $m \ge 2$, there exists a relatively prime to m, with $1 \le a < m$, such that $a/m = [a_1, a, 2 \ldots a_r]$ with all partial quotients $a_j \le 5$. The strong version of this conjecture would be that even if we restrict to $a_j \le 2$, all sufficiently large m admit of an a so that a/m has such a continued fraction expansion.

Fractions a/m for which all the partial quotients are small can be useful for numerical integration. The list $Z[a,m] := ((j/m, \{ja/m\}) \colon 0 \le j < m)$ forms part of a lattice in $\mathbb{R}^2$ which is well distributed in the plane. Moreover, in any segment of the list, the second entry is reasonably evenly distributed over the interval $(0,1)$. Fractions of this form are also relevant to the construction of linear-congruential pseudo-random numbers. [Ni4],[Ni5].

But, do such fractions exist? If we get to pick m, asking only that m lie between x and $3x$ say, then the answer is trivially yes. We need only choose a list $\mathbf{v}$ of 1's and 2's at random, continuing until $|\mathbf{v}| > x$, and $|\mathbf{v}|$ will be our m, and $|\mathbf{v}_-|$ our a. We can, with some tinkering, hit a target far narrower than the interval $[x, 3x]$. There are also certain special types of m for which it is known that an appropriate a exists, and better yet, there is a fast and simple algorithm for a. Niederreiter has shown that

Theorem 3.1 *For all positive integers n, $K_{2^n} \le 3$.*

The proof of the theorem stated above rests on two lemmas that allow us to pass from a pair (a, m) to a new pair (a', m^2) or to $(a'', 2m^2)$ while preserving a certain structure in the corresponding list $\mathbf{v}$.

Lemma 3.1 *Let $a/m = [\mathbf{v}] = [a_1, \ldots a_r]$ with $\gcd(a,m) = 1$ and $m \ge 2$ and $v_r \ge 2$. Let $b \ge 2$ be an integer. Let $\mathbf{w}$ be the list*

$$\mathbf{w} = (a_1, \ldots a_{r-1}, -1+a_r, 1, b-1, a_r, a_{r-1}, \ldots a_1) = (\mathbf{v}^-, -1+a_r, 1, b-1, \mathbf{v}')$$

where $\mathbf{v}'$ denotes the reversed list $(a_r, a_{r-1}, \ldots a_1)$. Then $|\mathbf{w}| = b|\mathbf{v}|^2$.

Lemma 3.2 *Let $u = (\mathbf{v}^-, -1+a_r, 1+a_r, \mathbf{v}^{-'})$. Then $|\mathbf{u}| = |\mathbf{v}|^2$.*

Induction, of course, is the method. We defer the proof of the lemmas, and get on with the main business.

Proof. The inductive step takes as established that for some N, and for each n with $N \leq n < 2N$, there exists an odd a with $1 \leq a < 2^n$ such that $\mathbf{v}[a, 2^n]$ begins and ends with 2 or 3, and consists entirely of 1's, 2's and 3's, so that $K(a, 2^n) \leq 3$. The same then holds for the interval $[2N, 4N)$. If $2N \leq k < 4N$ and k is odd, then Lemma 3.1 applies with $b = 2$. The list $\mathbf{w}$ begins and ends with the a_1 of $\mathbf{v}$, which by hypothesis began and ended with entries 2 or 3. All interior entries are still 1 or 2 or 3. If $2N \leq k < 4N$ and k is even, Lemma 3.2 applies and again we get a $\mathbf{u}$ that begins and ends with 2 or 3 and has all interior entries consisting of 1, 2, or 3.

It is unusual in an inductive proof that there is any difficulty whatever in getting started, but on this occasion we must be careful. There is no list of the sort required for the induction, when $n = 5$, though there is a $\mathbf{v}$ of the sort required by the theorem in question. Thus we must first exhibit values of a suitable for the theorem when $m = 2, 4, 8, 16$, or 32, and then take $N = 6$, and exhibit corresponding values of a for $6 \leq n \leq 11$. Here they are: $1/2$, $3/4$, $5/8$, $7/16$, and $25/32$ will serve for the first task, while $23/64$, $47/128$, $95/256$, $223/512$, $367/1024$, and $791/2048$ have $K(a, m) = 2$ with a continued fraction expansion that begins and ends with 2. □

We now turn to establishing Lemma 3.1.

Proof. Assume that $\mathbf{v}$ is a list of length at least 2, ending with $a_r \geq 2$. We then calculate.

$$|\mathbf{w}| = |\mathbf{v}^-, a_r - 1, 1||b - 1, \mathbf{v}'| + |\mathbf{v}^-, a_r - 1||\mathbf{v}'|;$$
$$= (|\mathbf{v}^-, a_r - 1| + |\mathbf{v}^-|)((b-1)|\mathbf{v}| + |\mathbf{v}^-|) + |\mathbf{v}^-, a_r - 1||\mathbf{v}|.$$

Now $\mathbf{v}^- = (\mathbf{v}^{--}, a_{r-1})$, so $|\mathbf{v}^-, a_r - 1| = (a_r - 1)|\mathbf{v}^-| + |\mathbf{v}^{--}|$, and $|\mathbf{v}^-, a_r - 1| + |\mathbf{v}^-| = |\mathbf{v}^-, a_r| = |\mathbf{v}|$. Thus, using our identities,

$$|\mathbf{w}| = |\mathbf{v}^-, a_r - 1, 1||\mathbf{v}, b - 1| + |\mathbf{v}^-, a_r - 1||\mathbf{v}|$$
$$= |\mathbf{v}|((b-1)|\mathbf{v}| + |\mathbf{v}^-|) + ((a_r - 1)|\mathbf{v}^-| + |\mathbf{v}^{--}|)|\mathbf{v}|$$
$$= |\mathbf{v}|((b-1)|\mathbf{v}| + |\mathbf{v}^-| + |\mathbf{v}| - |\mathbf{v}^-|) = b|\mathbf{v}|^2.$$

□

The proof of the second lemma is similar:

Proof. We have

$$|\mathbf{u}| = |\mathbf{v}^-, a_r - 1||\mathbf{v}^-, a_r + 1| + |\mathbf{v}^-|^2$$
$$= ((a_r - 1)|\mathbf{v}^-| + |\mathbf{v}^{--}|)((a_r + 1)|\mathbf{v}^-| + |\mathbf{v}^{--}|) + |\mathbf{v}^-|^2$$
$$= (a_r^2 - 1)|\mathbf{v}^-|^2 + 2a_r|\mathbf{v}^-||\mathbf{v}^{--}| + |\mathbf{v}^{--}|^2 + |\mathbf{v}^-|^2$$
$$= a_r^2|\mathbf{v}^-|^2 + 2a_r|\mathbf{v}^-||\mathbf{v}^{--}| + |\mathbf{v}^{--}|^2 = (a_r|\mathbf{v}^-| + |\mathbf{v}^{--}|)^2 = |\mathbf{v}|^2.$$
$\square$

Niederreiter gives the additional detail that $[\mathbf{w}] = [\mathbf{v}] + \frac{(-1)^{r+1}}{b|\mathbf{v}|^2}$ and that $[\mathbf{u}] = [\mathbf{v}] + \frac{(-1)^{r+1}}{|\mathbf{v}|^2}$. He also proves similar results for the case of $m = 5^n$. Note that the constructions of the two lemmas provide us with the heart of an algorithm for a given $m = 2^n$. If n is small, we use our table. If not, we first work out, going backwards from n, the chain of doublings, and doublings plus 1, that will take us from some number n_0 in the range $[6, 11)$ to n. We then apply Lemma 3.1, with $b = 2$, in case a doubling plus 1 was used, and Lemma 3.2 if it was a doubling alone, to build the original $\mathbf{v}$ associated with n_0, into a suitable list associated with n.

For a general m, nothing so efficient is known, by which an a with minimal $K(a, m)$ may be found, or (a lesser task), by means of which an a with $K(a, m) = c$ may be found if one exists. We now give a method which is, at any rate, significantly better than trying $a = 1$, then $a = 2$, and so on until either success occurs, or we reach $a = m - 1$ without having hit upon a suitable a. This method, which lends itself to algorithmic implementation but is not presented in such detail as to merit calling it an algorithm, also gives us a proof that, for arbitrary integer $c \geq 2$, there are rather more numbers $m \leq x$ than guaranteed by a result of Sander [Sa], for which $K_m \leq c$. We shall require a result from the literature [He2]. Associated with each integer $c \geq 2$ there is a positive real number $\beta[c] > 1/2$ so that $\lim_{r \to \infty} \sum_{\mathbf{v} \in \{1,2,\ldots c\}^r} |\mathbf{v}|^{-2s} = 0$ if and only if $s > \beta[c]$. This $\beta[c]$ is known to be the Hausdorff dimension of a certain continued fraction Cantor set, but that is beside the point at the moment. What is important is that, for $c = 2$, we have $\beta = 0.53128\ldots$, and as $c \to \infty$, $\beta = 1 - \frac{6}{\pi^2}c^{-2} + o(c^{-2})$.

The basic idea behind our approach is that real numbers which are especially near a simple rational number have a continued fraction expansion which tracks that of the nearby rational until that is completed (ending either in 1 or in some larger a_r), and then continues with a large next a_{r+1}. Thus, any a so that a/m lies especially near any simple rational may be skipped over, in the search for an a for which $K(a, m) \leq c$. Since there are many simple rationals, there is a good chance that we can prescreen most of the available a's and save ourselves most of the effort of a direct search.

The pictures show the situation for $c = 2$. Pairs (a, m) each cast a 'shadow', into which no further pairs fall. The largest gap, down the middle of the graphics, is the shadow of the pair $(1, 2)$.

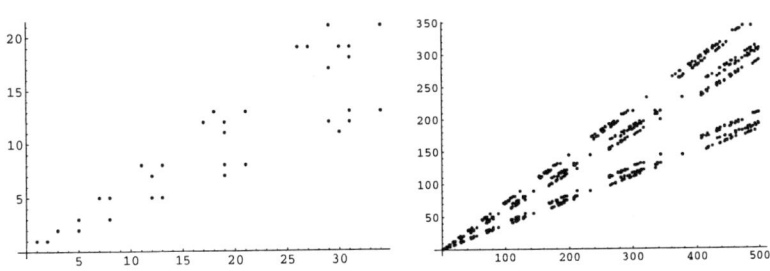

Fig. 3.1   Pairs with partial quotients one and two only.

At first blush, it may seem that this is an idea which, however well it works, can only go to proving an upper bound for the number of $m \leq x$ for which $K_m \leq c$. But there is a twist. For fixed $\epsilon > 0$, the number of $\mathbf{v}$ for which $x \leq |\mathbf{v}| \leq (1 + \epsilon)x$ is known to be comparable to $x^{2\beta[c]}$.[He2] The fewer the number of a potentially associated with a particular m, the greater the number of m for which at least one such a must exist. (If the peanut butter is spread thin, much of the toast must be covered.)

We now turn to the details. Let $V[c]$ denote the set of all finite lists $\mathbf{v} = (a_1, a_2, \ldots, a_r)$, of length $r \geq 1$, for which each entry is no greater than c. Let $\lambda_c = [c, 1, c, 1, \ldots]$, so that $\lambda_c = 1/(c+1/(1+\lambda_c))$. Let $I[\mathbf{v}, c]$ denote the real interval about $[\mathbf{v}]$ with endpoints $[\mathbf{v} + \lambda_c]$ and $[\mathbf{v}^-, a_r - 1, 1 + \lambda_c]$. Which endpoint is the lesser depends on the parity of r.

The intervals corresponding to distinct $\mathbf{v} \in V[c]$ are disjoint, and their union is all of $[0, 1]$ save for a Cantor dust of measure zero. (And Hausdorff dimension $\beta[c]$, though we do not need that fact at the moment.) The length of $I[\mathbf{v}, c]$ is comparable to $1/(c^2|\mathbf{v}|^2)$. The number of integers a so that $a/m \in I[\mathbf{v}, c]$ is thus $O(1 + mc^{-2}|\mathbf{v}|^{-2})$.

Consider, then, a procedure in which first, all $\mathbf{v} \in V[c]$ with $|\mathbf{v}| \leq \sqrt{x/c}$ are listed. There will be on the order of $(x/c)^{\beta[c]}$ of these. The real interval $[0, m]$ is dissected into that many intervals in which no a for which $K(a, m) \leq c$ is to be found, and as many intervals, give or take 1, in which we must search case by case. The aggregate length of the intervals which must be searched individually, is comparable to $\sum_{\mathbf{v}\in V[c],|\mathbf{v}|>\sqrt{x/c}} c^{-2}|\mathbf{v}|^{-2}$. But using [He2], this quantity is comparable to

$$\int_{t=\sqrt{x/c}}^{\infty} c^{-2}t^{-2} \cdot t^{2\beta[c]-1} \, dt = \frac{1}{2-2\beta[c]} \left(\frac{x}{c}\right)^{\beta[c]-1}.$$

For large c, this is comparable to $c^{3-\beta[c]}x^{\beta[c]-1}$.

For $x \leq m \leq (1+\epsilon x)$, then, the number $R[m]$ of a which must be examined individually is thus bounded by

$$R[m] = O\left((x/c)^{\beta[c]} + m \sum_{\mathbf{v}\in V[c],|\mathbf{v}|>\sqrt{x/c}} c^{-2}|\mathbf{v}|^{-2}\right) = O_c\left(x^{\beta[c]}\right).$$

The total number of times we shall have to either calculate a continued fraction expansion of a fraction a/m, or the corresponding fraction for a list $\mathbf{v}$, is thus $O(x^{\beta[c]})$. This improves upon the naive approach of trying cases one by one, from 1 up to m, by a factor of $x^{1-\beta[c]}$, and is thus most effective when c is small. On the other hand, if the Zaremba conjecture is correct, c need never be greater than five.

Our bound on the number of cases we should have to check individually is of course an outright upper bound on the number of $a \leq m$ for which $K(a,m) \leq c$. Since there are, for $x \leq m \leq (1+\epsilon)x$, on the order of $x^{2\beta[c]}$ suitable pairs, and since no single m gives rise to more than $O(x^{\beta[c]})$ such pairs, we have the following result:

Theorem 3.2 *There are at least on the order of $x^{\beta[c]}$ integers m between x and $(1+\epsilon)x$ for which $K_m \leq c$.*

There are enough rationals a/q with continued fraction expansion of the form $[a_1, a_2, \ldots a_r]$ with $r \geq 1$, $q \leq x$ and $a_i \leq 2$ for $1 \leq i \leq r$ to allow for one to each $q \leq x$, for as we shall see in Chapter 9, there are in fact on the order of $x^{1.062561}$ such pairs (a,q) with $x/2 < q < x$. This is leeway sufficient that probabilistic modelling of the question corroborates the stronger conjecture: a process choosing on the order of $x^{1.06256}$ integers, with reasonably uniform distribution, in an interval of length on the order of x will for large x hit every integer at least once.

The actual distribution of these points may have irregularities or patterns, but at any rate it does not favor particular congruence class pairs. That is, for any modulus m, and any a', b' so that gcd[gcd[a', m], gcd[b', m]) = 1, one has with asymptotically equal frequency $(a \equiv a' \mod m, b \equiv b' \mod m)$.

The evidence in favor of this stronger conjecture is on its face not overwhelming. In the interval $1 \leq m \leq 1000$, the minimum, over $1 \leq a < m$ and relatively prime to m, of the maximum partial quotient in that continued fraction expansion of a/m not ending in one, is more often 3 or 4, than 2, while for $m = 6$, 54 and 150, it is five. Sampling among larger values of m indicates that we cannot expect 'sufficiently large' in this conjecture, to be at all small.

3.1 The Sequence ($\{n\alpha\}$) of Multiples of a Number

There are several results in the literature on distribution of sequences which involve continued fractions, or throw light on continued fractions.

Here we are concerned with particular sequences of the form

$$(n\alpha - \lfloor n\alpha \rfloor) = (\{n\alpha\}), n \geq 1.$$

Without meaningful loss of generality we require $0 < \alpha < 1/2$. When $\alpha = a/b$ is rational with a and b relatively prime, $(\{n\alpha\})$ is periodic with period b, and the sequence cycles through a permutation of $(1/b)(0, 1, \ldots, b-1)$.

When α is irrational, the sequence $(\{n\alpha\}), n \geq 1$ is dense in $(0, 1)$ and asymptotically uniformly distributed. Of course, for any finite N the list $(\{n\alpha\}), 1 \leq n \leq N$ is discrete and the distribution is necessarily somewhat lumpy. One of the basic issues in the theory of distribution of sequences is to determine the extent to which sequences can approximate a uniform distribution. If N is fixed, then the list $(0, 1/N, 2/N, \ldots, (N-1)/N)$, or any translate thereof, affords a trivial best possible approximation to uniform distribution. But what one really wants of a sequence (x_j) is that it provides 'streaming' uniform distribution: any initial segment $(x_1, \ldots, x_N)$ should give not too far from best-possible discrepancy. The best that is possible along these lines is that the discrepancy, as a function of N, be $O(\log N/N)$. [KuN]. Taking α to be an irrational number with bounded partial quotients gives a sequence $(x_j) = \{j\alpha\}$ that achieves this best-possible streaming nearness to uniform distribution.

3.1.1 The sequence of increments of $\lfloor n\alpha \rfloor$

For any positive irrational α, consider the sequence $s_k(\alpha) = \lfloor (k+1)\alpha \rfloor - \lfloor k\alpha \rfloor$, $(k \geq 1)$. This sequence is a string of $\lfloor \alpha \rfloor$'s and $\lceil \alpha \rceil$'s, and so is effectively a bit sequence. The sequence is highly ordered, and almost periodic in this sense: most finite subsequences of this sequence occur infinitely often in the sequence, and with positive density. There is a sequence of integers $j_k \geq 2$, and pairs of *words*, that is, finite strings of $\lfloor \alpha \rfloor$'s and $\lceil \alpha \rceil$'s, (A_k, B_k), so that beyond an initial string W_k of $\lfloor \alpha \rfloor$'s and $\lceil \alpha \rceil$'s with length comparable to either of the words, the rest of the sequence consists of concatenations of one or the other of these words, strung together so that between each word B_k there are either $j_k - 1$, or j_k, consecutive copies of A_k. The pattern of the A_k and B_k and j_k, in that part of the sequence $s_n(\alpha)$ beyond its initial segment W_k, is itself an exact copy of the pattern of a sequence $s_n(\alpha_k)$, but with words A_k and B_k (in one order or the other) in place of individual integers $\lfloor \alpha_k \rfloor$'s and $\lceil \alpha_k \rceil$'s. The α_k come out of the centered continued fraction expansion of α.

The centered continued fraction expansion of a number α writes

$$\alpha = a_0 \pm \cfrac{1}{a_1 \pm \cfrac{1}{a_2 \pm \cfrac{1}{a_3 \pm \dots}}},$$

where each a_k is a positive integer other than 1. The expansion is calculated by means of the following modification of the standard continued fraction expansion:

> Input α, and a depth n to which the calculation is to be taken. Local variables C, k, and α'. Initialize the list C with the single entry $\lfloor \alpha + \tfrac{1}{2} \rfloor$, and set $k = 0$ and $\alpha' = \alpha$. While $k \leq n$, execute the Loop subroutine with input α':
>
> > Loop Step: Given input α', let $\sigma = 1$ if $\{\alpha'\} < 1/2$, else $\sigma = -1$. Then reset α' with the update rule
> >
> > $$\alpha' \leftarrow \frac{1}{\alpha' - \lfloor \alpha' + \tfrac{1}{2} \rfloor}.$$
> >
> > Increment k. Append to C the (new) pair (σ, α'). Return$[(\sigma, \alpha')]$.
>
> Output C.

The resulting list begins with a single integer a_0, and continues with pairs of the form (σ_k, a_{k+1}); these give the continued fraction expansion described, using σ_k to determine the sign of the $\pm$ after a_k.

The conventional continued fraction expansion of a number given in the centered continued fraction may be easily obtained by a simple substitution algorithm:

> Input the sequences (a_k) and (σ_k) to depth n. Pass through this list, and in each case in which $\sigma_k = -1$, decrement a_k, and replace a_{k+1} with 1 followed by $a_{k+1} - 1$. Go to the next case in which $\sigma = -1$ until the entire list has been processed. Output the modified list of a's.

Note that this substitution procedure can produce a 0 in the output list if it is given consecutive -2's in the input. However, consecutive -2's do not occur in the output of the centered continued fraction expansion.

We now describe the connection between $(n\alpha)$, $s_n(\alpha)$, and the centered continued fraction expansion of α, when α is irrational. The sequence $(s_n(\alpha))$ has the basic property that there are two entries: $\lfloor \alpha \rfloor$ and $1 + \lfloor \alpha \rfloor$. The pattern of these is then a mix of strings of the more frequent entry, call it A_0, punctuated by single instances of the less frequent entry, B_0.

We define a *word* to be a finite list of entries A_0 and B_0. Given two words A, B, and a positive integer j, let $A^j B$ denote the word got by concatenation of j copies of A followed by a single B.

We associate with α two auxiliary sequences of irrational numbers and two auxiliary sequence of integers. Let $\alpha_0 = \alpha$. For $k \geq 0$, let β_k be the unsigned distance from α_k to the nearest integer. Let $j_k = \lfloor 1/\beta_k \rfloor$, and let $\sigma_k = 0$ if $\{1\beta_k\} < 1/2$, else $\sigma_k = 1$. The connection to continued fractions is that the α_k in this algorithm are the same as the α_k in the centered continued fraction algorithm. The following result cannot be new, but the author has not been able to determine the original discoverer.

Theorem 3.3 *If α is an irrational real number, and $s_n(\alpha) = \lfloor (n+1)\alpha \rfloor - \lfloor n\alpha \rfloor$, then there are sequences $(A_k), (B_k)$ and (C_k) of words, and sequences (α_k) and (β_k) of irrational numbers and (j_k) of positive integers, such that*

(1) *If $\{\alpha\} < 1/2$, then $A_0 = \lfloor \alpha \rfloor$ and $B_0 = 1 + A_0$, else $B_0 = \lfloor \alpha \rfloor$ and $A_0 = 1 + B_0$. C_{-1} is the empty word, and $C_0 = A_0^{j_0-1} B_0$.*

(2) For $k \geq 1$, $A_k = A_{k-1}^{j_{k-1}-1+\sigma_{k-1}} B_{k-1}$, $B_k = A_{k-1}^{j_{k-1}-\sigma_{k-1}} B_{k-1}$, and $C_k = C_{k-1} A_k^{j_k-1} B_k$.

(3) For each $k \geq 0$, the sequence $s_n(\alpha)$ consists of an initial word C_{k-1} followed by a string of words A_k or B_k, and the rth such A or B is in fact A_k, if and only if $s_r(\beta_k) = 0$; it is B_k if and only if $s_r(\beta_k) = 1$.

Proof. We first establish a lemma.

Lemma 3.3 *Suppose $0 < \beta < 1/2$ is irrational and let γ be the unsigned distance from $1/\beta$ to the nearest integer. Let $j = \lfloor 1/\beta \rfloor$. Then*

(1) *The first j entries of $s_n(\beta)$ are the word C consisting of $n - 1$ 0's followed by a single 1;*

(2) *If $0 < \{1/\beta\} < 1/2$ then the rest of $s_n(\beta)$ after the initial word C consists of strings of the word $A = 0^{j-1}1$ broken by single instances of the word $B = 0^j 1$, and the rth such A or B is A if $s_r(\gamma) = 0$, B if $s_r(\gamma) = 1$; and*

(3) *If $1/2 < \{1/\beta\} < 1$ then $A = 0^j 1$ and $B = 0^{j-1} 1$ but the rest is the same.*

Proof. For $1 \leq n < j$, $(n+1)\beta < 1$ so $s_n(\beta) = 0$. For $n = j$, then $j\beta < 1 < (j+1)\beta$ so $s_j(\beta) = 1$. Thus $(s_n(\beta))$ begins with the string $0^{j-1}1 = (0, 0, \ldots 0, 1)$ as claimed. Now there are two cases. If $\{1/\beta\} < 1/2$, then $\gamma = \{1/\beta\}$. Consider the sequence (n_k) of successive positive integers at which $s_{n_k}(\beta) = 1$. Observe that

$$s_n(\beta) = 1 \Leftrightarrow 1 - \beta < \lfloor n\beta \rfloor < 1.$$

Let

$$t_k = \{n_k \beta\} \in (1-\beta, 1), u_k = \frac{1}{\beta}(1 - t_k), \text{ and let } v_k = \{k\gamma\}.$$

Then we claim $u_k = v_k$ for all $k \geq 1$.

If $k = 1$ then $t_1 = j\beta$, $u_1 = \frac{1}{\beta}(1 - j\beta) = \gamma$, and $v_1 = \gamma$. Now suppose $u_k = v_k$ for some $k \geq 1$ and consider t_{k+1}, u_{k+1}, and v_{k+1}. We have

$$1 - \beta < n_k \beta - \lfloor n_k \beta \rfloor < 1,$$
$$0 \leq (n_k + r)\beta - (1 + \lfloor n_k \beta \rfloor) < 1 - \beta \text{ for } 1 \leq r < n_{k+1} - n_k, \text{ and}$$
$$1 - \beta < n_{k+1}\beta - (1 + \lfloor n_k \beta \rfloor) < 1.$$

Thus

$$n_{k+1} - n_k = j \text{ if and only if } 1 - \beta < (j\beta - 1) + t_k < 1,$$

or equivalently,

$$n_{k+1} - n_k = j \Leftrightarrow -\beta < (j\beta - 1) + (t_k - 1) < 0$$
$$n_{k+1} - n_k = j \Leftrightarrow 0 < u_k + \gamma < 1$$
$$n_{k+1} = n_k + j \Leftrightarrow 0 < u_k < 1 - \gamma$$

so that in this case,

$$t_{k+1} = n_{k+1}\beta - \lfloor n_{k+1}\beta \rfloor = t_k + (j\beta - 1)$$

and

$$u_{k+1} = u_k + \gamma.$$

Now

$$v_k = u_k < 1 - \gamma \Rightarrow v_{k+1} = v_k + \gamma,$$

which proves that $v_{k+1} = u_{k+1}$.

On the other hand, $n_{k+1} - n_k = j+1$ if and only if $(j\beta-1)+t_k < 1-\beta$, so that in this case $u_k > 1 - \gamma$. But then

$$t_{k+1} = n_k\beta - \lfloor n_k\beta \rfloor + (j+1)\beta - 1 = t_k + (j+1)\beta - 1$$

so that

$$u_{k+1} = u_k + \frac{1}{\beta} - (j+1) = u_k + \gamma - 1.$$

Since $u_k = v_k > 1 - \gamma$, $v_{k+1} = v_k + \gamma - 1 = u_k + \gamma - 1 = u_{k+1}$. Thus again, $v_{k+1} = u_{k+1}$. It now follows that $n_{k+1} = n_k + j$ if and only if $s_k(\gamma) = 0$, while $n_{k+1} = n_k + j + 1$ if and only if $s_k(\gamma) = 1$. As the kth string of either $0^{j-1}1$ or 0^j1 after the initial string in $(s_n(\beta))$ is $0^{j-1}1$ exactly when $n_{k+1} = j + n_k$, this completes the proof of the lemma for the case $\{1/\beta\} < 1/2$. When $\{1/\beta\} > 1/2$ the proof is similar. We use the same n_k's and t_k's. But this time, $\gamma - (j+1) - 1/\beta$, and $u_k = \frac{1}{\beta}(t_k - (1-\beta))$. Thus $u_1 = \frac{1}{\beta}(l_1 - (1-\beta)) = \gamma$, and u_k moves by steps of γ or $\gamma - 1$. Again $u_k = v_k$ for all k, but the consequence for $s_n(\beta)$ is that the nth word after the initial $0^{j-1}1$ is 0^j1 when $s_n(\gamma) = 0$ and is $0^{j-1}1$ when $s_n(\gamma) = 1$. $\square$

We now turn to completing the proof of the theorem. Recursive use of the lemma gives us that

$$C_k = C_{k-1} A_k^{j_k - 1} B_k$$

where the words A_k and B_k are determined by reference to α_k. $\square$

Example. If $\alpha = \sqrt{2}$ then for $k \geq 1$, $\alpha_k = \sqrt{2}+1$ while for $k \geq 0$, $\beta_k = \sqrt{2}-1$. Thus $C_0 = (1,2)$, $A_0 = (1,2)$ and $B_0 = (1,1,2)$, so that $C_1 = (1,2,1,2,1,1,2)$, $A_1 = A_0 B_0 = (1,2,1,1,2)$, $B_1 = A_0^2 B_0 = (1,2,1,2,1,1,2)$, and $C_2 = C_1 A_1 B_1$ and so on.

This particular sequence has come in for a lot of attention. It is also generated by the replacement rules $1 \leftarrow 12$, $2 \leftarrow 121$, starting with the list (1). The list is featured as sequence number A006337 in Sloane's Encyclopedia of Integer Sequences:

http://www.research.att.com/~njas/sequences

Let $\alpha > 0$ be given. We want a good way to determine long initial segments of the sequence $(\lfloor (n+1)\alpha \rfloor - \lfloor n\alpha \rfloor)$. The basic algorithm we now describe requires irrational inputs, so that the dichotomy $0 < \{\beta\} < 1/2$ or $1/2 < \{\beta\} < 1$ will not be violated during execution. If α is rational, however, the sequence is purely periodic with length equal to the denominator of α. Thus it is only necessary to determine the initial sequence of some sufficiently close but larger α^* to appropriate depth. One workable and simple choice if $\alpha = a/b$ is $(ab+1)/b^2$.

Input α, and a depth d to which the calculation is to be taken. Working variables are A, B, C, j, β, d', and α'. Let $A_0 = \lfloor \alpha \rfloor$ and $B_0 = 1 + \lfloor \alpha \rfloor$ if $\{\alpha\} < 1/2$, while if $\geq 1/2$, set $A_0 = 1 + \lfloor \alpha \rfloor$ and $B_0 = \{\alpha\}$.

If α is an integer then $s_n(\alpha)$ is constant at that integer. Exit the algorithm and return a message to that effect. If α is rational and equal to a/b with a and b relatively prime and $b > 1$, let $\alpha^* = (ab+1)/b^2$ and let $\beta = \alpha^* - \lfloor \alpha^* + 1/2 \rfloor$. Otherwise α is irrational; let $\beta = \alpha - \lfloor \alpha + 1/2 \rfloor$. Now let $\alpha' = 1/\beta$. Let $j = \lfloor 1/\beta \rfloor$. Let $C = A_0^{j-1} B_0$ be the finite list in which the first $j-1$ entries are A_0 while the last entry is B_0. Let $d' = 0$. While $d' < d$, (and while Length$[C] < b$ if $\alpha = a/b$ is rational), execute the following loop:

> Let $\beta = \alpha' - \lfloor \alpha' + 1/2 \rfloor$. Let $j = \lfloor 1/\beta \rfloor$. Update A and B to $A^{j-1}B$ and $A^j B$ respectively if $\{1/\beta\} \leq 1/2$, or to $A^j B$ and $A^{j-1} B$ if $\{1/\beta\} > 1/2$. Update C to $CA^{j-1}B$. Update α' to $1/\beta$. Increment d'.

Output C, α', A, and B.

Remark 3.1 *There is a recreational mathematics side to this. Larry Tessler has given a fetching metaphor: "A tank rolls forever along a perfectly straight infinite highway whose center line is studded with reflectors spaced exactly 1 meter apart. The center of the tread rides the highway's center line and leaves marks on the pavement that are spaced exactly r meters apart, where r is an irrational number. At the beginning of the trip, the initial mark falls directly on a reflector." For the rest of this story, see*

http://www.nikora.com/puzzles/
HowDoTheTankTreadsRollOnTheReflectors.htm#Answers

3.2 Discrepancy

There are various kinds of discrepancy, but in the one dimensional case the distinctions are minor. The definition we choose is this: the *discrepancy* $D_N^*(x)$ of a sequence $x = (x_1, x_2, \ldots)$ of real numbers in $[0, 1)$ is

$$D_N^*(x) = \max_{0 \leq t \leq 1} \left| N^{-1} \#\{k : 1 \leq k \leq N \text{ and } 0 \leq x_k < t\} - t \right|.$$

Clearly $D_N^*(x)$ is invariant under permutations of $[1, 2, \ldots, N]$, and is between $1/N$ and 1. But, no sequence consistently yields this lower bound for all N. In fact, there exists $C > 0$ ($C = 0.04667$ will do, according to theorem 1.51 of [DT]) so that for any sequence, $D_N^*(x)$ exceeds $C \log N/N$ infinitely often.

For sequences of the special form $x_n := n\alpha \mod 1$ the continued fraction expansion of α gives reasonably complete information about the discrepancy. Suppose a positive integer N is given. Let m be the largest integer so that $q_m \leq N$, and choose further positive integers b_j using the greedy algorithm: take $b_m = \lfloor N/q_m \rfloor$, $N_m = N$, and then once $b_m, \ldots b_{j+1}$ and $N_m, \ldots, N_j$ have been determined, set $b_j = \lfloor N_j/q_j \rfloor$ and $N_{j-1} = N_j - b_j q_j$. With this algorithm we have $N = \sum_{j=0}^m b_j q_j$, and $0 \leq b_j \leq a_{j+1}$. The maximum value of $N' D_{N'}^*(\alpha n)$, over $1 \leq N' \leq N$, is known to be comparable to $b_m + \sum_0^m a_j$.[DT] (Corollary 1.64 page 52.)

If a number α has only small partial quotients, as we have seen, then the discrepancy of $(k\alpha), 1 \leq \alpha \leq n$ is about as small as we could hope for. If $\alpha = p/q$ is rational, then this will hold so long as $n \leq q$.

It should be noted that there are other sequences that give low discrepancy, and that in higher dimensions, generalizing the tactic of picking a number and looking at the fractional part of its successive integer multiples, is not the best approach. Instead, one wants things called Halton or Hammersley sequences, or, a more recent development, Niederreiter's (t, m, s) nets. See [En] for starters and some great eye candy, and then see [Ni6]. The field is an active one, and has various practical applications, so much so that there is even a demand for commercial software implementing these ideas: http://www.mathdirect.com/.

3.3 The Sum of $\{n\alpha\}$ from 1 to N

This, like the discrepancy, depends essentially on the continued fraction expansion of α to that depth so that $q_m \leq N < q_{m+1}$. The basic result is given in [DT], who in turn credit Schoißengeier. [S] Let N_j and b_j be defined as before, so that $N = \sum_0^m b_j N_j$. By convention, set $N_{-1} = 0$. Next, let

$$A_j = N_{j-1}(\alpha - \frac{p_j}{q_j}) + \sum_{t=j}^{m} b_t(q_t\alpha - p_t)$$

for $0 \leq j \leq m$. Then (p 63)

$$2\sum_{n=1}^{N}\{n\alpha\} - N = A_0 - \sum_{j=0}^{m} b_j((-1)^j - q_j A_j).$$

In connection with the estimate (p 50)

$$q_j A_j = b_j(-1)^j a_{j+1} + O(1/a_{j+1}),$$

this allows us to conclude immediately that if α has bounded partial quotients a_j, then $2\sum_{n=1}^{N}\{n\alpha\} - N = O(\log N)$, a result which also follows from the bounds on the discrepancy of the sequence.

We close with some illustrations of these sums. There are strong hints of a fractal structure to these sums.

The last shows the sequence of partial sums of a discretized version of this last sum: $\sum_{k=1}^{n} \text{sign}(\{k\sqrt{2}\} - 1/2)$:

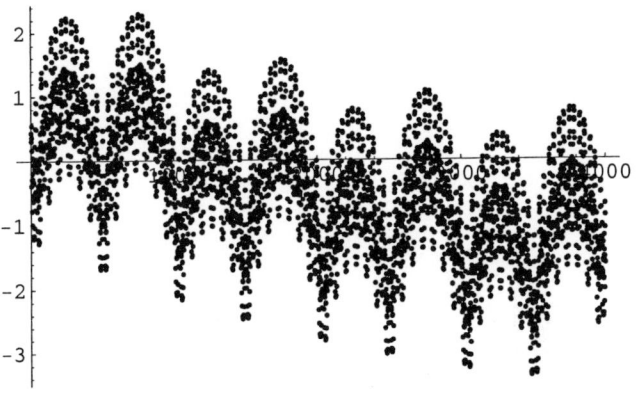

Fig. 3.2 Partial sums of $\{ne\}$ to 4000.

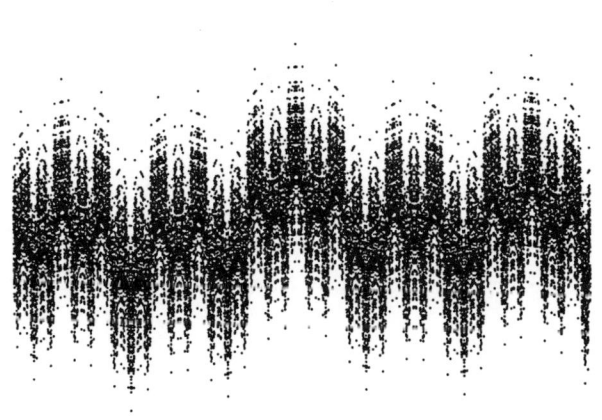

Fig. 3.3 Partial sums of $\{n\sqrt{2}\}$ to 34000.

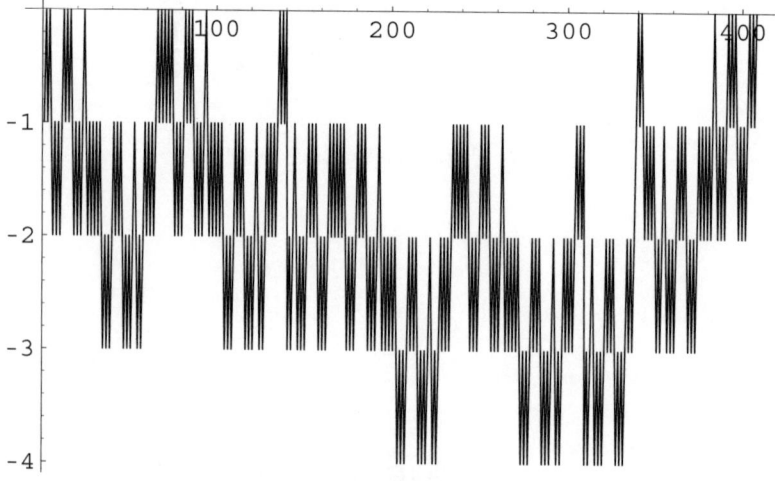

Fig. 3.4 Running total of parity of $\{k\sqrt{2} - 1/2\}$.

Chapter 4

Ergodic Theory

4.1 Ergodic Maps

Ergodic theory is based on the intuitive idea that a process which shuffles points around may well cause most points to have an orbit that visits every neighborhood of the same size equally often. More exactly, suppose $(X, \mathcal{B}, \mu)$ is a probability space. The points are the elements of X, $\mathcal{B}$ is the σ-algebra set of measurable subsets of X, and μ is the function from $\mathcal{B}$ to $[0, 1]$ that assigns measure to measurable sets. An excellent reference is the recent book by Dajani and Kraiikamp [DK2] *Ergodic Theory of Numbers*.

There is a considerable literature connecting ergodic theory to the metric theory of continued fractions. Here, we consider one particular theorem in this vein, and extend it to the case of continued fractions with restricted partial quotients.

Consider the distribution of the normalized approximation error $\theta_n(x) = q_n|p_n - xq_n|$, where p_n and q_n are the numerator and denominator respectively of the nth continued fraction convergent to x. That is, if we make a scatterplot of $\theta_n(x)$ for $1 \leq n \leq N$, what will it look like?

This varies with N and x. For $\phi = (\sqrt{5} - 1)/2$, $\theta_n(\phi) \to \phi/(1 + \phi^2)$, while $\theta_n(e)$ gives a scatterplot approaching two point masses: two-thirds of the values lie near $1/2$, while the rest lie near 0. But these distributions are not typical.

H. Lenstra thought about the normal distribution of $\theta_n(x)$ and conjectured the answer. The theorem of Bosma, Jager and Wiedijk [BJW] consists of the affirmation of this conjecture:

Theorem 4.1 *Let*

$$f(z) = \begin{cases} z/\log 2, & \text{if } 0 \le z \le 1/2; \\ \frac{1}{\log 2}(1 - z + \log(2z)), & \text{if } 1/2 \le z \le 1; \\ 1, & \text{if } z > 1. \end{cases}$$

Then for almost all $x \in [0,1]$,

$$\lim_{N \to \infty} \frac{1}{N} \# \left\{1 \le n \le N \colon \theta_n(x) \le z\right\} = f(z).$$

In [Na], along with a number of other theorems, Nair extends this to show that for a certain class of increasing sequences (n_j), one has a.e. that

$$\lim_{N \to \infty} \frac{1}{N} \# \left\{1 \le j \le N \colon \theta_{n_j}(x) \le z\right\} = f(z).$$

In [He8] the restriction on the sequences is removed; any strictly increasing sequence will do.

Here, we give Nair's proof as it applies to the basic sequence $(n_j = j)$ and show how it can be extended in another direction to give the distribution of $\theta_n(x)$ when x is taken from the set E_M consisting of all numbers in $[0,1]$ with infinite continued fractions for which all partial quotients come from M.

There is one catch: not just any set M will do. There are certain sets M for which the analysis falls through. We defer the discussion of all this for later, and proceed now with setting the stage for Nair's proof as it applies to the simplest case, Lenstra's original conjecture.

4.2 Terminology

A *dynamical system* consists of a set X, a σ-algebra $\mathcal{B}$ of subsets of X, a probability measure μ on $(X, \mathcal{B})$, and a map $T \colon X \to X$ such that T is measurable and measure-preserving with respect to μ. That is, $\mu(T^{-1}A) = \mu(A)$ for all $A \in \mathcal{B}$.

The map T is called *ergodic* if T has no invariant measurable subsets of measure other than 0 or 1, that is, T has the property that if A is a measurable set and $TA \subset A$ then $\mu(A) = 0$ or $\mu(A) = 1$. A condition equivalent to ergodicity is that if f is an integrable function from X to $\mathbb{R}$ so that $f(Tx) = f(x)$ μ a.e. then f is a.e. equal to a constant.

A stronger condition which implies ergodicity is *mixing*: T is mixing, or 2-mixing, if for any measurable sets A and B,

$$\lim_{n\to\infty} \mu(A \cap T^{-n}B) = \mu(A)\mu(B).$$

The transformation $Tx = x + \sqrt{2}$ (mod 1) on the unit interval equipped with the usual measure, is ergodic but not mixing.

Just about as useful in practice is *weak mixing*: T is weak mixing if for all measurable sets A and B,

$$\lim_{r\to\infty} \frac{1}{r} \sum_{j=1}^{r} \left|\mu(T^{-j}B \cap A) - \mu(A)\mu(B)\right| = 0.$$

Even weak mixing is stronger than mere ergodicity. For a discussion of the nuances of these distinctions, and some related theorems, see [Hl].

To study the continued fraction transformation $T\colon x \to 1/x - \lfloor 1/x \rfloor$ on the measure space $(0,1)$ we need the right measure; T is not measure preserving with respect to Lebesgue measure. As we might have expected, though, it is measure preserving with respect to the Gauss measure μ which assigns to $(0,x)$ mass $\int_0^x \frac{dt}{(1+t)\log 2}$. The verification of this is left to the reader.

Nair's proof, in the special case of the basic BJW theorem, depends upon two main points. The first point is a general one about sequences and dynamical systems. A sequence (n_j) of nonnegative integers is termed L^p-*good universal* if it satisfies certain technical properties for some $p \geq 1$. He obtains a theorem to this effect: If, moreover, the sequence satisfies the further condition that for all irrational numbers α, $(n_j\alpha)$ is uniformly distributed mod 1, and if $(X, \mathcal{B}, \mu, T)$ is weak mixing, then a.e. on X with respect to μ,

$$\lim_{N\to\infty} \sum_{j=1}^{N} f(T^{n_j}x) - \int_X f\, d\mu.$$

In the case we are interested in here, $n_j = j$ for all n, and the premises of Nair's good-universal theorem are satisfied. But here, we can obtain the conclusion of his theorem from scratch, as it were. The *individual ergodic theorem*[Hl] says that if T is ergodic on X and if $f \in L^1(X)$ then the same conclusion as above holds.

The second point is that taking $X = [0,1]$ and T to be the usual operator $Tx = 1/x - \lfloor 1/x \rfloor$ is not the optimal choice of dynamical system for this game. Instead, we should consider the related dynamical system in which $X = \Omega = ([0,1]\backslash\mathbb{Q}) \times [0,1]$, and $\mathcal{T}: \Omega \to \Omega$ is given by

$$\mathcal{T}(x,y) = \left(\frac{1}{x} - \left\lfloor\frac{1}{x}\right\rfloor, \frac{1}{\lfloor\frac{1}{x}\rfloor + y}\right).$$

Then we take ω to be the measure on Ω with density $\frac{1}{\log 2}\frac{1}{(1+xy)^2}$. We take $\mathcal{B}$ to be the Lebesgue measurable sets of Ω. This $\mathcal{T}$ is one-to-one on Ω, and is *weak mixing*. The first coordinate of $\mathcal{T}^n(x,0)$ is $T^n x$. We omit the proof that $(\Omega, \mathcal{B}, \mathcal{T}, \omega)$ is weak mixing, as we shall later be showing, in greater generality, that $(\mathcal{T}, \omega_M)$ is outright mixing, whenever M is a regular set (defined in section 4.4) and ω_M is the measure appropriate to M.

4.3 Nair's Proof

Let $\Omega_c = \{(s,t) \in \Omega: \frac{1}{s} + t \geq c\}$. Note that $\omega(\Omega(1/z)) = f(z)$. Let

$$\theta_n(x) = \frac{1}{\left(\frac{1}{T^n x} + \frac{q_{n-1}}{q_n}\right)},$$

so $\theta_n(x) \leq z$ if and only if $\mathcal{T}^n(x,0) \in \Omega(1/z)$.

Now, fix $\epsilon > 0$. For n sufficiently large, we have

$$y \in [0,1], \mathcal{T}^n(x,y) \in \Omega\left(\tfrac{1}{z} + \epsilon\right) \Rightarrow \mathcal{T}^n(x,0) \in \Omega(1/z), \quad \text{and}$$
$$\mathcal{T}^n(x,0) \in \Omega(1/z) \Rightarrow \forall y \in [0,1], \mathcal{T}^n(x,y) \in \Omega\left(\tfrac{1}{z} - \epsilon\right).$$

Thus, a.e in Ω, we have

$$\lim_{N\to\infty} \frac{1}{N}\#\left\{n: 1 \leq n \leq N, \mathcal{T}^n(x,y) \in \Omega\left(\tfrac{1}{z}+\epsilon\right)\right\}$$
$$\leq \lim_{N\to\infty} \frac{1}{N}\#\{n: 1\leq n \leq N \text{ and } \mathcal{T}^n(x,0) \in \Omega(1/z)\}$$
$$\leq \overline{\lim}_{N\to\infty} \frac{1}{N}\#\{1\leq n \leq N: \mathcal{T}^n(x,0) \in \Omega(1/z)\}$$
$$\leq \lim_{N\to\infty} \frac{1}{N}\#\left\{1\leq n \leq N: \mathcal{T}^n(x,y) \in \Omega\left(\tfrac{1}{z}-\epsilon\right)\right\}.$$

But by the individual ergodic theorem, the first and last limits here exist for almost all $(x,y) \in \Omega$, and are equal to $\omega(\Omega(1/z+\epsilon))$ and $\omega(\Omega(1/z-\epsilon))$, respectively. This completes the proof.

4.4 Generalization to E_M

We now show how all this can be extended to the case in which the choices of x are restricted to x taken from some E_M. To make sense of this, we must have a suitable measure on E_M, and to define that measure, we need M to be *regular*.

Consider

$$\lambda_M(s) = \lim_{r \to \infty} (\sum_{\mathbf{v} \in M^{(2^r)}} |\mathbf{v}|^{-s})^{1/2^r}.$$

Since for lists $\mathbf{u}$ and $\mathbf{v}$ of positive integers, $|\mathbf{uv}| = |\mathbf{u}||\mathbf{v}|(1 + \{\mathbf{u}\}[\mathbf{v}])$, it follows that $(\sum_{\mathbf{v} \in M^{(2^r)}} |\mathbf{v}|^{-s})^{1/2^r}$ is decreasing in r, so if $\sum_{k \in M} k^{-s}$ is finite, the limit exists. (If, on the other hand, $\sum_{k \in M} k^{-s} = \infty$, then $\lambda_M(s) = \infty$ as well.) A set M is *regular* if there exists an $s > 0$ such that $\lambda_M(s) = 1$. Since for $\mathbf{u} \in M^r$, $|\mathbf{u}| \gg ((1 + \sqrt{5})/2)^r$, it follows that if $\lambda_M(s) < \infty$ and $\epsilon > 0$, then $\lambda_M(s + \epsilon) \leq ((1 + \sqrt{5})/2)^{-\epsilon} \lambda_M(s)$. Thus, there can be at most one s such that $\lambda_M(s) = 1$. When $M = \mathbb{Z}^+$, this value of s is $s = 2$, and it is never more than 2.

If M is empty or has just one element, then it is not regular because $\lambda_M(s) < 1$ for all $s > 0$. If M is finite and has at least two elements, it is regular. $\mathbb{Z}^+$ is regular. But there do exist (infinite) subsets of $\mathbb{Z}^+$ that are not regular. One such set is the set of integers that occur in the sequence (a_n), where a_n is the integer nearest $1 + 100n \log^2 n$. What happens in such cases is that $\lambda_M(s)$ leaps from below 1 straight to ∞ as s slides below some critical value.

Whenever $\lambda_M(s)$ is finite, there exists a positive, decreasing function $g_{M,s}(t)$, with $g_{M,s}(0) = 1$, such that $\sum_{k \in M} (k+t)^{-s} g_{M,s}(t) = \lambda_M(s) g_{M,s}(t)$ for $0 \leq t \leq 1$ (see Chapter 9). That is, $g_{M,s}$ is the eigenfunction corresponding to the leading eigenvector $\lambda_{M,s}$ of the operator $H_{M,s}$ taking a function f defined on $[0,1]$ to $\sum_{k \in M} (k+t)^{-s} f(1/(k+t))$.

So suppose M is regular, and let β_M be the unique positive real number s such that $\lambda_M(s) = 1$. Let g_M be the eigenfunction corresponding to this dominant eigenvalue 1 for the linear operator $H_M = H_{M,\beta}$ given by $H_M f = \sum_{k \in M} (k+t)^{-\beta} f(1/(k+t))$, normalized by the condition that $g_M(0) = 1$. Let $\mu = \mu_M$ be the probability measure on $[0,1]$, and supported on E_M, determined by the condition that for all $r \geq 0$ and $\mathbf{v} \in M^r$, if $I_\mathbf{v} = \{[\mathbf{v} + t] : 0 \leq t \leq 1\}$, then

$$\mu(I_\mathbf{v}) = |\mathbf{v}|^{-\beta_M} g_M(\{\mathbf{v}\}).$$

This collection of requirements involves some overlapping conditions, for instance, that $I_\mathbf{v} = \cup_{k \in M} I_{\mathbf{v}k}$. Thus, we must check that they are consistent. They are, because

$$\sum_{k \in M} \mu(I_{\mathbf{v}k}) = \sum_{k \in M} |\mathbf{v}k|^{-\beta} g(\{\mathbf{v}k\}) = |\mathbf{v}|^{-\beta} \sum_{k \in M} (k + \{\mathbf{v}\})^{-\beta} g(1/(k + \{\mathbf{v}\}))$$
$$= |\mathbf{v}|^{-\beta} g(\{\mathbf{v}\}) = \mu(I_\mathbf{v}).$$

Now $g(0) = 1$, so $\mu([0,1]) = 1$. We now show that μ is supported on E_M. Consider an open interval J disjoint from E_M. There exists r such that $\cup_{\mathbf{v} \in M^r} I_\mathbf{v}$ is also disjoint from J, while

$$\mu\left(\cup_{\mathbf{v} \in M^r} I_\mathbf{v}\right) = \sum_{\mathbf{v} \in M^r} \mu(I_\mathbf{v}) = \sum_{\mathbf{v} \in M^r} |\mathbf{v}|^{-\beta} g(\{\mathbf{v}\})$$
$$= \sum_{\mathbf{w} \in M^{r-1}} \sum_{k \in M} |\mathbf{w}k|^{-\beta} g(\{\mathbf{w}k\})$$
$$= \sum_{\mathbf{w} \in M^{r-1}} |\mathbf{w}|^{-\beta} \sum_{k \in M} (k + \{\mathbf{w}\})^{-\beta} g(1/(k + \{\mathbf{w}\}))$$
$$= \sum_{\mathbf{w} \in M^{r-1}} |\mathbf{w}|^{-\beta} g(\{\mathbf{w}\}) = \mu\left(\cup_{\mathbf{w} \in M^{r-1}} I_\mathbf{w}\right).$$

By induction, this boils down to $\mu([0,1]) = 1$. Thus there is no mass left over to be assigned to J.

Let $\gamma = \gamma_M = 1/\int_0^1 g_m \, d\mu_M$. This will be a normalizing constant for our next measure. We need some information about the connections between μ_M, g_M, and the linear functional Λ_M given by $\lambda_M f = c_f$ when $\lim_{r \to \infty} H_M^r f = c_f g_M$. We fix the space B of functions on which H_M is acting to be the Banach space

$$\mathcal{H} := \{f \colon [0,1] \to \mathbb{R} \colon f, f' \text{ are continuous on } [0,1]\}$$
$$\text{with } \|f\|_A := \sup_{0 \le t \le 1} |f(t)| + \sup_{0 \le t \le 1} |f'(t)|.$$

This space is discussed in Chapter 9, section 9.2.1.

Proposition 4.1 *If $M \subset \mathbb{Z}^+$ is regular, then for $f \in \mathcal{H}$ and $x \in [0,1]$,*

$$\int_0^1 H_M f \, d\mu_M = \int_0^1 f \, d\mu_M$$

$$\Lambda_M f = \gamma_M \int_0^1 f \, d\mu_M$$

$$g_M(x) = \int_{\theta=0}^1 (1+\theta x)^{-\beta_M} \, d\mu_M(\theta)$$

$$\lim_{r \to \infty} H_M^r (1+x\theta)^{-\beta_M} = \gamma_M g_M(\theta) g_M(x).$$

Proof. For the first claim, suppose $f \in \mathcal{H}$. Fix M, and abbreviate μ_M as μ, β_M as β, and so on. We calculate:

$$\int_0^1 H_M f \, d\mu = \int_0^1 \sum_{k \in M} (k+t)^{-\beta} f\left(\tfrac{1}{k+t}\right) d\mu(t)$$

$$= \lim_{r \to \infty} \sum_{\mathbf{v} \in M^r} \sum_{k \in M} \int_{I_{k\mathbf{v}}} (k+t)^{-\beta} f\left(\tfrac{1}{k+t}\right) d\mu(t)$$

$$= \lim_{r \to \infty} \sum_{\mathbf{v} \in M^r} \sum_{k \in M} (k+[\mathbf{v}])^{-\beta} f\left(\tfrac{1}{k+[\mathbf{v}]}\right) |\mathbf{v}|^{-\beta} g(\{\mathbf{v}\})$$

$$= \lim_{r \to \infty} \sum_{\mathbf{u} \in M^{r+1}} f([\mathbf{u}]) |\mathbf{u}|^{-\beta} g(\{\mathbf{u}\}) = \int_0^1 f \, d\mu.$$

Here we used continuity of f on $[0,1]$, the fact that the lengths of the intervals $I_{\mathbf{v}}$ tended to zero, and the fact that for long $\mathbf{v}$, the front and back continued fractions $[\mathbf{v}]$ and $\{\mathbf{v}\}$ are close to the front and back fractions of their initial and end segments respectively.

For the second part, consider $h = f - \gamma_M \left(\int_0^1 f \, d\mu\right) g$. By construction, $\int_0^1 h \, d\mu = 0$, so for all $r \geq 1$, by the first part, $\int_0^1 H_M^r h \, d\mu = 0$. Thus, $\Lambda_M h = 0$, and since $\Lambda_M g = 1$, it follows that

$$0 = \Lambda_M \left(f - \gamma_M (\int_0^1 f \, d\mu) g\right) = \Lambda_M f - \gamma_M \int_0^1 f \, d\mu$$

as claimed.

For the third part, we have

$$H_M \int_0^1 (1+\theta x)^{-\beta}\, d\mu(\theta) = \sum_{k \in M} (k+x)^{-\beta} \int_0^1 \left(1 + \frac{\theta}{k+x}\right)^{-\beta} d\mu(\theta)$$

$$= \lim_{r\to\infty} \sum_{k \in M} (k+x)^{-\beta} \sum_{\mathbf{v} \in M^r} \int_{\theta \in I_{\mathbf{v}}} \left(1 + \frac{\theta}{k+x}\right)^{-\beta} d\mu(\theta)$$

$$= \lim_{r\to\infty} \sum_{k \in M} (k+x)^{-\beta} \sum_{\mathbf{v} \in M^r} (k + [\mathbf{v}] + x)^{-\beta} |\mathbf{v}|^{-\beta} g(\{\mathbf{v}\})$$

(passing from an integral to a Riemann sum). Using the identity for g and returning from Riemann sum to integral, this simplifies to

$$\lim_{r\to\infty} \sum_{\mathbf{u} \in M^{r+1}} (1 + x[\mathbf{u}])^{-\beta} |\mathbf{u}|^{-\beta} g(\{\mathbf{u}\}) = \int_0^1 (1+\theta x)^{-\beta}\, d\mu(\theta).$$

Thus, $\int_0^1 (1+\theta x)^{-\beta} d\mu(\theta)$ is a positive function of x on $[0,1]$, is invariant under H_M, and evaluates to 1 at $x = 0$. Since the eigenfunction of H_M corresponding to the leading eigenvalue 1 is unique up to scaling, it follows that $g = \int_0^1 (1+\theta x)^{-\beta} d\mu(\theta)$ as claimed.

For the final claim, we first use the two previous, and now established, parts of the proposition. This gives $\Lambda_M[(1+\theta x)^{-\beta})] = \gamma_M \int_0^1 (1+\theta t)^{-\beta} d\mu_M(t) = \gamma_M g_M(\theta)$. But Λ_M gives the coefficient for the limiting behavior of H_M^r applied to the function $x \to (1+\theta x)^{-\beta}$, so we are done. This completes the proof of the proposition. □

We now define a related measure $\nu = \nu_M$ on E_M, determined by the condition that $\nu(I_{\mathbf{v}}) = \gamma_M \int_{I_{\mathbf{v}}} g(x)\, d\mu(x)$. Equivalently, for any continuous f on $[0,1]$, $\int_0^1 f\, d\nu = \gamma_M \int_0^1 f(x) g_M(x)\, d\mu$. Like μ, ν is a probability measure supported on E_M; if μ is like the uniform measure, but for E_M, then ν is like the Gauss measure, but for E_M.

What exactly does this mean? The Gauss measure is invariant under T: For any measurable set A, $m(T^{-1}A) = m(A)$. And the same holds for ν.

Proposition 4.2 *The probability measure ν_M on E_M is invariant under T: $x \to 1/x - \lfloor 1/x \rfloor$; $\nu_M(T^{-1}A) = \nu_M(A)$ for all μ-measurable subsets A of $[0,1]$.*

Proof. It is sufficient to show that for $r \geq 1$ and $\mathbf{v} \in M^r$, $\nu_M(T^{-1}I_{\mathbf{v}}) = \nu_M(I_{\mathbf{v}})$. Since each $I_{\mathbf{v}}$ can be written as a disjoint union, apart from common endpoints, (which can occur if M is infinite), of the form $I_\mathbf{v} = \cup_{\mathbf{w} \in M^r} I_{\mathbf{vw}}$, it will suffice to show that as $r \to \infty$, and for $\mathbf{v} \in M^r$,

$$\nu(T^{-1}I_\mathbf{v}) = (1 + O(\phi^r))\nu(I_\mathbf{v}).$$

But if $\mathbf{v} \in M^r$, then

$$\nu(T^{-1}I_\mathbf{v}) = \nu\left(\cup_{k \in M} I_{k\mathbf{v}}\right) = \sum_{k \in M} \nu(I_{k\mathbf{v}}) = \gamma_M \sum_{k \in M} \int_{I_{k\mathbf{v}}} g_M(z) \, d\mu(z)$$

$$= (1 + O(\phi^r))\gamma_M \sum_{k \in M} g_M\left(\tfrac{1}{k+[\mathbf{v}]}\right) |k\mathbf{v}|^{-\beta} g_M(\{k\mathbf{v}\})$$

$$= (1 + O(\phi^r))\gamma_M g_M(\{\mathbf{v}\})|\mathbf{v}|^{-\beta} \sum_{k \in M} (k + [\mathbf{v}])^{-\beta} g_M\left(\tfrac{1}{k+[\mathbf{v}]}\right)$$

$$= (1 + O(\phi^r))\gamma_M g_M(\{\mathbf{v}\})g_M([\mathbf{v}])|\mathbf{v}|^{-\beta} = (1 + O(\phi^r))\nu_M(I_\mathbf{v}).$$

This proves the proposition. □

We need a stronger statement along the same lines. Not only is ν an invariant measure for T, T is ergodic with respect to ν, and even mixing.

Proposition 4.3 (T, ν_M) *is mixing on* E_M.

Proof. What is required is that for all measurable A and B,

$$\lim_{r \to \infty} \nu(T^{-r}A \cap B) = \nu(A)\nu(B).$$

Again we show that this holds to within arbitrarily tight tolerances by looking at what happens with the canonical intervals $I_\mathbf{v}$.

Any measurable A and B can be approximated to arbitrary closeness by unions of such intervals, so we assume $A = \cup_{\mathbf{u} \in S_A} I_\mathbf{u}$, $B = \cup_{\mathbf{u} \in S_B} I_\mathbf{u}$, where $S_A, S_B \subset M^n$ for some n. We need to show that $\nu(T^{-r}A \cap B) \to \nu(A)\nu(B)$ as $r \to \infty$. We abbreviate ν_M to ν, g_M to g, and so on, and calculate:

$$\nu(T^{-r}A \cap B) = \sum_{\mathbf{u} \in S_A} \sum_{\mathbf{v} \in M^{r-n}} \sum_{\mathbf{w} \in S_B} \nu(I_{\mathbf{wvu}})$$

$$= (1 + O(\phi^{r-n}))\gamma_M \sum_{\mathbf{u} \in S_A} \sum_{\mathbf{w} \in S_B} |\mathbf{w}|^{-\beta}|\mathbf{u}|^{-\beta} g([\mathbf{w}])g(\{\mathbf{u}\})$$

$$\cdot \sum_{\mathbf{v} \in M^{r-n}} |\mathbf{v}|^{-\beta}(1 + \{\mathbf{w}\}[\mathbf{v}])^{-\beta}(1 + \{\mathbf{v}\}[\mathbf{u}])^{-\beta}.$$

Now the inner sum here has the form

$$\sum_{\mathbf{v} \in M^{r-n}} |\mathbf{v}|^{-\beta}(1 + [\mathbf{v}]x)^{-\beta}(1 + \{\mathbf{v}\}t)^{-\beta}$$

with $x = \{\mathbf{w}\}$ and $t = [\mathbf{u}]$. But

$$\sum_{\mathbf{v} \in M^{r-n}} |\mathbf{v}|^{-\beta}(1+[\mathbf{v}]x)^{-\beta}(1+\{\mathbf{v}\}t)^{-\beta}$$

$$= (1 + O(\phi^{r-n})) \sum_{\mathbf{v} \in M^{r-n}} |\mathbf{v}|^{-\beta}(1+[\mathbf{v}]x)^{-\beta}(1+\{x+\mathbf{v}\})^{-\beta}$$

$$= (1 + O(\phi^{r-n})) H_M^{r-n}[(1+xz)^{-\beta}]|_{z=t}$$

$$= (1 + o(1))\gamma_M g_M(x) g_M(t),$$

so that as $r \to \infty$,

$$\nu(T^{-r}A \cap B) = (1 + o(1))\gamma_M \sum_{\mathbf{u} \in S_A, \mathbf{w} \in S_B} |\mathbf{w}|^{-\beta}|\mathbf{u}|^{-\beta} \cdot$$

$$g([\mathbf{w}])g(\{\mathbf{u}\})\gamma_M g(\{\mathbf{w}\})g([\mathbf{u}])$$

$$= (1 + o(1))(1 + O(\phi^n)) \sum_{\mathbf{u} \in S_A, \mathbf{w} \in S_B} \nu(I_\mathbf{u})\nu(I_\mathbf{w})$$

$$= (1 + o(1))(1 + O(\phi^n))\nu(A)\nu(B).$$

Thus, on sending both n and $r - n$ to ∞, we conclude that $\lim_{r \to \infty} \nu(T^{-r}A \cap B) = \nu(A)\nu(B)$. □

4.5 A Natural Extension of the Dynamic System (E_M, μ, T)

Recall that $T \colon \Omega = [0,1]\backslash \mathbb{Q} \times [0,1]$, $T(x,y) = (1/x - k, 1/(y+k))$, with $k = \lfloor 1/x \rfloor$. Let $\Omega_M = E_M \times [0,1]$. Then T is a one to one mapping on Ω that carries Ω_M into itself.

As always in this chapter, we assume M is regular. Let $\nu' = \nu'_M$ be the measure on Ω, and supported on Ω_M, determined by

$$\int_{\Omega_M} f \, d\nu'_M = \gamma_M \int_0^1 \int_0^1 f(x,y)(1+xy)^{-\beta} \, d\mu_M(x) d\mu_M(y).$$

Then ν'_M is a probability measure, since the measure of Ω is given by

$$\gamma_M \int_0^1 \int_0^1 (1+xy)^{-\beta} \, d\mu_M(x) \, d\mu_M(y) = \gamma_M \int_0^1 g_M(y) \, d\mu_M(y) = 1.$$

If $f \colon \Omega \to \mathbb{R}$ depends on x alone, say $f(x,y) = \hat{f}(x)$ for $y \in [0,1]$, then $\int_{\Omega_M} f \, d\nu' = \int_{E_M} \hat{f} \, d\nu$. We claim that ν'_M is invariant under T.

Proof. As usual, we may approximate a measurable set $A \subseteq \Omega$ by a union of subsets of Ω of the form $I_\mathbf{u} \times I_\mathbf{v}$, with $\mathbf{u}, \mathbf{v} \in M^r$, r large. It will suffice, then, to show that for all $\mathbf{u}, \mathbf{v} \in M^r$,

$$\nu'(T^{-1}(I_\mathbf{u} \times I_\mathbf{v})) = (1 + O(\phi^r))\nu'(I_\mathbf{u} \times I_\mathbf{v}).$$

So suppose $\mathbf{u} \in M^r$ and $\mathbf{v} \in M^r$, with $\mathbf{v} = (v_1, v_2, \ldots, v_r)$. As we have done before, let $\mathbf{v}_- = (v_2, \ldots, v_r)$. Then

$$\begin{aligned}
&\nu'(T^{-1}(I_\mathbf{u} \times I_\mathbf{v})) \\
&= \nu'(I_{v_1\mathbf{u}} \times I_{\mathbf{v}_-}) \\
&= (1 + O(\phi^r))\gamma_M(1 + [v_1\mathbf{u}][\mathbf{v}_-])^{-\beta}|v_1\mathbf{u}|^{-\beta}g_M(\{v_1\mathbf{u}\})|\mathbf{v}_-|^{-\beta}g_M(\{\mathbf{v}_-\}) \\
&= (1 + O(\phi^r))\gamma_M(1 + [\mathbf{v}_-]/(v_1 + [\mathbf{u}]))^{-\beta}(v_1 + [\mathbf{u}])^{-\beta} \\
&\quad \cdot |\mathbf{u}|^{-\beta}|\mathbf{v}_-|^{-\beta}g_M(\{\mathbf{u}\})g_M(\{\mathbf{v}\}) \\
&= ((1 + O(\phi^r))\gamma_M(1 + [\mathbf{u}]/(v_1 + [\mathbf{v}_-]))^{-\beta}|\mathbf{u}|^{-\beta}|\mathbf{v}|^{-\beta}g_M(\{\mathbf{u}\})g_M(\{\mathbf{v}\}) \\
&= (1 + O(\phi^r))\gamma_M(1 + [\mathbf{u}][\mathbf{v}])^{-\beta}|\mathbf{u}|^{-\beta}|\mathbf{v}|^{-\beta}g_M(\{\mathbf{u}\})g_M(\{\mathbf{v}\}) \\
&= (1 + O(\phi^r))\nu'(I_\mathbf{u} \times I_\mathbf{v}).
\end{aligned}$$

$\square$

As in the case of T, $\mathcal{T}$ is not merely measure preserving, but mixing. The proof is not entirely routine, though many of the techniques will by now be familiar.

First note that if $\mathbf{u}, \mathbf{v} \in M^s$ and $r > 4s$, then

$$\mathcal{T}^{-r}(I_\mathbf{u} \times I_\mathbf{v}) = \cup_{\mathbf{w} \in M^{r-s}} I_{\mathbf{w}\mathbf{v}'\mathbf{u}} \times [0,1]$$

where $\mathbf{v}' = (v_s, \ldots, v_1)$ is the reversal of $\mathbf{v}$.

Now consider measurable sets $A, B \subseteq \Omega$. Since ν' is supported on $E_M \times E_M$, and since $\mathcal{T}(E_M \times E_M) \subseteq E_M \times E_M$, for purposes of discussing $\nu'(\mathcal{T}^{-r}A \cap B)$, we may restrict attention to $A, B \subseteq E_M \times E_M$. Fix $\epsilon > 0$. Approximate A and B by open sets $A^* = \cup_{(\mathbf{u},\mathbf{v}) \in S_A} I_\mathbf{u} \times I_\mathbf{v}$ and $B^* = \cup_{(\mathbf{w},\mathbf{z}) \in S_B} I_\mathbf{w} \times I_\mathbf{z}$ so that $A \subseteq A^*$, $B \subseteq B^*$, $\nu(A^* \setminus A) < \epsilon$, $\nu(B^* \setminus B) < \epsilon$, and for all $(\mathbf{u}, \mathbf{v}) \in S_A$ and $(\mathbf{w}, \mathbf{z}) \in S_B$, $\mathbf{u}, \mathbf{v}, \mathbf{w}$, and $\mathbf{z}$, which are finite lists of integers belonging to M, have the same length s.

With this in place, we have

$$|\nu'(\mathcal{T}^{-r}A \cap B) - \nu'(\mathcal{T}^{-r}A^* \cap B^*)| \leq 2\epsilon.$$

Thus, it will suffice to show that as $r \to \infty$, $\nu'(\mathcal{T}^{-r}A^* \cap B^*) \to$

$\nu'(A^*)\nu'(B^*)$. We therefore calculate:

$$\nu'(T^{-r}A^* \cap B^*) = \sum_{(\mathbf{u},\mathbf{v})\in S_A} \sum_{(\mathbf{w},\mathbf{z})\in S_B} \nu'(T^{-r}(I_\mathbf{u} \times I_\mathbf{v}) \cap (I_\mathbf{w} \times I_\mathbf{z}))$$

$$= \sum_{(\mathbf{u},\mathbf{v})\in S_A} \sum_{(\mathbf{w},\mathbf{v})\in S_B} \sum_{\mathbf{y}\in M^{r-s}} \nu'((I_{\mathbf{yv'u}} \times E_M) \cap (I_\mathbf{w} \times I_\mathbf{z}))$$

$$= \sum_{(\mathbf{u},\mathbf{v})\in S_A} \sum_{(\mathbf{w},\mathbf{z})\in S_B} \sum_{\mathbf{x}\in M^{r-2s}} \nu'(I_{\mathbf{wxv'u}} \times I_\mathbf{z})$$

$$= (1 + O(\phi^s))\gamma_M \sum_{(\mathbf{u},\mathbf{v})\in S_A} \sum_{(\mathbf{w},\mathbf{z})\in S_B} \sum_{\mathbf{x}\in M^{r-2s}}$$

$$\cdot |\mathbf{wxv'u}|^{-\beta}|\mathbf{z}|^{-\beta}(1 + [\mathbf{w}][\mathbf{z}])^{-\beta} g(\{\mathbf{u}\})g(\{\mathbf{z}\})$$

$$= (1 + O(\phi^s))\gamma_M \sum_{\mathbf{u},\mathbf{v},\mathbf{w},\mathbf{z}} g(\{\mathbf{u}\})g(\{\mathbf{z}\})|\mathbf{z}|^{-\beta}|\mathbf{w}|^{-\beta}|\mathbf{v}|^{-\beta}|\mathbf{u}|^{-\beta}$$

$$\cdot (1 + [\mathbf{u}][\mathbf{v}])^{-\beta}(1 + [\mathbf{w}][\mathbf{z}])^{-\beta}$$

$$\cdot \sum_{\mathbf{x}\in M^{r-2s}} |\mathbf{x}|^{-\beta}(1 + \{\mathbf{w}\}[\mathbf{x}])^{-\beta}(1 + \{\mathbf{v}\}\{\mathbf{x}\})^{-\beta}.$$

The innermost sum here has the form

$$\sum_{\mathbf{x}\in M^{r-2s}} |\mathbf{x}|^{-\beta}(1 + \sigma[\mathbf{x}])^{-\beta}(1 + \tau\{\mathbf{x}\})^{-\beta}$$

with $\sigma = \{\mathbf{w}\}$ and $\tau = \{\mathbf{v}\}$. But this is equal, to within the tolerance of the error factor of $(1 + O(\phi^s))$, to

$$(1 + O(\phi^s)) \sum_{\mathbf{x}\in M^{r-2s}} |\mathbf{x}|^{-\beta}(1 + \sigma[\mathbf{x} + \tau])^{-\beta}(1 + \tau\{\mathbf{x}\})^{-\beta}$$

which in turn is equal (exactly) to

$$H_M^{r-2s}(1 + \sigma\zeta)^{-\beta}\Big|_{\zeta=\tau}.$$

In view of Proposition 4.1, last item, this is, again to within the tolerances allowed, simply $(1 + O(\alpha^{r-2s}))\gamma_M g(\{\mathbf{w}\})g(\{\mathbf{v}\})$. We conclude that

$$\nu'(T^{-r}A^* \cap B^*) = (1 + O(\phi^{r-2s}))\gamma_M^2 \sum_{\mathbf{u},\mathbf{v},\mathbf{w},\mathbf{z}} g(\{\mathbf{u}\})g(\{\mathbf{z}\})g(\{\mathbf{w}\})g(\{\mathbf{v}\})$$

$$\cdot (1 + [\mathbf{u}][\mathbf{v}])^{-\beta}(1 + [\mathbf{w}][\mathbf{z}])^{-\beta}$$

$$= (1 + O(\phi^{r-2s}))\nu'(A^*)\nu'(B^*).$$

This completes the proof that T is mixing with respect to ν' on Ω_M.

With this mixing established, we may plug in directly to the arguments of theorem 2.2 of [Na] and conclude the following, bearing in mind that the measure ν', supported on E_M, depends on E_M:

Theorem 4.2 *With respect to ν', for almost every $x \in E_M$ the fraction over $1 \leq n \leq N$ of n so that $\theta_n(x) = q_n^2 |x - p_n/q_n| \leq z$ tends to $F_M(z) = \nu'(\{(s,t): 0 \leq t \leq z - 1/s\})$.*

4.5.1 *Variants on the theme*

There are many variants of continued fraction algorithms to which the same circle of ideas may be applied. The Knopfmacher CF-algorithm takes as input a pair of polynomials or power series over some field, say f and g. The algorithm provides an expansion of $(1+zf)/(1+zg)$ of the form

$$\frac{1+zf}{1+zg} = 1 + \cfrac{C_1 z^{k_1}}{1 + \cfrac{C_2 z^{k_2}}{1+C_3 z^{k_3}/\ldots}}.$$

We single out the particular case when the field is $\{0,1\}$. The expansion then takes the form

$$\frac{1+zf}{1+zg} = 1 + \cfrac{z^{k_1}}{1 + \cfrac{z^{k_2}}{1+z^{k_3}/\ldots}}.$$

The algorithm proceeds by steps, keeping a *working pair* (f_j, g_j) and outputting the successive positive integers $k_1, k_2, k_3 \ldots$ as it goes. Given a pair (f, g), let $K[(f,g)] := (g, ((f+g)z^{1-k} + 1)z^{-1})$, where $k-1$ is the number of consecutive terms, starting with the constant term, at which f and g agree. The CF algorithm terminates when, in the next working pair, $f = g$. Example: We execute the CF algorithm on the pair $(1 + z + z^3, z)$ to get an expansion of $(1 + z(1 + z + z^3))/(1 + z^2)$. The initial working pair, with f and g viewed as a bit sequences instead of polynomials, is $((1,1,0,1),(0,1))$. The sequence of values of k spun off by the algorithm is $(1,2,1,1,1)$. The second working pair is $((0,1),(0,0,1))$, the third is $((0,0,1),(1))$, the fourth is $((1),(0,1))$, the fifth is $((0,1),(1))$, and the last is $((1),(1))$. Because the two entries in this final working pair are equal, the algorithm halts.

Let P denote the set of all formal power series $f = \epsilon_0 + \epsilon_1 z + \epsilon_2 z^2 + \ldots$ with all $\epsilon_j \in \{0,1\}$. Let $B(f, g, n) := (f + O(z^n)) \times (g + O(z^n))$ be a 'box' about (f, g). Let μ be the usual product measure on $\Omega = P \times P$, so that $\mu B(f, g, n) = 2^{-2n}$. We equip P with the 2-adic metric, so that if

$f = z^k(1+zg) \in P$ then $\|f\| = 2^{-k}$, and we equip the product space Ω with the sup norm, that is, $\|(f,g)\| = \max[\|f\|, \|g\|]$.

Now K is a measure preserving map on Ω. Let K_j^{-1} be the *function* given by $K_j^{-1}(f,g) := (f + z^{j-1} + z^j g, f)$. Then

$$K_j^{-1}(B(f,g,n)) = \{(u,v): \; v \in f + O(z^n), u \in v + z^{j-1} + z^j g + O(z^n)\}.$$

Thus,

$$K^{-1}B(f,g,n) = \bigcup_{j=1}^{\infty} K_j^{-1}B(f,g,n).$$

Now any open $U \subseteq \Omega$ can be written as the disjoint union of boxes: $U = \bigcup_{m=1}^{\infty} B_m(f_m, g_m, n_m)$. Thus,

$$K^{-1}U = \bigcup_{j=1}^{\infty} \bigcup_{m=1}^{\infty} K_j^{-1} B_m(f_m, g_m, n_m).$$

We claim that the components in this representation are again disjoint. Consider two components, corresponding to (j_1, m_1) and (j_2, m_2). Without loss of generality we assume $j_1 \leq j_2$.

If $m_1 = m_2 = m$ and $j_1 < j_2$, and if $(u,v) \in K_{j_1}^{-1}U_m \cap K_{j_2}^{-1}U_m$, then $u \in v + z^{j_1-1} + z^{j_1}g_m + O(z^{j_1+n_m}) \cap v + z^{j_2-1} + z^{j_2}g_m + O(z^{j_2+n_m})$. This is impossible since contradictory values of the z^{j_1-1} coefficient are specified.

If $m_1 \neq m_2$ then $U_{m_1} \cap U_{m_2} = \emptyset$. If $j_1 = j_2 = j$, then

$$K_{j_1}^{-1} U_{m_1} \cap K_j^{-1} U_{m_2} = \emptyset$$

because the inverse images of disjoint sets, under the function K_j, are disjoint. Finally, if $m_1 \neq m_2$ and $j_1 < j_2$, then again the coefficients required on z^{j_1-1} of first entry u in a (u,v) belonging to the intersection, would be different.

We now show that K is measure preserving with respect μ. Under the circumstances, it will suffice to show that for any *open* set U, $\mu K^{-1}U = \mu U$. This is because Ω is a compact set, so that by theorems 2.14 and 2.17 of [Ru], for any measurable $E \subseteq \Omega$, and for any $\epsilon > 0$, there exists an open set U such that $E \subseteq U$ and $\mu(U \setminus E) < \epsilon$. Thus

$$\mu K^{-1}E \leq \mu K^{-1}U = \mu U \leq \mu E + \epsilon.$$

In the other direction, $U \setminus E$ is measurable, so there exists an open V such that $U \setminus E \subseteq V$ and $\mu V < 2\epsilon$. Now $U \setminus V \subseteq E$ so $K^{-1}(U \setminus V) \subseteq K^{-1}E$, so $\mu K^{-1}E \geq \mu K^{-1}(U \setminus V) \geq \mu K^{-1}U - \mu K^{-1}V = \mu(U) - \mu(V) \geq \mu E - 2\epsilon$.

As these inequalities hold for all $\epsilon > 0$, $\mu K^{-1}E = \mu E$. That is, K is measure preserving on Ω. This is of course the prelude to the main result, that K is in fact ergodic.

Theorem 4.3 *K is ergodic on P equipped with the measure μ.*

Proof. We must show that there is no measurable set in Ω, other than sets of measure 0 or 1, that is invariant under K. So suppose E is some set of measure δ with $0 < \delta < 1$ and suppose $K^{-1}E \subseteq E$. This will lead to a contradiction.

We use the *transfer operator* G associated with K: If ρ is bounded and measurable on Ω, then $G\rho$ is the function given by

$$G\rho(u,v) = \sum_{n=1}^{\infty} 2^{-n}\rho(u + z^{n-1} + z^n v, u).$$

This transfer operator has the key property that if D is a measurable subset of Ω, and if ρ is a bounded measurable function on Ω, then

$$\int_{K^{-1}D} \rho = \int_D G\rho.$$

This is because $K^{-1}D$ can be written as a disjoint union $\cup_1^\infty K_n^{-1}D$, and the Jacobian of K_n^{-1} is 2^{-n}. Thus,

$$\int_{K^{-1}D} \rho \, d\mu = \sum_{n=1}^{\infty} \int_{K_n^{-1}D} \rho \, d\mu = \sum_{n=1}^{\infty} \int_D (\rho \circ K_n^{-1}) \, d\mu = \int_D G\rho \, d\mu.$$

Now for $n \geq 1$, let $Z_n = \{(r,s) \in \Omega \mid r + s \in z^{n-1} + O(z^n)\}$. If we identify a point $(r,s) \in \Omega$ with the real number pair obtained from (r,s) by setting $z = 1/2$, we can depict these zones. (The remaining on-diagonal squares comprise $Z_k, k \geq 4$.)

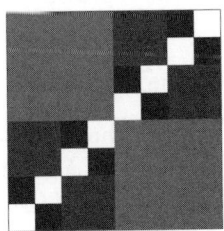

Fig. 4.1 Zones for the CF-algorithm.

By a suitable change of variables,
$$2^{-n} \int_{(u,v)\in\Omega} \rho(u + z^{n-1} + z^n v, u) = \int_{(r,s)\in Z_n} \rho(r,s).$$
Summing over n, we see that $\int_\Omega \rho = \int_\Omega G\rho$. Now we are going to take two perspectives on what happens, as $n \to \infty$, to $G^n \rho_E$ when
$$\rho_E(x) = \begin{cases} 1/\mu(E), & \text{if } x \in E; \\ 0 & \text{if not.} \end{cases}$$
On the one hand, since $K^{-1}E \subseteq E$, and since $\mu(K^{-1}E) = \mu(E)$, we must have
$$1 = \int_E \rho_E = \int_{K^{-n}E} \rho_E = \int_E G^n \rho_E.$$
On the other hand, perhaps if we look at the 'fourier series' representation of ρ_E, we shall see that the nonconstant terms decay under iteration of G.

A direct realization of this ambition will encounter certain difficulties, unfortunately. E need only be a generic measurable set of measure strictly between 0 and 1. Convergence of any sort of fourier expansion is doubtful. We thus assume that $4\epsilon < \mu(E) < 1 - 4\epsilon$, and consider an open set E' with $\mu(E' \Delta E) < \epsilon^4$, of the form $E' = \bigcup_{1 \le m \le M} B(f_m, g_m, n_m)$. Now $\mu(K^{-n} E' \Delta E') < \epsilon^4$, so $\int_{E'} G^n \rho_{E'} = 1 + O(\epsilon^3)$.

The other 'hand' here is the fourier approach. For $S, T \subseteq \mathbb{N}$, $u = \sum a_k z^k$, and $v = \sum b_k z^k$, let $\psi_{S,T} \colon \Omega \to \{-1, 1\}$ be given by
$$\psi_{S,T}(u,v) = \prod_{j \in S} \prod_{k \in T} (-1)^{(a_j + b_k)}.$$
Then the $\psi_{S,T}$ form an orthonormal set of functions on Ω with respect to the inner product $\langle \psi, \phi \rangle = \int_\Omega \psi \phi \, d\mu$. Furthermore, if $B = B(u,v,n)$, then $\rho_B(f,g) = 2^{2n} I[(f,g) \in B]$ has the representation
$$\rho_B = \sum_{S,T \subseteq \{0,1,\ldots,n-1\}} c_{S,T} \psi_{S,T}$$
where $c_{S,T} = \psi_{S,T}(u,v)$. In general, the coefficient of $\psi_{S,T}$ in the representation of some measurable, L_2 function ϕ on Ω, is $\int_\Omega \phi \psi_{S,T} \, d\mu$. Now consider the finite-dimensional subspace F_N of (trivially measurable, bounded, L_1 and L_2) functions on Ω which are constant on all boxes of the form $B(u,v,N)$. Setting $N = 1 + \max\{n_m \colon 1 \le m \le M\}$, we have $\rho_{E'} \in F_N$. Let $P_N = \{(S,T) \colon (S,T) \subseteq \{0,1,\ldots,N-1\}^2\}$.

Now suppose $\phi \in F_N$, $\phi = \sum_{(S,T) \in P_N} c_{S,T} \psi_{S,T}$. Then

$$(G\phi)(u,v) = \sum_{P_N} \sum_{n=1}^{\infty} 2^{-n} c_{S,T} \psi_T(u) \psi_S(u + z^{n-1} + z^n v).$$

Now the coefficient of $1 = \psi_{\emptyset,\emptyset}$ is invariant under G, but as we shall now show, G acts (almost) like a contraction on the orthogonal complement $F_N^\perp$ of 1 in F_N.

For $\gamma \in F_N$, $\gamma = \sum_{(S,T) \in P_N} c_{(S,T)} \psi_{(S,T)}$, let $\|\gamma\| = \sum_{(S,T) \in P_N} |c_{(S,T)}|$. This is a norm on F_N.

Lemma 4.1 *For $\gamma \in F_N^\perp$, $\|G^2 \gamma\| \le (1 - 2^{-N}) \|\gamma\|$.*

Proof. First, we note that $G\psi_{\emptyset,T} = \psi_{T,\emptyset}$. Thus, it will be sufficient to show that for $S \ne \emptyset$, $\|G\psi_{(S,T)}\| \le 1 - 2^{-N}$. For $n \in \mathbb{N}$, let $S - n = \{s - n : s \in S\} \cap \mathbb{N}$. Then

$$G\psi_{(S,T)} = \sum_{n=1}^{\max S} 2^{-n} (-1)^{|S \cap \{n-1\}|} \psi_{S \triangle T, S-n}.$$

Thus $\|G\psi_{(S,T)}\| = 1 - 2^{-\max S} < 1 - 2^{-N}$. □

From this lemma, it now follows that $G^n \rho_{E'} \to 1$ uniformly on Ω as $n \to \infty$. But then $\lim_{n \to \infty} \int_{E'} G^n \rho_{E'} = \mu(E')$. This contradicts our earlier calculation, contingent on the hypothetical existence of an E with measure other than 0 or 1, yet invariant under K, that $\int_{E'} G^n \rho_{E'} = 1 + O(\epsilon^3)$. Therefore, there is no such E after all, and K is ergodic on Ω. □

Chapter 5
Complex Continued Fractions

5.1 The Schmidt Regular Chains Algorithm

A *complex continued fraction algorithm* is, broadly speaking, an algorithm that provides approximations by ratios of Gaussian integers, or integers of another complex number field, to a given complex number. We are faced with an immediate choice: does speed matter, or does approximation quality trump everything else?

If quality of approximation is paramount, then the algorithm of choice is due to Asmus Schmidt[Sch]. The basic idea is to partition (we ignore issues concerning breaking ties at the boundaries) the complex plane into pieces bounded by line segments and arcs of circles, and to successively refine this partition. The partition component to which the target ξ belongs shrinks down to ξ as this process is iterated.

Each component of the partition is associated with a linear fractional map $z \to (az+b)/(cz+d)$, where the coefficients a, b, c, d are Gaussian integers, as well as with the matrix $\begin{pmatrix} a & b \\ c & d \end{pmatrix}$.

The Schmidt algorithm resembles the Lagarias multidimensional continued fraction algorithm (discussed in Chapter 6) in that both generate, for each ξ, a sequence $\langle M_n(\xi) \rangle$ of matrices with 'integer' entries, whose inverses also have 'integer' entries. The required Diophantine approximation to the target ξ can then be read off from $M_n(\xi)$.

It produces a tesselation of its space of possible targets, with the property that each set of ξ associated to a given M_n splits into some number of shards, each associated with a specific one of the possible successors $M_n T$. We show the fourth-generation partition of that portion of $\mathbb{C}$ bounded by the circle with diameter $(1, 1+i)$:

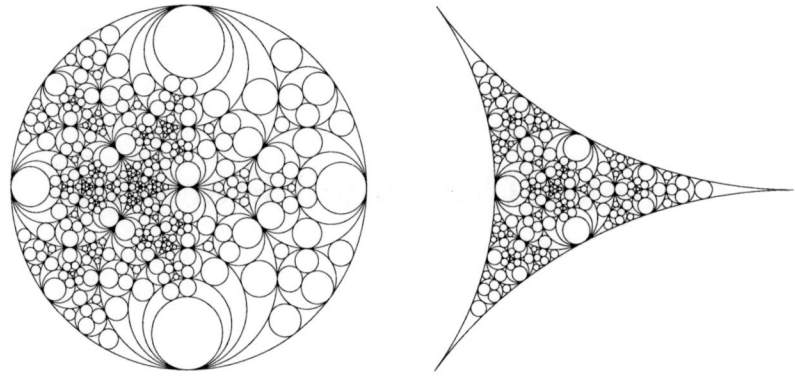

Fig. 5.1 Fourth generation and (zoom to left-center) fifth.

On average, the Schmidt algorithm would appear to require only polynomial time in the binary length of the numerator and denominator of Gaussian-rational inputs. The worst-case is a different matter. Given a real input, or one equivalent at an early stage to a real number, the algorithm can be quite slow.

The Schmidt *regular chains* algorithm [Sch] takes as input a complex number ξ and returns a sequence $\langle T \rangle = (T_1, T_2, \ldots)$ of 2×2 matrices in $GL_2(\mathbb{Z}[i])$ drawn from a certain finite set, or alphabet, $\mathcal{S} = \{V_1, V_2, V_3, E_1, E_2, E_3, C\}$. Associated with this sequence of 'letters', we have a sequence of products $\langle M_n \rangle$, where $M_n = T_1 \ldots T_n$, a sequence of candidate approximations $p_n^{(j)}/q_n^{(j)}$ given by

$$\begin{pmatrix} p_n^{(1)} & p_n^{(2)} & p_n^{(3)} \\ q_n^{(1)} & q_n^{(2)} & q_n^{(3)} \end{pmatrix} = M_n \begin{pmatrix} 1 & 0 & 1 \\ 0 & 1 & 1 \end{pmatrix}$$

and finally, a sequence of approximations $\langle r_n \rangle$, where r_n is whichever of the three fractions $p_n^{(j)}/q_n^{(j)}$ is nearest ξ. The building block T's are

$$V_1 = \begin{pmatrix} 1 & 0 \\ 1-i & 1 \end{pmatrix}, \quad V_2 = \begin{pmatrix} 1 & 0 \\ -i & 1 \end{pmatrix}, \quad V_3 = \begin{pmatrix} 1-i & i \\ -i & 1+i \end{pmatrix},$$

$$E_1 = \begin{pmatrix} 1 & 0 \\ 1-i & i \end{pmatrix}, \quad E_2 = \begin{pmatrix} 1 & -1+i \\ 0 & i \end{pmatrix}, \quad E_3 = \begin{pmatrix} i & 0 \\ 0 & 1 \end{pmatrix},$$

and
$$C = \begin{pmatrix} 1 & -1+i \\ 1-i & i \end{pmatrix}.$$

We take M_0 to be the identity matrix. Associated with each product $M_n = \begin{pmatrix} a & b \\ c & d \end{pmatrix}$ we have also a linear fractional map m_n on $\mathbb{C} \cup \infty$, given by $m_n(z) = (az+b)/(cz+d)$, and a region $\mathcal{F}_n$ that contain ξ and is delimited and determined by a/c, b/d, and $(a+b)/(c+d)$.

Given three noncollinear points $z_1, z_2, z_3 \in \mathbb{C}$, let $C = C(z_1, z_2, z_3)$ be the circle through the z_j, D the closed disk bounded by C, and T the connected closed subset of D bounded by arcs of circles perpendicular to C, running inside D from each z_j to the other two.

When $\det M_n$ is real, (± 1), $\mathcal{F}_n$ is the disk $D(a/c, b/d, (a+b)/(c+d))$. When instead the determinant is imaginary ($\pm i$), $\mathcal{F}_n$ is the 'triangle' (each vertex has interior angle of zero) $T(a/c, b/d, (a+b)/(c+d))$ with vertices $(r_1, r_2, r_3) = (a/c, b/d, (a+b)/(c+d))$; the 'edges' of this triangle are arcs, from each of the vertices to the other two, of circles orthogonal to the circle or line through all three vertices.

A *regular chain* is a sequence $(T_1, T_2, \ldots)$ of matrices in $\mathcal{S}$ such that

(1) If $\det M_{n-1} = \pm 1$ then $T_n \in \{V_1, V_2, V_3, C\}$,
(2) There is no $n_0 \in \mathbb{Z}^+$ and $j \in \{1, 2, 3\}$ such that for $n \geq n_0$, $T_n = V_j$.

There is a connection between m_n and $\mathcal{F}_n$. Let $\mathcal{J} = \{z \mid \Im z \geq 0\}$ and $\mathcal{J}^* = \{z \mid 0 \leq \Re z \leq 1, \Im z \geq 0, \text{ and } |z - 1/2| \geq 1/2\}$. If $\det(M_n) = \pm 1$, then $\mathcal{F}_n$ is whichever of $m_n(\mathcal{J})$ or $m_n(\overline{\mathcal{J}})$ forms the interior of a disk, while if $\det M_n = \pm i$, then $\mathcal{F}_n$ is whichever of $m_n(\mathcal{J}^*)$ or $m_n(\overline{\mathcal{J}^*})$ is the bounded, 'triangular' set. Thus, another description of which T is to be used next is this: take that T such that the new $\mathcal{F}$ will contain ξ.

Computationally, one may implement this algorithm on Gaussian rational inputs, breaking ties ad hoc by having the algorithm return the first permissible choice of T. Testing whether a 'triangular' region contains ξ may be accomplished by mapping that region onto $\mathcal{J}^*$ or its conjugate with an appropriately chosen linear fractional mapping ϕ and checking whether $\phi(\xi)$ belongs. The algorithm terminates at ξ, and all steps can be carried out with exact arithmetic on Gaussian rationals. The approximations r_n have the form $r_n = p/q$, $p, q \in \mathbb{G}, q \neq 0$, and are in each case the best of the three $p_n^{(j)}/q_n^{(j)}$. They satisfy (page 16 of [Sch])

$$\left|\xi - \frac{p}{q}\right| \leq \frac{1}{\sqrt{2}|q|^2},$$

and conversely, (theorem 2.5 of [Sch]) if $p,q \in \mathbb{G}, q \neq 0$, and if $|\xi - p/q| \leq \sqrt{2}/((\sqrt{2}+1)|q|^2)$, then p/q is one of the r_n found by the Schmidt algorithm.

We illustrate the algorithm by executing it on input $\xi = 7/11 + 14i/31$. We have $T_1 = V_3$ because ξ belongs to the circle through the points $(1+i, i/(1+i), 1)$ determined by V_3. The next few M_n are

$$M_2 = \begin{pmatrix} 2-i & i \\ 1-2i & 1+i \end{pmatrix}, \quad M_3 = \begin{pmatrix} 3-i & i \\ 2-3i & 1+i \end{pmatrix}, \quad M_4 = \begin{pmatrix} 4 & -1 \\ 4-3i & -1+i \end{pmatrix},$$

$$M_5 = \begin{pmatrix} 3+i & -4+3i \\ 4-i & -2+6i \end{pmatrix}, \quad M_6 = \begin{pmatrix} 2+8i & -3-4i \\ 8+7i & -6-2i \end{pmatrix},$$

$$M_7 = \begin{pmatrix} -5+7i & -6-9i \\ 11i & -13-5i \end{pmatrix}, \quad M_8 = \begin{pmatrix} -5+7i & -13-14i \\ 11i & -24-5i \end{pmatrix}.$$

The data can be presented as well in visual form: The regions are the successive $\mathcal{F}_n$, together with some neighboring candidates for those regions which were not selected because ξ lay outside them. In the first figure, we show the individual region as well.

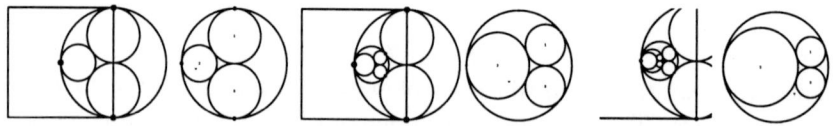

Fig. 5.2 First few iterations of the Schmidt algorithm.

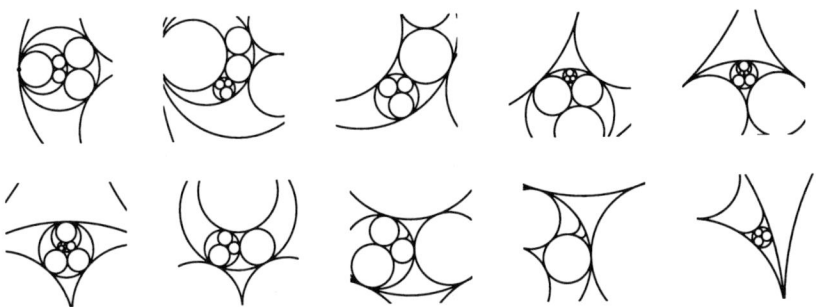

Fig. 5.3 Multiple zooms.

Schmidt also announces explicit chain expansions for $e^{1/(a-ib)}$ and $e^{1/(-ib)}$ when $a, b \in \mathbb{Z}^+$. In particular, the chain for e^i is

$$e^i \sim C^2 V_3 C^2 V_3^3 C^2 V_3^5 C^2 V_3^7 C^2 V_3^9 C^2 V_3^{11} C^2 \ldots.$$

5.2 The Hurwitz Complex Continued Fraction

If we relax the priority we set on quality of approximation, what do we get in exchange? Any complex continued fraction algorithm should take as input a complex number $z = z_0$, and return as output a sequence of approximations (p_n/q_n) to z, where $p_n, q_n \in G = \{a+ib : a \in \mathbb{Z}, b \in \mathbb{Z}\}$ are relatively prime Gaussian integers. The following properties are desirable:

(1) The sequence $(|q_n|)$ increases at least exponentially.
(2) The sequence (p_n/q_n) terminates in p/q if the input z is a Gaussian rational.
(3) All approximations are of quality at least comparable to that which is known to exist by virtue of the Dirichlet pigeonhole principle:

$$\left| \frac{p_n}{q_n} - z \right| \ll |q_n|^{-2}$$

uniformly over n and over all inputs z.

It would be nice if the algorithm were also easily executed, provided that basic arithmetic operations and truncations present no particular difficulty. (If the input is given only as the limit of a sequence of decimal approximations, for instance, it may be impossible to resolve a truncation operation.) We might hope that the p_n/q_n take the form

$$\cfrac{1}{a_1 + \cfrac{1}{a_2 + \cfrac{\cdots + \cfrac{1}{a_n}}{}}}.$$

We might also hope that every reasonably good Gaussian rational approximation p/q to z, of quality comparable to or better than that vouchsafed by the Dirichlet pigeonhole principle, was either a continued fraction convergent, or simply related to the two convergents whose denominators bracketed q in absolute value.

Finally, we might hope to understand both what typically happens for a 'random' initial z, and what happens in selected special cases such as inputs that satisfy a quadratic polynomial with Gaussian integer coefficients.

The classical algorithm due to A. Hurwitz [Hur] meets most of these desiderata. It takes as input a complex number z, and outputs a sequence $\langle a_n \rangle$ of Gaussian integers, from which further sequences $\langle p_n \rangle$ and $\langle q_n \rangle$ of Gaussian integers can be computed exactly as in the case of the classical continued fraction. The Gaussian rationals p_n/q_n then come to us in reduced form, and they furnish decent approximations to z. The arithmetic needed to decide on the next step is decidedly simpler than with the Schmidt algorithm, while the approximations are comparable if not always quite as good.

We denote by $[z]$ the Gaussian integer *nearest* z, rounding down, in both the real and imaginary components, to break ties. The Hurwitz complex continued fraction algorithm, like the near-classical *centered* continued fraction algorithm for real numbers, is simplest when restricted to inputs inside the fundamental domain of the truncation function. Here, that domain is $B = \{x + iy \mid -1/2 \leq x < 1/2, -1/2 \leq y < 1/2\}$. The algorithm proceeds by steps of the form $z_{n+1} = 1/z_n - [1/z_n]$. If $z = z_0 \in \mathbb{Q}(i)$, the algorithm terminates when, as must eventually occur, $z_n = 0$, and the final finite-depth continued fraction gives a reduced fraction p_n/q_n equal to z. The next-to-last convergent pair gives a solution to $uq_n - vp_n = 1$.

If initially $z \notin \mathbb{Q}(i)$ then the algorithm continues indefinitely, or in practice, until some other terminating condition is met.

Once $(a_0, a_1, \ldots a_n)$ have been computed, we compute the *convergents* p_k/q_k using the usual formula. (Details are provided in the next section.)

As in the case of the classical algorithm,

$$\begin{vmatrix} p_{n-1} & p_n \\ q_{n-1} & q_n \end{vmatrix} = (-1)^n.$$

Here, we show that this algorithm has excellent convergence properties, and we study some surprising cases in which algebraic inputs z lead to expansions that exhibit behavior that is neither periodic (along the lines of the classical continued fraction expansion of $\sqrt{n}$), nor 'typical'. We close by working out the details of what that 'typical' behavior is, and give an analog, for the case of the Hurwitz algorithm, to the classical Gauss-Kuz'min theorem, itself discussed in Chapter 9.

5.3 Notation

It will be convenient to introduce some additional notation. We assume that $z = z_0$ is given, with $z_0 \in B$, and that $p_{-1} = q_0 = 1$ while $p_0 = q_{-1} = 0$. For $n \geq 1$ let $a_n = [1/z_{n-1}]$, $p_n = a_n p_{n-1} + p_{n-2}$, and $q_n = a_n q_{n-1} + q_{n-2}$. Let $x_n = 1/z_{n-1}$, and let $w_n = q_{n-1}/q_n$. Then

$$z_{n+1} = \frac{1}{z_n} - a_{n+1}$$

$$w_{n+1} = \frac{1}{a_{n+1} + w_n}.$$

Let $1/B$ denote the set of reciprocals of the nonzero elements of B; $1/B$ is bounded by arcs of circles of radius 1 about ± 1 and $\pm i$, and these arcs pass through $\pm 1 \pm i$ and through ± 2 and $\pm 2i$. Let G' denote $G\backslash\{0, \pm 1, \pm i\}$.

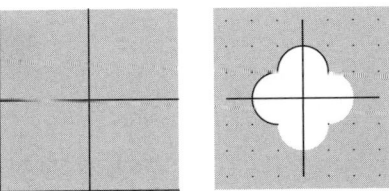

Fig. 5.4 B and $1/B$.

From the figure, it is evident that only elements of G' can occur as *partial quotients* a_n. On the other hand, it is not so simple to say which sequences can actually occur. There is a variant on this algorithm, due to Julius Hurwitz[JH] in which the a_n are restricted further to be multiples of $(1+i)$, and in this variant, it was apparently possible to ferret out the details of which sequences of (a_n) can occur.

For the algorithm now under investigation it is possible, at any rate, to work out that certain finite combinations of consecutive a_k cannot occur. The following table gives the set of all possible pairs (a, b) of consecutive a_j. The reason for these constraints is a matter of geometry, and we explain two of the cases in some detail. If $a_j = 1 + i$, then x_j belongs to the intersection of $1/B$ and the square $1/2 \leq \Re x < 3/2$, $1/2 \leq \Re y < 3/2$, so that $z_j = x_j - (1+i)$ belongs to the intersection of $1/B - (1+i)$ and B. The set of reciprocals of complex numbers in this set, in turn, is the intersection of the regions $\Re[z] \geq -1/2$, $\Im[z] \leq 1/2$, $|z+1| > 1$, and $|z+i| > 1$.

If $a_j = 2 + i$, then z_j belongs to that part of B lying on or outside the circle of radius 1 about $-1-i$, so $1/z_j$ lies in that part of $1/B$ lying on or outside the circle of radius 1 about $-1+i$, with the result that all of G' apart from $-1+i$ is a possibility for a_{j+1}.

The dots indicate Gaussian integers that are available for a_{j+1}. For

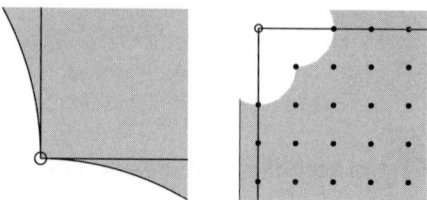

Fig. 5.5 $a_j = 1 + i$: Possible z_j,   Possible a_{j+1}.

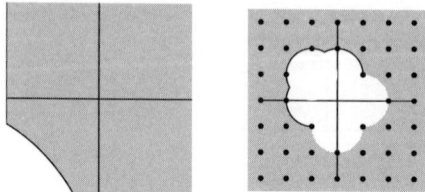

Fig. 5.6 $a_j = 2 + i$: Possible z_j,   Possible a_{j+1}.

other values of a_j, we have the following little table; negating or taking the conjugate of a_j negates or conjugates the set of possible a_{j+1}, while if $|a_j| \geq 2\sqrt{2}$, any element of G' is a possibility for a_{j+1}.

a_j	possible a_{j+1}
$1+i$	$x-iy, x+y \geq 2$
$1-i$	$x+iy, x+y \geq 2$
2	$x+iy, x \geq 0, |x|+|y| \geq 2$
$2+i$	$G'\setminus\{-1+i\}$
$1+2i$	$G'\setminus\{-1+i\}$

5.4 Growth of $|q_n|$ and the Quality of the Hurwitz Approximations

Hurwitz continued fraction convergents to a target z_0 are good approximations. With the notation established above, we have

$$z_0 - \frac{p_n}{q_n} = \frac{(-1)^n z_n}{q_n^2(1+z_n w_n)}.$$

Since $|z_n| < 1/\sqrt{2}$, and $|w_n| < 1$, it follows immediately that $|z_0 - p_n/q_n| < 2\sqrt{2}|z_n|/|q_n|^2$. When z_0 is a Gaussian quadratic irrational, such as $\sqrt{2+i}-1$ or, for that matter, $\sqrt{2}-1$, the Hurwitz continued fraction expansion is (eventually) periodic, and $|z_n|$ is bounded below by a positive constant. On the other hand, a simple computation with (algebraic) conjugates shows that when the target z_0 is a Gaussian quadratic irrational, there exists $\epsilon > 0$, depending on z_0, such that for all Gaussian integers p, q with $q \neq 0$, $|z_0 - p/q| > \epsilon|q|^{-2}$. Thus for such targets, the Hurwitz continued fraction gives approximations that are essentially best possible. Whenever z_n is small, we get particularly good approximations, but this alone does not exclude the possibility that there might be particularly good approximations that the Hurwitz algorithm misses. The algorithm does, however, achieve approximations that are essentially best possible, in the following sense:

Theorem 5.1 *Suppose $\alpha \in \mathbb{C}$ has a Hurwitz algorithm sequence of convergents to depth at least n, and suppose (p_{n-1}, q_{n-1}) and (p_n, q_n) are the numerators and denominators of the $n-1$th and nth convergents. Suppose*

$q \in G$ with $|q_{n-1}| < |q| \le |q_n|$, $p \in G$, and $p/q \ne p_n/q_n$. Then

$$\left|\frac{p}{q} - \alpha\right| \ge \frac{1}{5}\left|\frac{p_n}{q_n} - \alpha\right| \cdot \left|\frac{q_n}{q}\right|.$$

We defer the proof as we shall need some of the machinery developed below.

We now discuss the growth of the continuants, that is, the denominators q_n. The first result along these lines is due to Hurwitz, who showed that for all initial $z_0 \in B$, and all n such that the expansion has not yet terminated, $|q_n| > |q_{n-1}|$. His proof is based on the observations given above.

To this, we now add

Theorem 5.2 *If $z \in B$ has a Hurwitz continued fraction expansion to depth $n + 2$, then*

$$\left|\frac{q_{n+2}}{q_n}\right| \ge 3/2.$$

Proof. Recall that

$$q_{n+1} = a_{n+1}q_n + q_{n-1}$$

$$z_{n+1} = \frac{1}{z_n} - a_{n+1}$$

$$w_{n+1} = \frac{1}{a_{n+1} + w_n}$$

$$a_{n+1} = \left[\frac{1}{z_n}\right]$$

$$\frac{q_n}{q_{n+1}} = w_{n+1}.$$

Let G_a denote the set of possible values in G' for the successor a_{j+1} to a_j if $a_j = a$. We shall need some estimates for w_n.

Lemma 5.1 *Suppose $z \in B$, $n \ge 1$, and z has a Hurwitz continued fraction to depth $n + 2$. If $|w_n| \ge 2/3$, then $|w_{n+1}| < 2/3$. Furthermore, either $|w_n| < 2/3$, or $\frac{2}{3} \le |w_n| < 1$ and one of the following, or its negative or complex-conjugate counterpart, holds:*

$$\left|w_n - \frac{9}{14}(1-i)\right| < \frac{3}{7} \text{ and } a_n = 1 + i$$

$$\left|w_n - \frac{9}{16}\right| < \frac{3}{16} \text{ and } a_n = 2.$$

The proof of the lemma is inductive and begins with the observation that the claim is true at the outset because $w_0 = 0$. We use a fact about reciprocals of disks: If D is a disk in $\mathbb{C}$ with center z and radius r, and if $|z| > r$, then with the notation $D(s, r) := \{z \in \mathbb{C} : |z - s| < r\}$,

$$\frac{1}{D(s,r)} = D\left(\frac{\overline{s}}{|s|^2 - r^2}, \frac{r}{|s|^2 - r^2}\right).$$

Now let D_0 be the disk about 0 with radius $2/3$, and let $D_1 = 1/(1+i+D_0) = D((9/14)(1-i), 3/7)$, $D_2 = 1/(-1+i+D_0)$, $D_3 = -1/(1+i+D_0)$, $D_4 = 1/(1-i+D_0)$, $D_5 = 1/(2+D_0)$, $D_6 = 1/(2i+D_0)$, $D_7 = -1/(2+D_0)$, and $D_8 = 1/(-2i+D_0)$. Assuming the claim to be true for $k \leq n$, we have either that $|w_n| \in D_0$, or that w_n lies in the intersection of one of 8 other disks with the open unit disk, and a_n takes a corresponding specific value, either $\pm 1 \pm i$ or ± 2 or $\pm 2i$.

Now if $|a| \geq \sqrt{5}$, then $1/(a + D_0) \subseteq D_0$, while if a is one of the 8 Gaussian integers in G' nearest the origin, $1/(a + D_0)$ is one of $D_1, \ldots, D_8$. Thus the inductive step cannot fail in the case that $w_n \in D_0$. On the other hand, for $1 \leq k \leq 8$, one readily checks that for any eligible successor a to a_n, $1/(D_k + a) \subseteq D_0$. Thus, for instance, when $w_n \in D_1 \setminus D_0$, we have $a_n = 1 + i$ so that $a_{n+1} \in G_{1+i} = \{u - iv : u \geq 0, v \geq 0, u + v \geq 2\}$. For $a \in G_{1+i}$, though, if $|a| \geq 2\sqrt{2}$ and $|w| < 1$ then $|1/(a+w)| < 2/3$, while if $a \in \{2, 2-i, 1-i, 1-2i, -2i\}$ then, from case by case calculation, $1/(a + D_1) \subset D_0$. For instance, $1/(1 - i + D_1) = D((23/73)(1+i), 6/73)$, while $1/(2 - i + D_1) = D(37/133 + 23i/133, 6/133)$. Thus the inductive step cannot fail in any of these other cases either, as $|w_{n+1}| < 2/3$. We now note that we have shown that whenever $|w_n| \geq 2/3$, $|w_{n+1}| < 2/3$.

At this point the proof of the theorem falls right out: $|w_n||w_{n+1}| < 2/3$ because both factors are less than 1 and one of them is less than $2/3$, and so $|q_{n+2}/q_n| = 1/|w_n w_{n+1}| > 3/2$. $\square$

We now give the proof of Theorem 5.1. We first write $(p, q) = s(p_{n-1}, q_{n-1}) + t(p_n, q_n)$. If $s = 0$ the estimate is immediate. We now break the question into two main cases: $|s| = 1$ and $|s| > 1$.

If $|s| = 1$, then we may multiply s and t by the same unit and so take $s = 1$. Now $|\alpha - p_n/q_n| = |z_n|/(|q_n(q_n + z_n q_{n-1})|)$. Thus we must show that if $q \neq q_n$ then

$$\left|\frac{p_n + z_n p_{n-1}}{q_n + z_n q_{n-1}} - \frac{tp_n + p_{n-1}}{tq_n + q_{n-1}}\right| > \frac{1}{5}\frac{|z_n|}{|q(q_n + z_n q_{n-1})|},$$

or equivalently, $|t - 1/z_n| > 1/5$. Now $q = q_{n+1}$ if and only if $t = [1/z_n]$. But $|q| \leq |q_n| < |q_{n+1}|$, so $t \neq [1/z_n]$, so $|t - 1/z_n| \geq 1/2$. This completes the proof for the case $|s| = 1$.

We break the case $|s| > 1$ down into subcases depending on the value of a_n. There is sufficient symmetry that rotations and reflections are of no consequence, so these cases reduce to the following: $|a_n| \geq 3$, $a_n = 2 + 2i$, $a_n = 2 + i$, $a_n = 2$, and $a_n = 1 + i$. We now assume that p/q gives a counterexample at depth n. The premise $|q_{n-1}| < |q| \leq |q_n|$ gives $|w_n| < |sw_n + t| \leq 1$, while our assumption that p/q is a good approximation to α gives $|s/z_n - t| \leq (1/5)$. Equivalently,

$$\left|\frac{1}{z_n} - \frac{t}{s}\right| \leq \frac{1}{5|s|}, \quad \left|w_n + \frac{t}{s}\right| \leq \frac{1}{|s|}.$$

From these it follows that $|w_n + 1/z_n| < 6/5|s|)$.

Now the value of a_n constrains both w_n and z_n. It constrains w_n because $w_n = 1/(a_n + w_{n-1})$ and $|w_{n-1}| < 1$. Thus, w_n belongs to the disk $D(\overline{a_n}/(|a_n|^2 - 1), 1/(|a_n|^2 - 1))$ where $D(c, r)$ denotes the disk $\{z \in \mathbb{C} : |z - c| < r\}$. The effect of a_n upon the possible values of z_n comes from the fact that $z_n \in D(a_n, 1) \cap B$, which becomes important only when $|a_n| \leq \sqrt{5}$.

For the case $|a_n| \geq 3$, we have $|w_n| < 1/2$ while $|1/z_n| \geq \sqrt{2}$, so $|w_n + 1/z_n| > \sqrt{2} - 1/2 > 6/5\sqrt{2}$. If $a_n = 2 + 2i$ then $w_n \in D((2-2i)/7, 1/7)$. Thus it will suffice to complete this case, that $D((-2+2i)/7, (1/7 + 6/(5\sqrt{2}))$ sits inside the union of the disks about ± 1 and $\pm i$ of radius 1. It does, since $|-(1+i)+(2-2i)/7| = 5\sqrt{2}/7 > 6/(5\sqrt{2}) + 1/7$.

If $a_n = 2 + i$ then $z_n \subseteq B \backslash D(-1-i, 1)$, so $1/z_n$ lies outside the union of the disks of radius 1 about ± 1, $\pm i$, and $-1+i$. On the other hand, $w_n \in D((2-i)/4, 1/4))$. The disk about $-(2-i)/4$ of radius $(1/4 + 6/(5\sqrt{2}))$ fits comfortably within the region from which $1/z_n$ is excluded, which completes the proof for the case $a_n = 2 + i$.

If $a_n = 2$, then $w_n \in D(2/3, 1/3)$ while $z_n \in B \backslash D(-1, 1)$ so that $1/z_n$ belongs outside the union of our disks and the half-space $\Re z < -1/2$. The disk $D(-2/3, 1/3 + 6/(5\sqrt{2}))$ fits comfortably inside the region from which $1/z_n$ is excluded, which completes the proof for the case $a_n = 2$.

Finally, if $a_n = 1 + i$, then $w_n \in D(1 - i, 1)$ while $z_n \in B \backslash (D(-1, 1) \cup D(-i, 1))$. Thus $1/z_n$ is excluded from the union of our four unit disks about ± 1 and $\pm i$, and the half-planes $\Re z < -1/2$ and $\Im z > 1/2$. But $D(-1+i, 1+6/(5\sqrt{2}))$ fits comfortably within this excluded zone.

Fig. 5.7 Excluded values of $1/z_n$ and disks for $-w_n$.

5.5 Distribution of the Remainders

The sequence $\langle z_n \rangle$ of remainders arising out of execution of the Hurwitz continued fraction algorithm on input z_0 depends, of course, on the input z_0. If z_0 is rational, then it terminates. If z_0 is a Gaussian quadratic irrational, that is, if z_0 satisfies a quadratic polynomial over $\mathbb{Q}(i)$, then it is ultimately periodic. [Hur]. These things are no surprise in view of what is known about the classical continued fraction algorithm.

In the classical theory of continued fractions, we have the famous Gauss-Kuz'min theorem to the effect that if X is taken at random with uniform distribution in $[0,1]$, then the probability density function of $T^k X$ converges exponentially fast to the Gauss density $\frac{1}{\log 2}\frac{1}{1+x}$ on $[0,1]$. This density is characterized by the fact that it is absolutely continuous with respect to Lebesgue measure and is invariant under the map $x \to 1/x - \lfloor 1/x \rfloor$. Numerical experiments suggest that something in the same vein might be true for the Hurwitz complex continued fraction. We do obtain such a theorem, but for the moment, we just show a picture of the density that plays the rôle for the Hurwitz continued fraction that the Gauss density plays for the classical real continued fraction.

The invariant density ρ for the Hurwitz algorithm is, in its own way, rather pretty: we show here a false-color image of a somewhat crudely computed approximation to it. A simple expression for this invariant density is not known, but the topic is new and there may well be one. The figure suggests, and we later prove, that this density is real-analytic on the interiors of the 12 regions into which arcs of circles of radius 1 centered at ± 1, $\pm i$, and $\pm 1 \pm i$ cut B, and that it has the symmetry group of the rotations and reflections that map $\overline{B}$ to itself. A key fact is that T maps the union of these arcs, and the boundary of B, into that same union.

Knowing this density, we can determine the relative frequency of the partial quotients a_j. For instance, the limiting frequency with which $a_j = (1+i)$, as $j \to \infty$, is $\int_{E[1+i]} \rho$, where $E[a] = \{z \in B : \lceil 1/z \rceil = a\}$, so that

Fig. 5.8 Invariant density for the Hurwitz algorithm.

$E[1+i]$ is the region bounded by the line segments $1/2 + it$, $-1/2 \leq t \leq -\sqrt{3}/6$, $t - i/2$, $\sqrt{3}/6 \leq t \leq 1/2$, and by arcs of the circles $|z - 1/3| = 1/3$ and $|z + i/3| = 1/3$ joining $(1-i)/3$ and $(1-i)/2$. We can also make a well-founded conjecture as to the average speed with which the algorithm brings a random Gaussian-rational input to zero.

As in the case of the rational integers and the classical gcd algorithm, a continued fraction algorithm executed on rational inputs is, en passant, computing the gcd of the input pair, as well as auxiliary information. Consider an input $z_0 = s_0/t_0$. As execution of the Hurwitz algorithm proceeds, successive $z_n = s_n/t_n$ are computed according to the rule $z_{n+1} = 1/z_n - [1/z_n] = 1/z_n - a_{n+1}$. From this, we have $s_{n+1}t_{n+1} = s_n t_n (1 - a_{n+1} z_n) = s_n t_n z_{n+1} z_n$. We conjecture that for typical random rational inputs drawn from $|s_0| < R$, $|t_0| < R$ say, the distribution of (z_n) more or less conforms to ρ, except at the beginning, where it will be more uniform, or near the end when it will of necessity be concentrated on a few simple Gaussian rationals. From this it would follow that the average number of steps needed to execute the Gaussian gcd algorithm should be proportional to $\log |t_n|$, with the constant of proportionality being $1/\int_B \log|z| \rho(z)\,dm(z)$. This integral was estimated by Monte Carlo methods to be 1.092766. Numerical tests of the conjecture on 100 pseudorandom pairs with s, t chosen at random among Gaussian integers with real and imaginary parts of absolute value less than 10^{1000} gave an aggregate of 210863 steps to finish all 100 inputs, as against a predicted value of 210679.

It would be nice to know the asymptotic behavior of the growth of q_n. For this, it would seem at first that we would need the distribution of w_n; if we had a function ω along the lines of ρ for the distribution of the successive w_n, $n \int_D (-\log|z|\omega(z))$ would be the expected value of $\log|q_n|$. This hypothetical ω can be guessed at by amassing some statistics; here is a pictures of the result. The range of the picture is the square $|\Re(z)| \leq 1$, $|\Im(z)| \leq 1$. On grounds of Hurwitz's result, we know in advance that the picture sits within the unit circle tangent to the four edges of this frame. What is not clear in advance is that the set in which ω is positive is a fractal, and that within this domain, there are interior fractal boundaries delimiting regions of greater and lesser intensity.

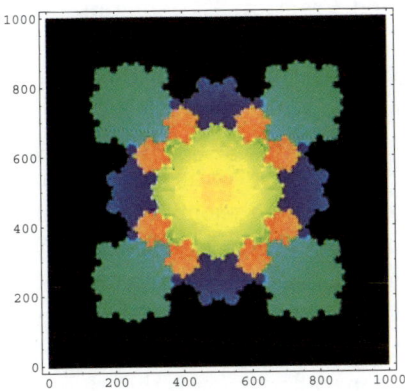

Fig. 5.9   Color representation of ω.

On the other hand, we can turn around the speculations about the rate at which the Hurwitz algorithm deals with random rational inputs. Suppose the algorithm is given s_0/t_0, and requires n steps to reduce this to zero. As the algorithm is executing it is at the same time building a fraction p_n/q_n equal to the input. In general, $\log|q_n|$ will be not much less than $|\log t_n|$ because it is rare that s_0 and t_0 will have a large common factor. Thus, we should expect the rate at which t_0 is eroded to zero to be equal to the rate at which q_k builds up, and if this rate is the same for typical random rational inputs as it is for the average complex input (taken at random from B with a uniform distribution), then we should expect that almost always, $\log|q_n| \approx 1.092766n$. A simple numerical experiment is not quite trivial, as it is not feasible to carry a computation to an appropriate depth with a pseudorandom number. What we can do is to take advantage of the rough and ready randomizing effect of rounding to a fixed number of digits as we go, which has the side benefit of restoring 'precision' to floating-point numbers that lose precision during extracting their Hurwitz expansion. The exact value one arrives at will depend on the details of the truncation and precision rules employed; the author got a rate of 1.09302 during 10^6 steps, on one run, and 1.09299 on another.

We defer the proof of the existence of this invariant density ρ for the moment, and take up a rather different topic.

5.6 A Class of Algebraic Approximants with Atypical Hurwitz Continued Fraction Expansions

Algebraic numbers may be expected to perhaps have atypical diophantine approximation properties. When we are dealing with real numbers, the situation is not all that well understood but certain things are well-known. Rational numbers have terminating continued fraction expansions. Real roots α of an irreducible quadratic polynomial over $\mathbb{Q}$ have ultimately periodic continued fraction expansions. As a result of this, the sequence of remainders is ultimately periodic, while the sequence of scaled errors $q_n^2(\alpha - p_n/q_n)$ converges to a periodic loop. For real α of degree greater than 2, there are results concerning simultaneous diophantine approximation of the powers of α, but concerning α alone, all we know is that the continued fraction expansion cannot involve extraordinarily large integers, such as would be typical of Liouville numbers. Thus, the theoretical frequency of $1, 2, \ldots, 10$ according to the Gauss density is $(0.415037, 0.169925, 0.0931094, 0.0588937, 0.040642, 0.0297473, 0.0227201, 0.0179219, 0.0144996, 0.0119726)$, while the actual frequency of these same partial quotients in the expansion of $2^{1/3}$ to a depth of 10000 is $(0.4173, 0.1675, 0.0946, 0.0636, 0.0421, 0.0295, 0.024, 0.0163, 0.0122, 0.0118)$.

In the case of the Hurwitz algorithm, at first it seems that we are in for more of the same. When the approximation target is in $\mathbb{Q}(i)$, the algorithm terminates, having found a reduced Gaussian rational equal to the original target. When the approximation target satisfies an irreducible quadratic polynomial over $\mathbb{Q}(i)$, the Hurwitz expansion is ultimately periodic. [Hur]. But if we look a bit further, we come upon a new behavior. Somewhat typical here is what happens with $z_0 = \sqrt{2} - 1 + i(\sqrt{5} - 2)$. This z_0 satisfies an irreducible quartic polynomial over $\mathbb{Q}(i)$, to wit, $x^4 + (4 + 8i)x^3 - (12 -$

$24i)x^2 - (32 - 16i)x + 24$. But the Hurwitz continued fraction expansion of z_0 is neither ultimately periodic, nor anything like the typical expansion. Instead, what we see is that

(1) The sequence of remainders is confined to a certain finite set of arcs of a circle. None of these arcs goes through the origin, and so the sequence of remainders is bounded away from the origin.
(2) The scaled errors (which we now define to be $|q_n|^2(z_0 - p_n/q_n)$), all lie very near one or another of a set of six parallel line segments cutting through B, equally spaced except that the one that would have gone through the origin is missing. The lines have equations of the form $\sqrt{2}x + \sqrt{5}y = n/2$ where n is an integer, $-3 \le n \le 3$.
(3) The ratios of successive q_n are confined as well. They are not distributed like the fractal we saw previously. Instead, they appear to lie very near one or another of a set of circular arcs, as with the remainders.

Fig. 5.10   Remainders for $\sqrt{2} + i\sqrt{5}$ and $\sqrt{2} + i\sqrt{3}$.

For most of these observations, we can provide reasons.

Theorem 5.3   *For arbitrary positive integers u, v such that u, v, and uv are not square, there is a finite set $\mathcal{A}$ of circles and lines in $\mathbb{C}$ such that for all $j \ge 1$, $T^j z_0 \in \cup_{A \in \mathcal{A}} A \cap B$.*

Proof.   Let $C(\gamma, r)$ denote the circle in $\mathbb{C}$ about γ of radius r. Fix positive integers u and v, neither square, and assume that uv is also not square. Let $a_0 = [\sqrt{u} + i\sqrt{v}]$, $m = u + v$, and $z_0 = \sqrt{u} + i\sqrt{v} - a_0 \in B$. As z_0 is not

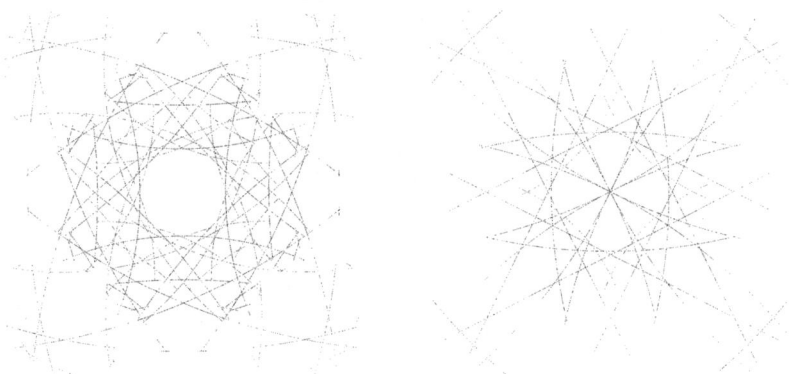

Fig. 5.11 Ratios of successive q_j for $\sqrt{2}+i\sqrt{5}$ and $\sqrt{2}+i\sqrt{3}$.

a Gaussian rational, the Hurwitz continued fraction expansion of z_0 is infinite. Let $\langle z_0, z_1, \ldots \rangle$ be the sequence of remainders, and $\langle a_0, a_1, \ldots \rangle$ the sequence of partial quotients, thus generated. Let $\langle p_0, p_1, \ldots \rangle$ and $\langle q_0, q_1, \ldots \rangle$ the sequence of numerators and denominators, generated by the Hurwitz algorithm on input $\sqrt{u}+i\sqrt{v}$. By convention, $p_{-1}=1, q_{-1}=0, p_0=a_0$, and $q_0=1$. Let $w_0=0$, and for $j\geq 1$, let $w_j=q_{j-1}/q_j$. Equivalently, for $j\geq 1$, $w_j=1/(a_j+w_{j-1})$.

We now consider a variant of the Hurwitz algorithm in which we keep track not only of the points $z_n = T^n z_0$, but of particular circles or lines on which z_n must lie. Our variant algorithm starts at an initial state $(z_0, C(-a_0, \sqrt{m}))$. All states have the form (x, K), where K is either a circle passing through B of the form $(C(\gamma/d, \sqrt{m}/|d|))$ with $\gamma \in G^*$, $d \in \mathbb{Z}^*$, and $d \mid (|\gamma|^2 - m)$, and with radius $r = \sqrt{m}/|d|$ satisfying $1/(2\sqrt{2}) \leq r \leq \sqrt{m}$, or a line passing through B of the form $L(k/(2\delta), i\bar{\delta})$ where $k \in \mathbb{Z}$ and $|\delta|^2 = m$.

Let $T(x, K) = (1/x - a, \phi_a(K))$ where $a = [1/x]$ and ϕ_a is the linear fractional map given by $\phi_a(z) = 1/z - a$.

Each statement below is readily verified and the details are left to the reader.

(1) If $x \in K \cap B$, K is a circle of the form $C(\gamma/d, \sqrt{m}/|d|)$ with $d \in \mathbb{Z}^*$, $d \mid (|\gamma|^2 - m) \neq 0$, and $\sqrt{m}/d \geq 1/(2\sqrt{2})$, then $T(x, K) = (x', K')$ where $x' \in K' \cap B$ and K' is another circle of the same form satisfying the same conditions.

(2) If $x \in K \cap B$, K is a circle of the form $C(\gamma/d, \sqrt{m}/d)$ with $d \in \mathbb{Z}^*$, $|\gamma|^2 = m$, and $\sqrt{m}/d \geq 1/(2\sqrt{2})$, then $T(x, K) = (x', K')$ with $x' \in B \cap K'$, and K' is the line $L((d - 2\Re(a\gamma))/(2\gamma), i\overline{\gamma})$, which passes through B.
(3) If $x \in K \cap B$ and K is a line of the form $L(0, i\overline{\delta})$ with $|\delta|^2 = m$, then $T(x, K) = (x', K')$ where $K' = L(-\Re(a\overline{\delta})/\overline{\delta}, i\delta)$.
(4) If $x \in K \cap B$ and K is a line of the form $L(k/2\delta, i\overline{\delta})$ with $k \in \mathbb{Z}^*$, and $|\delta|^2 = m$ then $T(x, K) = (x', K')$ where $x' \in K' \cap B$ and $K' = C\left(\frac{\overline{\delta}-ak}{k}, \frac{|\delta|}{|k|}\right)$, the new circle passes through 0 so that it has the required form, and the radius $|\delta|/|k| = \sqrt{m}/|k|$ is at least $1/\sqrt{2}$ because the diameter of K' is at least $\sqrt{2}$ because it is the reciprocal of the distance from 0 to the point on K nearest 0, and that nearest point lies within B and thus lies at a distance of less than $1/\sqrt{2}$ from 0.

There are only finitely many lines of the form $L(k/2\delta, i\overline{\delta})$ passing through B, with $|\delta|^2 = m$. There are only finitely many circles of the form $C(\gamma/d, \sqrt{m}/|d|)$ with $\gamma \in G$, $d \in \mathbb{Z}^*$, $d \mid (|\gamma|^2 - m) \neq 0$, and $\sqrt{m}/d \geq 1/(2\sqrt{2})$. We take $\mathcal{A}$ to be the set of all such lines and circles. □

We now turn to the topic of scaled errors. Let e_j be the scaled error for the jth convergent. Since $z_0 = (p_j + z_n p_{j-1})/(q_j + z_n q_{j-1})$, and since $\begin{vmatrix} p_{j-1} & p_j \\ q_{j-1} & q_j \end{vmatrix} = (-1)^j$, the scaled errors are also given by

$$e_j = \frac{(-1)^j \overline{q_j} z_j}{q_j + z_j q_{j-1}}.$$

Theorem 5.4 *Let u and v be positive integers. Let $\alpha = \sqrt{u} + i\sqrt{v}$. Let p, q be Gaussian integers with $|q|^2 \geq 9$, and assume $|p - q\alpha| \leq 2\sqrt{2}/|q|$. Let $e(q, p, \alpha)$ be the scaled error $e(q, p, \alpha) = \overline{q}(q\alpha - p)$. Then*

$$\Re(\overline{\alpha} e(q, p, \alpha)) = \frac{n}{2} + \delta$$

where $n = (|p|^2 - (u+v)|q|^2)$ is an integer satisfying $|n| \leq 8\sqrt{u+v}$ and $\delta \in \mathbb{C}$ with $|\delta| \leq 4/|q|^2$.

Remark 5.1 *In other words, the scaled errors lie almost, but not quite, on a finite set of parallel lines. The amount by which the scaled errors stand off from these lines is itself at most comparable to $|q|^{-2}$.*

Proof. Let $p = p_1 + ip_2$, $q = q_1 + iq_2$ with $p_1, p_2, q_1, q_2 \in \mathbb{Z}$. Let $c = p_1 q_1 + p_2 q_2$, $d = p_2 q_1 - p_1 q_2$, so that $c^2 + d^2 = |p|^2 |q|^2$. Let $\theta_1 + i\theta_2 = e(q, p, \alpha)$.

Then the given conditions are then equivalent to

$$(p_1 + ip_2)(q_1 - iq_2) = |q|^2(\sqrt{u} + i\sqrt{v}) + \theta_1 + i\theta_2,$$
$$e(q,p,\alpha) = (c - |q|^2\sqrt{u}) + i(d - |q|^2\sqrt{v}) = \theta_1 + i\theta_2,$$
$$\theta_1^2 + \theta_2^2 \leq 8.$$

Thus, $\bar{\alpha}e(q,p,\alpha) = (\theta_1\sqrt{u} + \theta_2\sqrt{v}) + i(\theta_2\sqrt{u} - \theta_1\sqrt{v})$, and

$$\Re(\bar{\alpha}e(q,p,\alpha)) = \theta_1\sqrt{u} + \theta_2\sqrt{v} = (c\sqrt{u} - |q|^2 u) + (d\sqrt{v} - |q|^2 v).$$

Turning this around, we have $c = |q|^2\sqrt{u} + \theta_1$, $d = |q|^2\sqrt{v} + \theta_2$, and so

$$|p|^2|q|^2 = c^2 + d^2 = |q|^4(u+v) + 2|q|^2(\theta_1\sqrt{u} + \theta_2\sqrt{v}) + (\theta_1^2 + \theta_2^2).$$

Hence,

$$(\theta_1\sqrt{u} + \theta_2\sqrt{v}) = \frac{1}{2}(|p|^2 - (u+v)|q|^2) - \frac{\theta_1^2 + \theta_2^2}{2|q|^2}.$$

With $n = |p|^2 - (u+v)|q|^2$, we note that

$$|n| \leq |p^2 - (\sqrt{u} + i\sqrt{v})^2 q^2| = |p - \alpha q||2\alpha q + (p - \alpha q)|$$
$$\leq \frac{2\sqrt{2}}{|q|}\left(2\sqrt{u+v}|q| + \frac{2\sqrt{2}}{|q|}\right) \leq (4\sqrt{2}\sqrt{u+v} + 8/|q|^2) \leq 8\sqrt{u+v}.$$

$\square$

This takes us back to the question of just what *is* typical of the orbits $T^n z$?

5.7 The Gauss-Kuz'min Density for the Hurwitz Algorithm

There can be more than one measure on a set that is invariant under a self-mapping T. There can, in principle, be more than one that is absolutely continuous with respect to Lebesgue measure.

We are certainly not exempt from the first possibility, as there are periodic orbits and uniformly weighted discrete measures on these are invariant under T. But the second pathology does not exist.

Theorem 5.5 *There is a unique measure μ on B that is absolutely continuous with respect to Lebesgue measure and invariant under T. This measure has a density function ρ; the density is continuous except perhaps along the arcs $|z \pm 1| = 1$, $|z \pm i| = 1$, and $|z \pm 1 \pm i| = 1$. It is real-analytic on an each of the 12 open regions into which B° is dissected by these arcs.*

The mapping T on B is mixing with respect to μ: for Lebesgue-measurable $A_1, A_2 \subseteq B$,

$$\lim_{n\to\infty} \mu(T^{-n}A_1 \cap A_2) = \mu(A_1)\mu(A_2).$$

The expected symmetry obtains: $\mu(iA) = \mu(A)$ and $\mu(\overline{A}) = \mu(A)$ for all measurable A.

The general plan of the proof is this: We show that if f is a continuous probability density function on B and X a random variable on B with density f, then there exists a positive integer n, depending on f, and $\epsilon > 0$, such that the density of $T^n X$ is greater than ϵ everywhere in $B°$.

The theory of positive operators [Kr] can then be applied. The result we use is an extension of the Perron-Frobenius theorem to certain positive operators. (The classical Perron-Frobenius theorem states that if M is a square matrix with nonnegative entries, and if there exists n such that all entries of M^n are positive, then the largest eigenvalue of M is positive and has multiplicity 1, and the corresponding eigenvector has positive entries.) What we first need is a Banach space $\mathcal{V}$ of functions on B and a positive compact linear operator L on $\mathcal{V}$ such that if a random variable X has density $f \in \mathcal{V}$ then the density of TX is Lf. Then we need an element z_0 of $\mathcal{V}$ with the property that for all nonzero positive $f \in \mathcal{V}$, there exists an n, and positive constants c_1 and c_2, such that $c_1 z_0 \leq L^n f \leq c_2 z_0$. The operator L is then said to be u_0-positive, and as a consequence, the operator L has a decomposition as $L = \lambda P + N$, where P is a positive projection of $\mathcal{V}$ onto a one-dimensional subspace of $\mathcal{V}$, λ a positive real number, and N an operator of spectral radius less than λ such that $PN = NP = 0$.

In our specific circumstances, some care must be given to the selection of the Banach space $\mathcal{B}$ in which our transfer operator will lie. We recall that $T: B\backslash\{0\} \to B$, with $Tz = 1/z - [1/z]$. We want $\mathcal{B}$ to consist of reasonably nice functions, and we want the transfer operator L_T for T to be compact.

Since the formula for the transfer operator is basically set in stone by the requirement that it *be* the transfer operator for T, our only latitude is in the selection of $\mathcal{B}$. We use the ideas of O. Bandtlow and O. Jenkinson from their recent work [BJ]. Their work features a single compact connected subset X of $\mathbb{R}$ and a map $T: X \to X$ that is "real-analytic full branch expanding with countably many branches." In our setting, it might at first seem as though we could take $X = B$ and use T as is, but unfortunately, the part about "full branch" fails. The conclusion they reach fails as well, at least on the numerical evidence, which should not be surprising as the premises were not met. The invariant density for our T is, on the numerical evidence, discontinuous along the arcs through B of the eight circles that dissect it into 12 distinct sectors.

Nevertheless, their work can be adapted to the situation at hand. The trick is to stitch together $\mathcal{B}$ as the Cartesian product of 12 distinct subsidiary Banach spaces $\mathcal{B}_1 \ldots \mathcal{B}_{12}$, and to break L_T up into a matrix of 144 compact operators carrying the various $\mathcal{B}_j$ to each other.

First, we describe the zones into which the numerical evidence prompts us to expect we must dissect B. Our zones B_j are connected, compact subsets of B with disjoint interiors, whose union is $\overline{B}$. We take the view that B is a subset of $\mathbb{R}^2$ rather than a subset of $\mathbb{C}$, we set $R : \mathbb{R}^2 \to \mathbb{R}^2$, $R(x,y) = (-y, x)$ (a counterclockwise rotation through an angle of $\pi/2$ radians), and we set

$$B_1 = \{(x,y) \mid 0 \leq x \leq 1/2, x^2 + (y-1)^2 \geq 1, \text{and } x^2 + (y+1)^2 \geq 1\},$$
$$B_2 = \{(x,y) \mid x^2 + (y-1)^2 \leq 1, (x-1)^2 + y^2 \leq 1, \text{and}$$
$$(x-1)^2 + (y-1)^2 \geq 1\},$$
$$B_3 = \{(x,y) \mid x \leq 1/2, y \leq 1/2, \text{ and } (x-1)^2 + (y-1)^2 \leq 1\},$$
$$B_4 = RB_1, \ B_5 = RB_2, \ B_6 = RB_3, \ B_7 = -B_1, \ B_8 = -B_2,$$
$$B_9 = -B_3, \ B_{10} = -B_4, \ B_{11} = -B_5, \ B_{12} = -B_6.$$

We take $U = \{(z_1, z_2) \in \mathbb{C}^2 \mid |z_1|^2 + |z_2|^2 < 1\}$ to be the open unit ball in $\mathbb{C}$, and, reserving the choice of ϵ but with $0 < \epsilon < 1/10$, we set $D_k = B_k + \epsilon U$, $1 \leq k \leq 12$. Thus, D_k is an ϵ-neighborhood of B_k, and all elements of D_k are close to a real pair $(x,y) \in B_k$.

Let $\psi : \mathbb{C}^2 \setminus \{(z, \pm iz) \mid z \in \mathbb{C}\} \to \mathbb{C}^2$ be given by

$$\psi(z_1, z_2) = \left(\frac{z_1}{z_1^2 + z_2^2}, \frac{-z_2}{z_1^2 + z_2^2} \right).$$

Remark 5.2 *Of course this is a thinly disguised translation of the mapping $1/z$ acting on $\mathbb{C}$. But we need a map on pairs of complex numbers that tracks what happens to a single complex number in terms of the real and imaginary parts.*

Let $G' \subset \mathbb{Z} + i\mathbb{Z} = \{a_1 + ia_2 \mid a_1, a_2 \in \mathbb{Z}, a_1 + ia_2 \neq 0, \pm 1, \pm i\}$, and for $1 \leq k \leq 12$, let $G_k = \{a \in G' \mid \psi(a + B_k) \subseteq \overline{B}.\}$. Then $G_1 = G' \setminus \{-1 \pm i, -2\}$, $G_2 = G' \setminus \{-1 \pm i, -2, -2i\}$, and $G_3 = G' \setminus \{-1 \pm i, 1 - i, -2, -2i, -2 - i, -1 - 2i\}$. Together with $G_{k+3} = iG_k$, this describes G_k, $1 \leq k \leq 12$.

Now T maps the arcs partitioning B into themselves or the boundary of B, so for each $a \in G'$, (abusing terminology by equating a with $(a_1, a_2) = (\Re a, \Im a)$,) either $\psi(a+B_k) \subseteq \mathbb{R}^2 \setminus B^\circ$, or for some j, $1 \leq j \leq 12$, $\psi(a+B_k) \subset B_j$.

Let $G_{j,k} = \{a \in G_k \mid \psi(a + B_k) \subset B_j\}$, and for $z \in D_k$, let $T_a(z) = \psi(a_1 + z_1, a_2 + z_2)$. At this point we need a lemma.

Lemma 5.2 *For ϵ sufficiently small, and $a \in G_{j,k}$, $T_a(B_k + \epsilon U) \subset B_j + \frac{4}{5}\epsilon U$.*

(That is, with an appropriate choice of ϵ, T_a maps D_k compactly into D_j, and T_a takes B_k into B_j.)

Proof. It will suffice to show that for $a \in G_k$ and $z \in B_k + \epsilon U$, $\|\psi'\| < 2/3$. Here, ψ' is the matrix derivative of ψ, and for $w \in \mathbb{C}^2$, $\|w\| = \sqrt{|w_1|^2 + |w_2|^2}$. Now for $a \in G_k$ and $z \in \epsilon U + B_k$, there exists $s \in a + B_k$ and $\zeta \in \epsilon U$ such that $z = s + \zeta$.

Consider $1/\overline{B}$ as a subset of $\mathbb{R}^2$. It is the exterior of the union of the open disks about $(\pm 1, 0)$ and $(0, \pm 1)$ of radius 1. Thus if $(a_1, a_2) \in G_k$ and $(s_1, s_2) \in (a_1, a_2) + B_k \subset 1/\overline{B}$, then $s_1^2 + s_2^2 \geq 2$.

It follows that for $(\zeta_1, \zeta_2) \in \epsilon U$ and $(s_1, s_2) \in 1/\overline{B}$,

$$\left|(s_1 + \zeta_1)^2 + (s_2 + \zeta_2)\right|^2 \geq (s_1^2 + s_2^2) - 2|s_1\zeta_1 + s_2\zeta_2| - \epsilon^2$$
$$\geq (s_1^2 + s_2^2) - 2\epsilon(s_1^2 + s_2^2)^{1/2} - \epsilon^2 \geq \|s\|^4(1 - 2\epsilon).$$

Now with $z = s + \zeta$ with $s \in 1/\overline{B}$ and $\zeta \in \epsilon U$, $\psi'(z) = (z_1^2 + z_2^2)^{-2} M_z$, where $|(z_1^2 + z_2^2)|^2 \geq (1 - 2\epsilon)\|s\|^4$ and where

$$M_z = \begin{pmatrix} z_2^2 - z_1^2 & -2z_1 z_2 \\ 2z_1 z_2 & z_2^2 - z_1^2 \end{pmatrix},$$

we have

$$M_z = M_s + 2\begin{pmatrix} s_2\zeta_2 - s_1\zeta_1 & s_1\zeta_2 - s_2\zeta_1 \\ s_1\zeta_2 + s_2\zeta_1 & s_2\zeta_2 - s_1\zeta_1 \end{pmatrix} + M_\zeta.$$

But the entries of $M_{s,\zeta}$ are each bounded in absolute value by $\epsilon(s_1^2 + s_2^2)^{1/2}$ by Cauchy's inequality, so $\|M_{s,\zeta}\| \leq 2\epsilon(s_1^2 + s_2^2)$ and similarly $\|M_\zeta\| \leq 2\epsilon^2$. From this it follows that for ϵ sufficiently small, $\|M_z\| \leq \|M_s\| + 3\epsilon\|s\|$. But by direct calculation, we see that $\|M_s\| = \|s\|^2$, so

$$\|M_z\| \leq \tfrac{4}{5}(1 - 2\epsilon)\|s\|^4 \leq \tfrac{4}{5}|z_1^2 + z_2^2|^2.$$

Thus $\|\psi'\| \leq 4/5$ for $z \in 1/\overline{B} + \epsilon U$. Since T_a mapped B_k into B_j, and since excursions from B_k in the input to T_a are damped by a factor of 4/5 or less in the output, the lemma is proved. □

Now consider the space $H^\infty(D_k)$ of bounded holomorphic functions from D_k into $\mathbb{C}$, equipped with the sup norm. Let $\mathcal{B}_k$ be the subspace of $H^\infty(D_k)$ consisting of elements that take B_k into $\mathbb{R}$, (equipped with the same norm.) Let $D''_j = \cup_{a \in G_{j,k}} T_a D_k$. As we have seen in our lemma above, D''_j fits inside a compact subset of D_j. We take D'_j to be a connected, open subset of D_j containing D'' and contained in a compact subset of D_j. Within $H^\infty(D'_j)$ we take $\mathcal{B}'_j$ to be the subspace consisting of all $f \in H^\infty(D'_j)$ such that f is real on $D'_j \cap B_j$.

For $a \in G_{j,k}$, $f \in H^\infty(D'_j)$ and $z \in D_k$ let

$$w_a(z) = ((a_1 + z_1)^2 + (a_2 + z_2)^2)^{-2},$$
$$L_a f = w_a(z) f(T_a(z)),$$
$$L_{j,k} = \sum_{a \in G_{j,k}} L_a.$$

Lemma 5.3 $L_{j,k}$ *is a bounded linear operator from $H^\infty D'_j$ into $H^\infty D_k$, it is compact, and it maps $\mathcal{B}'_j$ into $\mathcal{B}_k$.*

Proof. The proof of the first claim goes exactly as in Proposition 2.3 of [BJ], which (since this paper has not yet appeared) we quote with a few minor changes to adapt it to our slightly different circumstances. We take $\mathcal{L}$ to be $L_{j,k}$. In their work, a single domain serves both as D_k and D'_j, so all mention of either of these is our own adaptation; a verbatim quotation would have had D in place of both. We also use w_a in place of their w_i, and our set of subscripts is $G_{j,k}$ rather than $\mathbb{N}$.

Let $S := \sup_{z \in D_k} |w_a(z)| < \infty$. First we show that $\mathcal{L}$ maps $H^\infty(D'_j)$ to $H^\infty(D_k)$. Fix $f \in H^\infty(D'_j)$. If

$$g_k(z) = \sum_{|a| \leq k, a \in G_{j,k}} w_a(z) f(T_a(z))$$

for $k \in \mathbb{N}$, then $g_k \in H^\infty(D_k)$. Since

$$|g_k(z)| \leq \sum_{|a| \leq k, a \in G_{j,k}} |w_a(z)| \|f(T_a(z))\| \leq S \|f\|_{H^\infty(D'_j)} \quad (*)$$

for all $k \in \mathbb{N}$, we see that the sequence $\{g_k\}$ is uniformly bounded on D'_j. Moreover, $\lim_{k\to\infty} g_k(z) =: g(z)$ exists for every $z \in D_k$. By Vitali's convergence theorem (see e.g. [Nar], Chapter 1, Prop. 7) g_k thus converges uniformly on compact subsets of D_k. Hence, g is analytic on D_k. Moreover, by (*) we see that $|g(z)| \leq S\|f\|$ for any $z \in D_k$, so $g \in H^\infty(D_k)$. Thus, $\mathcal{L}f = g \in H^\infty(D_k)$.

In order to see that $\mathcal{L}$ is continuous we simply note that

$$\|\mathcal{L}f\|_{H^\infty(D_k)} = \sup_{z \in D_k} \left| \sum_{a \in G_{j,k}} w_a(z) f(T_a(z)) \right|$$
$$\leq \sup_{z \in D_k} \sum_{a \in G_{j,k}} |w_a(z)||f(T_a(z))|$$
$$\leq S\|f\|_{H^\infty(D'_j)}.$$

We next introduce their 'canonical embedding' operator J, defining J to be the operator which takes any element of $H^\infty(D_j)$ and regards it as an element of $H^\infty(D'_j)$. This, as they note, is, by Montel's theorem [Nar] Chapter 1, Prop. 6, a compact operator from $H^\infty(D'_j)$ into $H^\infty(D_j)$. But $L_{j,k} = \mathcal{L} \circ J$, and the composition of a bounded operator with a compact operator is compact. This proves the second claim. The third claim is obvious: if $z = (x, y) \in \mathbb{R}^2$, then $w_a(z) \in \mathbb{R}$ and $T_a(z) \in \mathbb{R}^2$, so for $f \in \mathcal{B}'_j$, $L_{j,k} f \in \mathcal{B}_k$. □

We are now at last in a position to define our overall transfer operator L. We take $\mathcal{B} := \mathcal{B}_1 \times \cdots \times \mathcal{B}_{12}$ to be our Banach space, equipped with the norm $\|f\| = \max_{1 \leq j \leq 12} \|f_j\|$, and we take L_T to be the operator on $\mathcal{B}$ given by

$$L_T \begin{pmatrix} f_1 \\ \vdots \\ f_{12} \end{pmatrix} = \begin{pmatrix} L_{1,1} & \cdots & L_{1,12} \\ \vdots & \ddots & \vdots \\ L_{12,1} & \cdots & L_{12,12} \end{pmatrix} \begin{pmatrix} f_1 \\ \vdots \\ f_{12} \end{pmatrix}.$$

This L_T is compact since its components are compact.

Remark 5.3 *One point worth noting is that for $z = (x, y) \in \mathbb{R}$, $w_a(z) = ((a_1 + x)^2 + (a_2 + y)^2)^{-2} = |(a_1 + ia_2) + (x + iy)|^{-4}$. Thus if we had defined w_a as a map on $\mathbb{C}$ rather than on $\mathbb{C}^2$, then it would not have been holomorphic, and the arguments of [BJ] could not have been used. Another point is that this formulation of L_T opens the way for effective computation of the invariant density ρ, which will be a suitably normalized multiple of the* one-dimensional eigenspace of L_T corresponding to the dominant eigenvalue 1. The operators $L_{j,k}$ will have matrices with respect to the expansion of f in a two-dimensional power series, and these matrices can be truncated with controllable errors, so that L_T becomes a computationally tractable object. A final point is that by taking full advantage of symmetry we can work with just $\mathcal{B}_1 \times \mathcal{B}_2 \times \mathcal{B}_3$.

At this point, we may as well regard L_T as an operator acting on a space of real-valued functions of $B \subseteq \mathbb{C}$, since the details of just which functions count as elements of our Banach space, and just how we coped with the difficulties presented by the boundaries of the B_k, are no longer relevant.

Having now constructed $\mathcal{B}$ and L_T, we next set about demonstrating that L_T, seen as an operator on $\mathcal{B}$, has the needed positivity properties.

What are these? We need the following, in the topology induced by our norm $\|\cdot\|$ on $\mathcal{B}$. There is a *positive cone* P with the following properties, some intrinsic to P and others related to L_T. We write $f \geq g$ if $f - g \in P$.

(1) P is closed under addition and scaling by positive numbers.
(2) $P^\circ \neq \emptyset$.
(3) P is a closed subset of $\mathcal{B}$.
(4) For all $f \in \mathcal{B}$, there exist $f_1, f_2 \in P$ such that $f = f_1 - f_2$. (P is reproducing.)
(5) L_T maps P into P.
(6) There exists $u_0 \in P^\circ$ such that if $f \neq 0$ and $f \in P$ then there exist $n \in \mathbb{Z}^+$ and $c_1, c_2 \in \mathbb{R}^+$ such that $c_1 u_0 \leq L_T^n f \leq c_2 u_0$. ($L_T$ is u_0-positive.)

We define P to be the set of all $f \in \mathcal{B}$ such that each of the f_k is nonnegative on B_k. Clearly P° is nonempty, and P is closed in $\mathcal{B}$. Any element of $\mathcal{B}$ can be written as $f_1 - f_2$ with $f_1 = f + C$ where $C = 2\max_{j=1}^{12} \|f\|$ and $f_2 = C$. Clearly f maps P into P. We claim that the element $f \in \mathcal{B}$ given by $f_k \equiv 1$ can serve as u_0, but this is far from clear. The reason it is *true* is that, given any open disk E in any of the B_k, there exists an open squarish (bounded by arcs of circles that meet at right angles) subset Q of E, and a positive integer n, such that the image under T^n of Q is B°. This we prove by demonstrating the existence of such a domain Q that does not, during iteration of T, touch the boundaries of B until the end, when it covers B. To this end, we first demonstrate that given $x \in \widetilde{B}$ and $\epsilon > 0$, there exists a Gaussian rational $r' \in \widetilde{B}$ and a neighborhood A of r', such that the orbit of r' under T stays inside $\widetilde{B}$ until the end, when, as it eventually must, it reaches zero and stops.

Using the existence of r', we then show that there exists also r, another Gaussian rational near x, a positive integer n, and a neighborhood A_r of r and contained within A, with the following properties:

(1) $T^k r \in \widetilde{B}$ for $1 \leq k < n$.
(2) $T^n r = 0$.
(3) If $a_1, a_2 \ldots$ are given by $a_k = [1/T^{k-1} r]$, and if ψ_1 is the linear fractional map $z \to 1/z - a_1$, and $\psi_k(z) = 1/\psi_{k-1}(z) - a_k$, then for $1 \leq k < n$, ψ_k takes A_r into the same one of the B_j to which $T_k r$ belongs, while the restriction of ψ_n to A_r is a conformal bijection from A_r to B°.

We begin our proof that such an r' exists by considering an auxiliary $s \in \widetilde{B}$ with $|s - x| < \epsilon/2$, and the histories of various elements of a small neighborhood of s under iteration of not only T, but all the variants of T got by truncating to the right, or top, of B instead of to the bottom and left as we have been in the habit of doing. We do this by means of a tree, which we shall call the *disk partition tree* of s, in which the vertices are ordered pairs $(P, \mathbf{a})$. The first entry is a 'triangle' in $\mathbb{C}$ with one vertex at s, bounded by arcs of a circle, and contained in the disk $|s - x| < \epsilon/2$. The second entry is a list $\mathbf{a}$ of Gaussian integers, all in G'. For vertices at distance j from the root of the tree, the triangles P are disjoint and fit together to cover an open disk $D_j \subset D_0$ centered on s, apart from their shared boundaries and from the perimeter of D_j. The lists $\mathbf{a}_P$ paired with these 'pizza slices' all have length j, and an edge extends down the tree from $(P, \mathbf{a}_P)$ to $(Q, \mathbf{a}_Q)$ if and only if $\mathbf{a}_Q$ is an extension of $\mathbf{a}_P$ and $Q \subseteq P$. The key point of the construction is that we arrange matters so that each slice P at depth j is carried intact to a region inside B° by all iterates of T of depth 1 through j. We now turn to the details.

Let $s_0 = s$. Since $s \in \widetilde{B}$, $1/s_0 \pmod{G} \in B^\circ$, though it could happen that $1/s_0 \pmod{G} \notin \widetilde{B}$. For however long it may be that this does not happen, we construct a list of further s_j, and a nested list D_j of open disks about s_0, starting with $D_0 = \{z \in \widetilde{B} : |z - s| < \epsilon/4\}$. We put $a_1 = [1/s_0]$ and $s_1 = 1/s_0 - a_1$, and we take D_1 to be a (possibly smaller) disk about s_0 such that $1/D_1 - a_1 \subseteq B^\circ$. We continue the construction taking $a_2 = [1/s_1]$, $s_2 = 1/s_1 - a_2$, and $D_2 \subseteq D_1$ so that

$$\frac{1}{\frac{1}{D_2} - a_1} - a_2 \subseteq B^\circ,$$

and so on. Eventually, (unless we are lucky) there will be a first integer k

such that $T^j s_0 \in \widetilde{B}$ for $1 \le j < k$ but $1/T^{k-1} s_0$ has either its real part, or its imaginary part, on the boundary of some translate of B by a Gaussian integer.

To track what now happens, we bring in some linear fractional mappings defined in terms of a list $\mathbf{a}$ of Gaussian integers. If $\mathbf{a} = (a_1, \ldots, a_k)$, let $\psi[\mathbf{a}, j]$ be the linear fractional given by

$$\psi[\mathbf{a}, j](z) = \begin{cases} \frac{1}{z} - a_1, & j = 1 \\ \frac{1}{\psi[\mathbf{a}, j-1](z)} - a_j, & 1 < j \le k \end{cases}.$$

Let $\psi[\mathbf{a}] = \psi[\mathbf{a}, k]$. Thus if we put $\psi_g := z \to 1/z - g$, then $\psi[\mathbf{a}, j] = \psi_{a_j} \circ \psi_{a_{j-1}} \cdots \circ \psi_{a_1}$. By our construction, if $\mathbf{a} = (a_1, \ldots, a_{k-1})$ then for $1 \le j \le k-1$, $\psi[\mathbf{a}, j]$ maps D_k conformally into B°, but $s_{k-1} = T^{k-1} s_0$ lies on the boundary of B. Let $a_k = [1/s_k]$, and consider the inverse image under $\psi[(a_1, \ldots, a_k), k]$ of the line (or, in the exceptional case that $s_k = (-1-i)/2$, two lines) in $\mathbb{C}$ and bounding B on which s_k lies. This line, (or these lines), will cut D_k into two or four pieces. We take the disks D_j

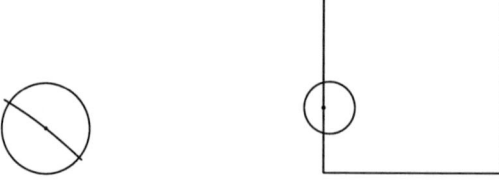

Fig. 5.12 D_{k-1} cut by ψ^{-1}(edge) and $\psi[\mathbf{a}](D_{k-1})$ cut by ∂B.

small enough that these circles, inverse images of one of the lines bounding B, meet each other only at the center, and so that they all exit all the D_j.

Now $\psi[(a_1, \ldots, a_k)]$ maps one of the parts into which this arc cuts D_k conformally into B°, and carries the other part to the exterior of B. But $\psi[(a_1, \ldots, a_{k-1}, 1+a_k)]$ carries that part of D_k conformally into B°. In the exceptional case that $T^k s_0 = -\frac{1}{2}(1+i)$, we would see instead a circle on the left about s_0, roughly quartered, mapped by ψ to a disk enclosing, (though typically not centered at) $-\frac{1}{2}(1+i)$. One of the quarters would map to inside B°, while the others would be mapped there by a modification of ψ corresponding to a change in the last entry in $\mathbf{a}$. Note that this exceptional case can only occur at stage n, because $T(-\frac{1}{2}(1+i)) = 0$.

Both in the main case, and in this special case, D_k has been partitioned into a finite number of 'slices', together with an equal number of bounding arcs each running along some circle from the center s_0 to the rim of D_k, and for each of the slices P, there is a corresponding list $\mathbf{a}_P$ of Gaussian integers, and a corresponding linear fractional map $\psi[\mathbf{a}_P]$, that maps $P \subset D_0$ conformally into $B°$. This begins the construction of a list of similar partitions $\mathcal{P}_j$ of neighborhoods D_j of s_0.

The vertices of the tree are pairs $(P, \mathbf{a})$. Edges join $(P, \mathbf{a})$ and $(Q, \mathbf{a}^*)$ if $\mathbf{a}^*$ is an extension of $\mathbf{a}$ to a list one entry longer, and in this case, $Q \subseteq P$.

Remark 5.4 *Normally, $Q \subset P$, as we shall have to shrink our neighborhoods of s_0 as construction proceeds. Any mapping $z \to 1/z - a$ is expanding when acting on B, and since we want the image of D_j under a composition of linear fractionals of this sort to be a subset of $B°$, we must make it small enough that the expansions to come do not cause it to overflow these bounds.*

The first $k - 1$ stages have an open disk D_j centered on s_0, associated with an integer list $\mathbf{a}_j$ of length j, each an extension of the previous, and with the property that for $1 \leq \ell \leq j$, $\psi[\mathbf{a}_j, \ell]$ maps D_j conformally into $B°$. Beginning with D_k, and continuing until such time as $T^n s_0 = 0$, we have disks D_j, and associated partitions $\mathcal{P}_j$ of D_j, such that for each slice $P \in \mathcal{P}_j$, there is a list $\mathbf{a}_P = (a_1, \ldots, a_j)$ of Gaussian integers such that for $1 \leq \ell \leq j$, $\psi[\mathbf{a}_P, \ell]$ maps P conformally into $B°$, is injective on D_j, and the inverse image under $\psi[\mathbf{a}_P, \ell]$ of the lines $\Re(z) = \pm\frac{1}{2}$ and $\Im(z) = \pm\frac{1}{2}$ bounding $B°$ either miss D_j entirely (if $T^j s_0 \notin \partial B$), or cut through D_j in a single arc through s_0 (if $T_j s_0$ lies on just one of these bounding lines), or quarter D_j, if $T_j s_0 = -\frac{1}{2}(1+i)$. In any case, we take D_j small enough that all these arcs enter and exit D_j, and none of them meets any of the others except at s_0.

Assuming we have carried this construction to depth m, with $k \leq m < n$, we now detail how it is extended to depth $m + 1$. Each $P \in \mathcal{P}_m$ is a 'triangle' bounded by three circular arcs. Two of these arcs have the form of the preimage under $\psi[\mathbf{a}_P, m]$ of a segment of one of the lines bounding $B°$, and s_0 lies at the intersection of these, while the third arc is part of the rim of D_m. For $1 \leq \ell \leq m$, if $z \in P$, then $T^\ell z = \psi[\mathbf{a}_P, \ell](z) \in B°$.

Remark 5.5 *Once we hit the first occasion on which $T^k s_0 \in \partial B$, the future orbits of s_0 under the various $\psi_P = \psi[\mathbf{a}_P]$ are confined to the union of the boundary of B and the internal boundaries of the B_k, as this set is closed under T and its variants that round to the right or up.*

Now consider the region

$$R_P = \frac{1}{\psi_P(P)} - \left[\frac{1}{\psi_P(s_0)}\right].$$

If the $T(\psi_P(s_0)) \in B^\circ$, then there is a new, possibly shrunken, neighborhood D_{m+1} of s_0 such that, with $a' = [1/\psi_P(s_0)]$, we have $(1/\psi_P(z)) - a' \in B^\circ$ for all $z \in D_{m+1} \cap P$. We won't need to split P, though we may have to trim it back towards its vertex s_0.

If, on the other hand, $b = 1/\psi_P(s_0) - a' \in \partial B$, we introduce a new arc cutting D_m through s_0. This arc is the inverse image under $z \to 1/\psi_P(z) - a'$ of the line bounding B° on which b lies. Say that an arc α *cuts* P if it meets P in any neighborhood of s_0. Our new arc may or may not cut P. If it does not cut P, we take a suitably shrunken portion of P about s_0 to be one of the pieces of $\mathcal{P}_{m+1}$, and we extend the Gaussian integer list **a** by appending a', to provide the pair $(P, \mathbf{a}')$ say. If it does cut P, we take suitably shrunken portions of the two sectors into which the portion of P near s_0 has been cut by our arc, to be the two pieces in $\mathcal{P}_{m+1}$ corresponding to P.

One of these pieces will be associated with the same extension of **a** we used before, while for the other, we extend instead by $a' + 1$, if b was on the left edge of B, or by $a' + i$, if b was on the bottom edge of B. We now further truncate our slices, if necessary, to a radius which is the minimum of the radii of the pieces we got during the construction of $\mathcal{P}_{m+1}$. The union of these pieces, together with their internal boundaries with each other, is now a pie-slice partition of an open disk D_{m+1} centered on s_0, and for each $P \in \mathcal{P}_{m+1}$, there is a corresponding Gaussian integer list **a**, an extension of the one that was associated with the parent P in $\mathcal{P}_m$, such that $\psi[\mathbf{a}, \ell]$ maps P into B° for $1 \le \ell \le m+1$.

We now claim that for all $P \in \mathcal{P}_{n-1}$, $1/\psi[\mathbf{a}_P](s_0) \in G'$. That is, the variants of the algorithm all run to completion on Gaussian rational input in the same number of steps. The reason for this is symmetry. The group S of rigid motions taking $\overline{B}$ onto $\overline{B}$ is generated by $\lambda_1(z) = \overline{z}$ and $\lambda_2(z) = iz$. We consider the equivalence relation $\sim$ on $\mathbb{C}$ determined by $u \sim v$ if $u = \lambda v$ for some $\lambda \in S$.

If $z \sim w$ then $1/z \sim 1/w$, and if $z, w \in \overline{B}$ and $z \sim w$, and if $a, b \in G'$ such that $1/z - a \in \overline{B}$ and $1/w - b \in \overline{B}$, then $1/z - a \sim 1/w - b$. Initially, $s_0 \sim s_0$. If $\psi[\mathbf{a}_1, \ell](s_0) = s_l \sim \psi[\mathbf{a}_2, \ell](s_0) = s'_l$ for all pairs $(\mathbf{a}_1, \mathbf{a}_2)$ of lists of length ℓ occurring in the disk partition tree of s_0, then $1/s_\ell \sim 1/s_{\ell'}$, and now for any of the choices a^* by which $\mathbf{a}_1$ may be extended, the various

$1/s_\ell - a^*$ are equivalent under $\sim$ to each other, and to the complex numbers that may arise out of $1/s_{\ell'}$ by extending $\mathbf{a}_2$ with a^{**}, say. (Another way to say this is to say that if we identified all elements of a given equivalence class under $\sim$, all the variants of T would condense to one.)

At the end, when $s_{n-1} \in 1/G'$, all other $s \sim s_{n-1}$ also belong to $1/G'$, so T maps them to zero as well. Therefore, all branches of our disk partition tree for s_0 have the same depth.

Now for each $(P, \mathbf{a}_P)$ of depth n in our disk partition tree,

$$\psi[\mathbf{a}, n](z) = \left(\frac{1}{z} - a_n\right) \circ \left(\frac{1}{z} - a_{n-1}\right) \cdots \circ \left(\frac{1}{z} - a_1\right)$$

takes s_0 to 0, and takes P on an orbit that at each stage remains inside B°.

Since these slices P partition the final disk D_n about s_0 into a finite number of slices, and since the sum of the angles at s_0 of the slices is 2π, there must be at least one slice that occupies a positive angle at s_0. Let P' be such a slice, and $\mathbf{a}'$ its Gaussian integer list. The image under $\psi[\mathbf{a}']$ of P is a 'triangle' in B° with its vertex $\psi[\mathbf{a}'](s_0) = 0$ with a positive angle at 0. For all $z \in P'$, $T^n z = \psi[\mathbf{a}'](z)$. Our disk partition tree has now served its purpose.

There are Gaussian integers $g \in G'$ such that $1/g \in P'$. Any neighborhood E of $1/g$ contained in P' will be taken conformally under $z \to 1/z - g$ to a neighborhood $\widetilde{E}$ of 0. We take another Gaussian integer h such that $1/(h + \overline{B}) \subseteq \widetilde{E}$, and such that $1/(h + \overline{B})$ is also a subset of some particular B_r. Let $\widetilde{\mathbf{a}}$ be the list got by appending g and then h to $\mathbf{a}'$. Now let $r = \psi[\mathbf{a}']^{-1}(1/(g + 1/h)) = \psi[\widetilde{\mathbf{a}}]^{-1}(0)$, and let

$$\widetilde{D} \subset D_n = \psi[\mathbf{a}']^{-1} \circ \left(\frac{1}{z+g}\right) \circ \left(\frac{1}{z+h}\right) B^\circ.$$

Iterating T takes $\widetilde{D}$ conformally to various squarish neighborhoods inside B°, until T^{n+1} takes it to the squarish region $(1/(h + B^\circ)) \subset \widetilde{E}$, and finally T^{n+2} carries it via $z \to 1/z - h$ onto B°.

A bit more is true. The orbit of $\widetilde{D}$ never meets any of the internal boundaries separating the various B_k until the last step, when it is splashed across B°. This is because if, at any earlier point any part of the orbit had touched an internal boundary, the subsequent iterations of T on that point would have also been in the union of these internal boundaries with the boundary of B itself, and yet, at the stage immediately prior to the final one, no point of $T^{n+1}\widetilde{D}$ belongs to any of these boundaries.

Now that we have our magic r, we are in a position to complete the proof that L_T is u_0-positive with $u_0 \equiv 1$. If $f \neq 0 \in \mathcal{V}$, and $f \geq 0$, then there exists $s \in \widetilde{B}$ such that $f(s) > 0$. We choose an open disk D_0 around s such that $D_0 \subset B_k$ where B_k is the one containing s, and such that for $z \in D_0$, $f(z) > \frac{1}{2}f(s)$. Now for $z \in \widetilde{B}$, $(L_T^{n+2}f)(z)$ is a sum of terms, among which (keeping in mind that $[c_1, c_2, \ldots, c_k] = 1/(c_1 + 1/(c_2 + \ldots + 1/c_k)))$ is the term

$$\prod_{j=1}^{n+2} |[a_j, \ldots, a_{n+2} + z]|^4 \, f([\langle a_{n+2}, a_{n+1}, \ldots, a_2, a_1 + z \rangle]).$$

But for arbitrary $z \in B^\circ$, the argument of f lies inside D_0, and the product is bounded below by a positive constant on B°, so this term is bounded below by a positive constant on $\widetilde{B} \subset B^\circ$. Since L_T is a bounded linear operator, we are almost done. We have proved most of the claims in our Gauss-Kuz'min theorem.

For the symmetry claims, we first note that $\rho(z) = \rho(\bar{z})$. Suppose not. Let $\rho_1(z) = \rho(z) - \rho(iz)$. Then $L_T\rho_1 = \rho_1$. But $\int_{\widetilde{B}} \rho_1 = 0$, so ρ_1 must be identically zero on $\widetilde{B}$. For mirror symmetry, let $\rho_2(z) = \rho(z) - \rho(\bar{z})$. Then

$$L_T\rho_2(z) = \sum_{g \in G'} |g + z|^{-4} \left(\rho(1/(g+z)) - \rho(1/(\overline{g+z})) \right)$$

$$= \rho(z) - L_T\rho(\bar{z}) = \rho(z) - \rho(\bar{z}) = \rho_2(z).$$

As before, $\int_{\widetilde{B}} \rho_2 = 0$, so ρ_2 is the zero element of $\mathcal{V}$. This completes the proof of our Gauss-Kuz'min theorem.

Chapter 6

Multidimensional Diophantine Approximation

The standard continued fraction expansion of a real number has several properties which we might hope to carry over into higher dimensions. Given a real number θ with $0 < \theta < 1$ as input, it returns a sequence of integer pairs (p_r, q_r) so that

(1)
$$\begin{pmatrix} p_{r-1} & q_{r-1} \\ p_r & q_r \end{pmatrix} = P_r \begin{pmatrix} p_{r-2} & q_{r-2} \\ p_{r-1} & q_{r-1} \end{pmatrix}$$

with

$$P_r = \begin{pmatrix} 0 & 1 \\ 1 & a_r \end{pmatrix}.$$

That is, the new integer pair is a positive integer combination of the previous two pairs, and the matrix P_r has integer entries and is invertible.

(2) The pairs generated are exactly the best-possible approximations in the following sense: For $1 \leq q' < q$, and for arbitrary integer p', if (p,q) is one of the (p_r, q_r) then $|p - q\theta| < |p' - q'\theta|$, and vice versa.

(3) The algorithm is fast. Each step is fast because the arithmetic needed to execute a given step is modest, amounting to a bounded number of multiplications, divisions, additions and subtractions, all involving integers no bigger than the inputs, if θ is a rational number, or floating point arithmetic to an accuracy no greater than the original accuracy with which θ is presented. Each two steps at least double the product of successive values of q.

It so happens that we cannot have all these properties in a multidimensional Diophantine approximation algorithm [Gr]. There are tradeoffs. The task of simultaneous Diophantine approximation is to take as input a list of real numbers $\theta = (\theta_1, \theta_2, \ldots \theta_n)$, which may as well all satisfy $|\theta_j| \leq 1/2$, and to return a sequence of integer-valued $(n+1)$-tuples of the form $(p_1, p_2, \ldots p_n, q)_r$ with increasing q's such that the vector $(p_1 - q\theta_1, p_2 - q\theta_2, \ldots p_n - q\theta_n)$ is "short". The output may as well be simply a sequence of integers q_r since the best choice of the various corresponding $p_{j,r}$ will be obvious.

The Dirichlet box principle gives some bounds on what "short" should mean, and points us to an algorithm which gets good approximations. For $\mathbf{x} = (x_1, x_2, \ldots, x_n) \in \mathbb{R}^n$, let $\{\mathbf{x}\} = (\{x_1\}, \{x_2\}, \ldots, \{x_n\})$. Given an integer $Q > 1$, partition the unit n-dimensional box into Q^n cubicles each of side $1/Q$. If $Q^* > Q^n$ then there exist $0 \leq j < k \leq Q^*$ so that $\{j\theta\}$ and $\{k\theta\}$ fall into the same cubicle. Thus the sup norm of $\{(k-j)\theta\}$ will be less than $1/Q$. From this it follows that there exists $1 \leq q \leq Q^n$ so that the approximation error vector $(\mathbf{p} - q\theta)$ will be no longer than $1/Q$ in the sup norm, or $\sqrt{n}/Q$ in Euclidean norm. The defect of this algorithm is that it requires inordinate time when Q is large.

J. Cassels [Ca] has proved that there exists $\epsilon > 0$ so that there are uncountably many vectors θ for which, for all q, $\|(\mathbf{p} - q\theta)\| > \epsilon q^{1/n}$, and W. Schmidt [Sc1] showed that the Hausdorff dimension of the set of badly approximable vectors in this sense is equal to n. M. Drmota and R. F. Tichy [DT] showed that for a real algebraic number α of degree n over $\mathbb{Q}$, the vector $(1, \alpha, \alpha^2, \ldots \alpha^{n-1})$ provides an explicit example of such a 'badly-approximable' vector.

The set of *badly approximable* elements of $\mathbb{R}^2$ has zero Lebesgue measure in $\mathbb{R}^2$, which complicates its visual representation. In the spirit of the striking pictures afforded by the stages enroute to the so-called *Sierpinski gasket* set, though, we offer a picture. The rays pointing into the first octant along badly approximable directions in $\mathbb{R}^3$ each cut the plane $x_1 + x_2 + x_3 = 1$ in a single point. If we fix a parameter $\epsilon > 0$ and consider just those directions $\mathbf{v}$ for which the angle between $\mathbf{v}$ and any integer vector $\mathbf{u}$ satisfies $\arg[\mathbf{u}, \mathbf{v}] > \epsilon \|\mathbf{u}\|^{-3/2}$, then the points in the cross-section are those which remain when the ellipses formed by the intersection of this plane with circular cones of the appropriate angle about each integer vector in the first octant. Fudging matters for computational simplicity, circles of radius $(1/9)\|\mathbf{u}\|^{-3/2}$ about the point $(u_1, u_2, u_3)/(u_1 + u_2 + u_3)$ were colored black, for all nonzero $\mathbf{u}$ with $0 \leq u_1, u_2, u_3 \leq 20$.

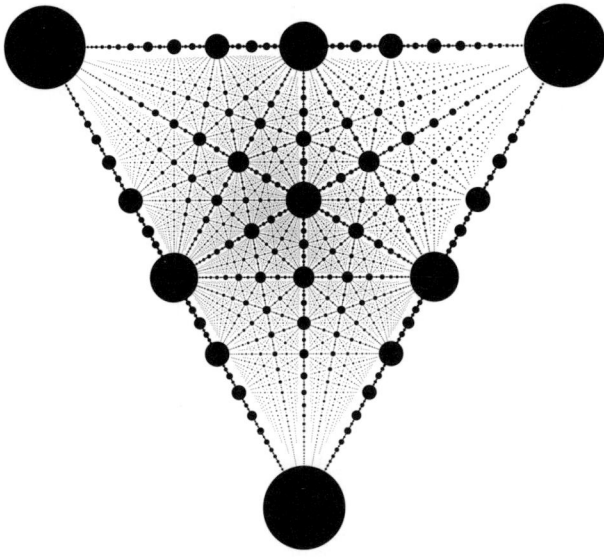

Fig. 6.1 A Sierpinski gasket of badly approximable pairs.

One of the first algorithms intended to provide simultaneous Diophantine approximations was the *Jacobi-Perron* algorithm. Here, we give a slightly simplified description of the algorithm, appropriate to the case of inputs that are linearly independent over $\mathbb{Q}$.

The algorithm takes as input a list θ of d positive real numbers. With $n = d+1$, it produces a sequence of $n \times n$ integer matrices M_k of determinant 1 and with all entries nonnegative. The first is the identity matrix $I_{n \times n}$. A *working vector* **w** is initialized to

$$(w_1, \ldots, w_n) = (1, \theta_1, \ldots, \theta_{n-1}).$$

With $b_k = \lfloor w_{k+1}/w_1 \rfloor$, the update rules below are applied repeatedly:

$$(w_1, \ldots, w_n) \leftarrow (w_2 - b_1 w_1, w_3 - b_2 w_1, \ldots, w_n - b_{n-1} w_1, w_1)$$
$$M_{k+1} = P_b M_k$$

where P_b has for its first $n-1$ rows, the second through nth rows of I_n, with its final row $(1, b_1, b_2, \ldots, b_{n-1})$. With this construction, at each stage,

$$(1, \theta_1, \ldots, \theta_d) = (w_1, w_2, \ldots, w_n)_k M_k.$$

Geometrically, the rows of M_k include $(1, \theta_1, \ldots, \theta_d)$ in their convex hull, and each update consists of rotating the rows $(m_2, \ldots, m_n)$ of the previous M, as in volleyball, into positions 1 through $n-1$, and generating a new last row (the 'server' m'_n), by the rule

$$m'_n = m_1 + \sum_2^n b_{k-1} m_k.$$

The b_k are the same as those described above, but they are also generated by choosing each b as large as possible so that the convex hull of the other vectors, together with the new final vector, shall include θ.

At the heart of modern algorithms for simultaneous Diophantine approximation lies the notion of lattice reduction.

For our purposes, a lattice Λ is a discrete additive subgroup of $\mathbb{R}^n$, equipped with some inner product. The dimension $D[\Lambda]$ of Λ is the dimension of the span of Λ; in this Chapter, D is always equal to n. A *lattice basis* is a sequence $(\mathbf{b}_1, \ldots, \mathbf{b}_n)$ of n nonzero elements of Λ such that, for all $\mathbf{v} \in \Lambda$, there exists a unique $\mathbf{a} \in \mathbb{Z}^n$ so that $\mathbf{v} = \sum_1^n a_k \mathbf{b}_k$. By a slight abuse of language, we shall conflate the *matrix* having rows $\mathbf{b}_1, \ldots, \mathbf{b}_n$ with the list of vectors, and call it, too, a lattice basis of Λ. Note that whether a list, or its associated matrix, is a basis of Λ is independent of the inner product in use.

Any lattice of dimension greater than one has infinitely many lattice bases. When working in a lattice, we cannot scale vectors by arbitrary ratios, so we cannot implement the Gram-Schmidt algorithm and transform some given initial lattice basis into an orthonormal lattice basis. We cannot even ensure that the vectors of our lattice basis are mutually orthogonal.

In two dimensions, there is really only one sensible definition of a reduced basis. A basis is *reduced* if its first vector is the shortest nonzero vector in the lattice, and its second vector is that vector most nearly orthogonal to the first, as well as the shortest vector which is not a multiple of the first. In higher dimensions, there are various ways to give concrete meaning to the intuitive notion of a reduced basis for a lattice. We use two such versions of lattice reduction: LLL reduction, and Minkowski reduction. We discuss LLL reduction immediately below, but defer a discussion of Minkowski reduction.

We have already seen one lattice reduction algorithm, the Gauss algorithm for two dimensional lattices. Given two vectors, this algorithm quickly yields an equivalent basis so that the first vector is a shortest nonzero vector of the lattice, and the second is shortest among those not parallel to the first. While one ought not feed this algorithm a pair of real numbers, or a pair of parallel vectors, instead of a pair of linearly independent vectors, if one does, the result is isomorphic to the classical continued fraction expansion of the ratio of the inputs, and to the classical Euclidean algorithm, if it be fed a pair of integers.

General lattice reduction algorithms aim to extend or approximate this sort of optimality in higher dimensions. One such algorithm, the famous Lenstra-Lenstra-Lovasz procedure [Co], known as LLL-reduction, rests on the well-known Gram-Schmidt algorithm. The Gram-Schmidt algorithm is not a lattice algorithm; it takes as input a list of vectors $(\mathbf{b}_1, \mathbf{b}_2, \ldots \mathbf{b}_N)$ and returns a list of mutually orthogonal vectors $(\mathbf{b}_1^* = \mathbf{b}_1, \mathbf{b}_2^*, \ldots \mathbf{b}_N^*)$ so that $\mathbf{b}_j^*$ is the component of $\mathbf{b}_j$ orthogonal to the span of $(\mathbf{b}_1, \mathbf{b}_2, \ldots \mathbf{b}_{j-1})$.

The LLL algorithm takes as input a list of N linearly independent vectors in $\mathbb{R}^N$. These form an ordered basis of a lattice. The algorithm returns a new basis for the same lattice, and if desired, an integer matrix P so that $P(\text{old basis}) = (\text{new basis})$. The new basis is 'LLL-reduced'. An ordered basis $(\mathbf{b}_1, \mathbf{b}_2, \ldots \mathbf{b}_N)$ is LLL reduced if the following holds, Let $\mu_{i,j} := (\mathbf{b}_i \cdot \mathbf{b}_j^*)/|\mathbf{b}_j^*|^2$. Then

(i) $|\mu_{i,j}| \leq 1/2$ for $i < j \leq N$,
(ii) $|\mathbf{b}_j^* + \mu_{j,j-1}\mathbf{b}_{j-1}^*|^2 \geq (3/4)|\mathbf{b}_{j-1}^*|^2$ for $1 < j \leq N$.

For a LLL-reduced basis of a lattice, the first vector $\mathbf{b}_1$ is not necessarily the shortest non-zero vector $\mathbf{x}$ in the lattice, but it does satisfy the condition $|\mathbf{b}_1| \leq 2^{(N-1)/2}|\mathbf{x}|$.

We can adapt the LLL algorithm to the task of simultaneous Diophantine approximation, and this is the best choice if speed is of paramount importance. The idea is (almost) this: start with the vectors $(1, 0, 0, \ldots 0)$, $(0, 1, 0, \ldots 0), \ldots (0, 0, \ldots 1)$ and $(\theta_1, \theta_2, \ldots \theta_{n-1})$, and reduce the resulting 'basis'. The snag is that the vectors are linearly dependent. This is fixed by a well-known trick which we shall be using throughout the rest of this Chapter. Hop up to the next higher dimension and insert a parameter which goes to zero. Start with the vectors $(1, 0, \ldots 0)$, $(0, 1, \ldots 0)$, $\ldots$ $(0, 0, \ldots 0, 1, 0)$ and finally $(\theta_1, \theta_2, \ldots \theta_{n-1}, t)$. For all positive t these n vectors are linearly independent, but for small t the matrix is ill-conditioned and the lattice generated by the given basis can be reduced substantially. That is, there is another basis, equivalent by way of a matrix $P \in GL(n, \mathbb{Z})$ so that the vectors in the new basis are short. Since the vectors all have the form

$(p_1 - \theta_1 q, p_2 - \theta_2 q, \ldots p_{n-1} - \theta_{n-1} q, qt)$, we will have found an approximation to θ.

How good will it be? The determinant of the lattice is t. A LLL-reduced basis of a lattice of dimension $n+1$ has a vector $\mathbf{x}$ which is no longer than $2^{n/2}$ times the shortest nonzero vector in the lattice. By Minkowski's theorem, a ball of radius R in real $(n+1)$ space, which has volume $(\sqrt{\pi}^{n+1}/\Gamma(1+(n+1)/2))R^{n+1}$, will contain a nonzero lattice point provided the volume is greater than $2^{n+1}t$. Thus, with R comparable to $\sqrt{n}$, or more accurately, R asymptotic to $\sqrt{2n/(\pi e)}t^{-1/(n+1)}$, we see that $\|\mathbf{x}\| \leq \sqrt{2n(1+\epsilon)/(\pi e)}t^{-1/(n+1)}$ and so the vector $\mathbf{b}$ found by LLL reduction will satisfy $\|\mathbf{b}\| \leq 2^{n/2}\sqrt{2n(1+\epsilon)/(\pi e)}t^{-1/(n+1)}$. By choosing $t_r := 2^{-r}$, then, we may obtain, quickly, a sequence of serviceable Diophantine approximations to θ. This approach has the drawback that none of the approximations is sure to be particularly good, and even if we chance upon some of the best approximations, we cannot hope to hit upon all of them.

There is a middle ground approach, due to Lagarias, which though of exponential complexity in the dimension n, gives us a reasonably dense subsequence of the Euclidean-norm best approximations and with tolerable rapidity for fixed and modest n. It so happens [La] that the sequence of q yielding these best-possible approximations increases at least exponentially, so in a sense we substantially match the ideal properties (ii) and (iii) of the one-dimensional algorithm. Unfortunately matters are not quite so simple, since the Lagarias algorithm works by updating a unimodular integer matrix and only updates that affect the first row yield new values of q.

We first adapt a changed perspective on lattice reduction. Instead of fixing the inner product and varying the basis $B_0[\theta, t]$ as $t \to 0$, we fix our initial basis, and thus the point set of our lattice, to be the set of all products of an integer row vector in $\mathbb{Z}^n$ with $B_0[\theta]$, which shall be the $n \times n$ matrix which is in its first $n-1$ rows, equal to the identity matrix, but which has for its last row, the entries $(-\theta_1, -\theta_2, \ldots, -\theta_{n-1}, 1)$. We then take, for $\alpha > 0$,

$$\langle \mathbf{u}, \mathbf{v} \rangle_\alpha = \sum_{1}^{n-1} u_k v_k + \alpha u_n v_n.$$

Associated with this inner product is the norm

$$\|\mathbf{x}\|_\alpha = \sqrt{\sum_1^{n-1} x_j^2 + \alpha x_n^2}.$$

The basis and the inner product, taken together, determine the lattice $\Lambda_\alpha[\theta]$, which is isomorphic to our earlier lattice with final row $(0,\ldots,0,t=\sqrt{\alpha})$. One advantage to this new perspective is that when the input vector is rational, the critical values of t in the first approach are square roots of rational numbers, while in this approach they are rational.

The Lagarias algorithm reduces, in the case $n = 2$ in which we have just one number θ to approximate, to an algorithm studied by Hermite. It is instructive to see how things go. We first note that the algorithm can be seen as a procedure for calculating through which cells in the picture presented here, the complex number $-\theta + it$ passes as t decreases from 1 to zero. Since the cells are 'triangles' with a foot meeting the real line at a rational number, we are getting, in effect, a series of rational approximations to θ.

Consider the complex domain $\overline{D}_0 := \{z \in \mathbb{C} : |z| \geq 1, \Im(z) > 0, \text{ and } -1/2 \leq \Re(z) \leq 1/2\}$. Let 'the scalloped region' be the union of all integer translates $D_0 + k$ of D_0. The various images of D_0 under the group Φ of linear fractionals generated by the three linear fractionals $z \mapsto -1/z$ and $z \mapsto z \pm 1$ tile the upper half plane. The tiling is essentially the

Fig. 6.2 Cells through which the Lagarias continued fraction algorithm passes, 1D case.

so-called modular tiling, with the difference that in the modular tiling, the original tile D_0 is split down the imaginary axis to make two tiles, and the group is correspondingly enlarged to include $z \pm 1/2$. (The modular tiling is described in John Stilwell's article [St].)

Our group Φ of linear fractionals consists of all functions of the form $z \mapsto \frac{az+b}{cz+d}$ for which $ad - bc = 1$ and $a, b, c, d \in \mathbb{Z}$. The points at which $-\theta + it$ pass from one tile to the next, along its descent toward the real axis, correspond to the t_r delimiting the intervals I_r on which $B_r(t)$ is reduced. They also correspond to the values of t at which the auxiliary quantity $z_r(t)$ enters and exits D_0.

The name 'geodesic' comes from the fact that there is a metric with respect to which the paths $z_r(t)$, and corresponding 'paths' taken by the matrices $B_r(t)$ in higher dimensions, are geodesics, either in the complex plane as here, or in a space of matrices. In the case at hand, we may think of the plane as a medium through which light passes. The local speed of light in this medium is $1/\Im(z)$. The 'distance' between two points is the time needed for a signal to reach from one to the other. For two points, or complex numbers $x + iy_1$ and $x + iy_2$ with the same real part, this distance is $|\log(y_2/y_1)|$. The metric is invariant under linear fractional transformations of H onto H. The geodesics are the vertical half lines, and all half-circles with diameter along the real line. All tiles are congruent in the sense that there is a $\phi \in \Phi$ taking one tile to the other. Tiles other than lateral translates $D_0 \pm n$, $n \in \mathbb{Z}$ have one corner, call it the foot, on the real line. If $D = \frac{az+b}{cz+d} D_0$ with $c \neq 0$ is such a tile, then a/c is the point at which the tile touches the real axis.

Fig. 6.3   Detail.

The algorithm begins with a matrix

$$B_0(t) := \begin{pmatrix} 1 & 0 \\ -\theta & t \end{pmatrix}$$

with $t_0 = 1$. A two-dimensional matrix is *reduced* if the two rows m_1, m_2 say satisfy $\|m_1\| \leq \|m_2\|$ and $\|m_2\| \leq \|m_2 + km_1\|$ for all integers k.

Initially, that is, for $t = 1$, $B_0(t)$ is reduced. As t decreases, the matrix passes from being reduced, to not reduced, as t hits $\sqrt{1-\theta^2}$. The idea of the Lagarias algorithm is that we keep track of a working basis $B_j(t)$ which is reduced in a certain interval $I_j = (t_{j+1}, t_j)$. The initial interval $I_0 = (\sqrt{1-\theta^2}, 1)$. At each juncture, as t hits the lower limit t_{j+1} of the interval I_j on which the working basis $B_j(t)$ is reduced, the working basis matrix $B_j(t)$ is updated so that the new basis will be reduced on some new interval (t_{j+2}, t_{j+1}).

6.1 The Hermite Approximations to a Real Number

One perspective on continued fractions is that we are given a real number α, and tasked to find the sequence of successive q with the property that for all $1 \leq q' < q$,

$$\min_{k \in \mathbb{Z}} |q'\alpha - k| < \min_{k \in \mathbb{Z}} |q\alpha - k|.$$

With the notation $\|x\| := \min_{k \in \mathbb{Z}} |x - k|$, we are finding a sequence of integers which in a certain sense provide best denominators for rational approximation to x.

Hermite had another way to think about what is a best denominator. A positive integer q is a Hermite denominator for α if there exist positive real numbers $a < b$ such that, for all $a < t < b$, and for all positive integers $q' \neq q$,

$$\|q\alpha\|^2 + tq^2 < \|q'\alpha\|^2 + tq'^2.$$

Every Hermite denominator for α is a continued fraction denominator, for the simple reason that if q satisfies the condition above, and if $q' < q$, then since also $\|q\alpha\|^2 + tq^2 < \|q'\alpha\|^2 + tq'^2$, we must have $\|q\alpha\| < \|q'\alpha\|$, and this property characterizes continued fraction denominators for α.

The Hermite denominators are not identical with the classical continued fraction denominators; it is possible for some of the latter to be skipped by the Hermite sequence for α. Any algorithm for finding Hermite denominators for general real numbers must somehow cope with the fact that inequalities which break one way or the other can hang on knife-edge while we accumulate further digits of our input α; our chosen method of coping is to punt and provide an algorithm only when the input α is rational, and further simplifying matters, to restrict attention to the case $0 < \alpha < 1/2$. If it is necessary to get an answer for an irrational number, one may compute upper and lower bound rational approximations to α and run the Hermite algorithm given here on both. Until the results diverge, they are also the results for α itself.

Our algorithm depends for its validity on an observation about a certain lattice. Consider the lattice $\Lambda[\alpha, t]$ generated by $((1,0), (-\alpha, 1))$ and equipped with the inner product associated with the norm $\|\mathbf{u}\|_t$ given by $\|(x,y)\|_t^2 = x^2 + ty^2$. That is, we fix α and the basis, and make the norm depend on the parameter t.

The Gauss lattice reduction algorithm terminates upon arriving at a lattice basis $(\mathbf{u}, \mathbf{v})$ with the property that $\|\mathbf{u}\| \le \|\mathbf{v}\|$ and that $\|\mathbf{v}\| \le \|\mathbf{v} + k\mathbf{u}\|$ for $k \in \mathbb{Z}, k \ne 0$. An ordered basis $(\mathbf{u}, \mathbf{v})$ of a two dimensional lattice is termed *Minkowski-reduced* if these inequalities hold.

The observation about Hermite denominators is that q is a Hermite denominator for α if and only if there is a positive real number t such that, when p is the integer nearest $q\alpha$, the vector $(p - q\alpha, q) \in \Lambda[\alpha, t]$ is, for that t, the unique shortest vector, up to multiplication by -1, in the lattice, and thus, the leading member of any Minkowski-reduced lattice basis of $\Lambda[\alpha, t]$.

The idea, then, is to generate a sequence of auxiliary variables, the successive t_k's, and successive second basis vectors, so that, as computation proceeds, we always have in hand a reduced basis $(\mathbf{u}_k, \mathbf{v}_k) = ((p_k - q_k\alpha, q_k), (r_k - s_k\alpha, s_k))$ and a parameter t_k, so that $\mathbf{u}_k$ will be the shortest vector of the lattice, for $t_{k+1} < t < t_k$. Since this topic is somewhat peripheral, we omit the somewhat pedestrian proof that the algorithm is correct, and just give the algorithm. We enumerate the key facts. At each transition, the new basis $(\mathbf{u}_k, \mathbf{v}_k) = ((p_k - q_k\alpha, q_k), (r_k - s_k\alpha, s_k)) = ((\beta_k, q_k), (\gamma_k, s_k))$, say, has these properties:

(1) For $k \ge 1$, $q_k, s_k \in \mathbb{Z}^+$, with $q_k > s_k$.
(2) p_k and r_k are the integers nearest $q_k\alpha$ and $s_k\alpha$ respectively.
(3) If $(p_k - q_k\alpha)(r_k - s_k\alpha) > 0$, then $q_k > 2s_k$.
(4) $\|q_k\alpha\|^2 + t_k q_k^2 = \|s_k\alpha\|^2 + t s_k^2$.
(5) For all $j \in \mathbb{Z}$, $\|\mathbf{v}_j + j\mathbf{u}_j\|_{t_k} \ge \|\mathbf{v}_j\|_{t_k}$.

(6) If $\sigma_k = \text{sign}(\beta_k \gamma_k)$, then $\mathbf{u}_{k+1} = \sigma_k \mathbf{v}_k + n_k \mathbf{u}_k$ where n_k is either $\lfloor |\gamma_k/\beta_k| \rfloor$ or $1 + \lfloor |\gamma_k/\beta_k| \rfloor$. If $\{\|\gamma_k/\beta_k|\} \leq 1/2$, n_k is the lesser choice. If $\{\|\gamma_k/\beta_k|\} > 1/2$, then which value n_k takes depends on whether the value of t at which, with $n = \lfloor |\gamma_k/\beta_k| \rfloor$, $\|\mathbf{u}_k\|_t = \|\sigma_k \mathbf{v}_k + n\mathbf{u}_k\|_t$ is greater than that at which $\|\sigma_k \mathbf{v}_k + n\mathbf{u}_k\|_t = \|\sigma_k \mathbf{v}_k + (n+1)\mathbf{u}_k\|_t$, or less.

(7) In all but certain exceptional cases, $\mathbf{v}_{k+1} = \mathbf{u}_k$. The exceptions occur when $\|\mathbf{u}_k\|_{t_{k+1}} = \|\sigma_k \mathbf{v}_k + n_k \mathbf{u}_k\|_{t_{k+1}} = \|\sigma_k \mathbf{v}_k + (n_k - 1)\mathbf{u}_k\|_{t_{k+1}}$. In that case, $\mathbf{v}_{k+1} = -\mathbf{v}_k + (n_k - 1)\mathbf{u}_k$.

A sample result: hermite$[41/99, \infty]$ gives

$$\begin{array}{cccccccc}
\frac{41}{99} & 1 & 1 & 0 & 0 & -1 & \frac{8120}{9801} \\
\frac{17}{99} & 2 & -\frac{41}{99} & 1 & 2 & -1 & \frac{464}{9801} \\
-\frac{7}{99} & 5 & \frac{17}{99} & 2 & 2 & -1 & \frac{80}{68607} \\
\frac{3}{99} & 12 & -\frac{7}{99} & 5 & 2 & -1 & \frac{40}{1166319} \\
-\frac{1}{99} & 29 & \frac{3}{99} & 12 & 2 & -1 & \frac{8}{6831297} \\
0 & 99 & -\frac{1}{99} & 29 & 3 & -1 & \frac{1}{87816960}
\end{array}$$

This gives the successive β_k, q_k, (in this case $(1, 2, 5, 12, 29, 99)$), γ_k, s_k, n_k, σ_k, and t_k that are encountered during the execution of the Hermite algorithm on input $41/99$.

There is a geometric interpretation of what is happening here. One keeps track of a point $p(t)$ moving inside the domain $D: -1/2 \leq x < 1/2$, $y > 0, x^2 + y^2 \geq 1$. Initially, the point is moving along the line $x = -41/99, y = t$, with t decreasing from infinity.

Any time the point hits the bottom scalloped arc boundary, it is sent off on a circular arc which is the negative reciprocal of the translation back into D_0 of the path it was on before. Thus, when t reaches $t_1 = \sqrt{1 - (41/99)^2}$, the point strikes the bottom boundary at $(-41/99, t_1)$. The negative reciprocal of this is $(41/99, t_1)$, and the new trajectory is is an arc along the circle through 0 and $99/41$, moving up and away from $(41/99, t_1)$. Each translation followed by reflection off the lower boundary corresponds to a new $\mathbf{u}_k$. Each transition from one zone to another in the scalloped region corresponds to a new reduced basis in which only the second vector has changed.

The doppelgänger of this sequence of movements within $\bigcup_k (D_0 \pm k)$ is a succession of milestones on the straight line path from $(-41/99 + i\infty)$ to $(-41/99)$. The images under appropriate linear fractional transformations, of D_0 and the various paths our moving point has taken through it, are conformal linear fractional copies of D_0, together with the appropriate segment of our straight line path. We illustrate with graphics showing one such packet of movement, corresponding to a circular arc along the scallop shaped top zone of the upper half plane. The arc runs from $(41/99, t_1) \in D_0$ across to $D_0 + 2$, where it crosses below the scalloped boundary and triggers the calculation of a new value of q. The zeroth and first segments of the straight line path are shown on the left, and the first corresponding arc on the right. The distorted 'strips' in the left picture converge at zero, which is the simplest p/q approximation to $-41/99$. In one dimension, a direct

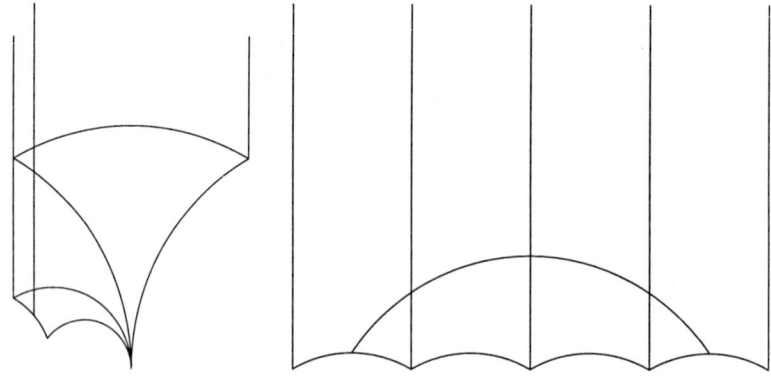

Fig. 6.4 Descent to zero through zones of the modular map.

implementation of the Lagarias algorithm resembles an additive variation on the algorithm presented above. The Lagarias algorithm pauses at each intermediate state corresponding to passages across a vertical line in the scalloped region. The expected value of any particular a_j in the classical continued fraction expansion of a random input $\theta \in (0, 1)$ is infinite, but one can in one dimension readily look ahead and avoid counting upward one by one–this is what we did above. In higher dimensions, matters are less likely to get stuck in this kind of a groove, and it becomes increasingly difficult in any case to 'see ahead'. Nevertheless, for any fixed dimension, there ought not in principle be any insuperable barrier to similar accelerations of the Lagarias algorithm.

The execution of the algorithm, in this case, can be expressed in terms of linear fractionals. Start with the identity linear fractional $\phi_0(z) := z$, and $z_0(t) := -\theta + it$. Let t decrease from 1 towards zero, and set $j = 0$.

Whenever $\phi_j(t)$ exits D_0, set t_{j+1} to be that value of t, and set ν_{j+1} to be that linear fractional, one of $-1/z$ or $z \pm 1$ as needed, so that $\nu_{j+1} \circ \phi_j(t) \in D_0$ for some interval $(t_{j+1} - \epsilon, t_{j+1})$. When t_j exits into $D_0 + 1$, $\nu_{j+1}(z) = z - 1$, and vice versa, while when t_j exits across the lower boundary of D_0, $\nu_{j+1}(z) = -1/z$. In the exceptional case that $\phi_j(t)$ exits D_0 through the right or left corner, $\nu_{j+1}(z)$ will instead be $-1/z \pm 1$. Set $z_{j+1}(t) = \nu_{j+1}(t)z_j(t) = \phi_{j+1}z_0(t)$. The map taking D_0 to the tile containing $-\theta + it$, when $t_{j+1} < t < t_j$, will be ϕ_j^{-1}.

All the continued fraction partial quotient pairs (p_j, q_j), as well as all the intermediate values $(p_{j-1} + kp_j, q_{j-1} + kq_j)$ with $k \leq a_{j+1}$, arise as $\pm(a, b)$ or $\pm(c, d)$ for one of the linear fractionals ϕ_j arising in the expansion. If θ is rational, we eventually arrive at a tile which has its foot at $-\theta$, and $-\theta + it$ descends through this tile to θ without exiting the tile. If θ is irrational, there will be infinitely many tiles, and t_j's and so on. The foot of the jth tile is the image of ∞ under ϕ_j^{-1}.

6.2 The Lagarias Algorithm in Higher Dimensions

Turning back to a discussion of the general case, there are several issues. First, we need just the right notion of a 'good' simultaneous Diophantine approximation to a list θ of $n - 1$ real numbers. Next, we must say what it means, for a lattice basis to be 'reduced', and we need to show that reducing the initial basis $B_0[\theta]$ gives rise to a good simultaneous Diophantine approximation of θ.

A *Euclidean-norm best denominator* for θ is a positive integer q so that for all positive integers $q' < q$,

$$\sum_{1}^{n-1}(p'_i - q'\theta_i)^2 > \sum_{1}^{n-1}(p_i - q\theta_i)^2,$$

where the p'_i and p_i are the integers nearest $q'\theta_i$ and $q\theta_i$ respectively. This very natural standard for judging the quality of Diophantine approximations turns out to not lend itself to the ideas we now wish to use.

A *Hermite denominator* for θ is a positive integer q so that there exist positive numbers $\beta < \alpha$ so that for *all* $q' > 0$ and all $\gamma \in (\beta, \alpha)$,

$$\sum_{1}^{n-1}(p'_i - q'\theta_i)^2 + \gamma q'^2 \geq \sum_{1}^{n-1}(p_i - q\theta_i)^2 + \gamma q^2.$$

Every Hermite denominator for θ is also a Euclidean-norm best denominator for θ, but the converse is not true. The sequence of Hermite denominators may skip some of the Euclidean-norm best denominators. On the other hand, it preserves the essential information about approximability of θ.

Let

$$C(\theta) = \lim_{q \to \infty} q^{2/(n-1)} \sum_{i=1}^{n-1} (p_i - q\theta_i)^2.$$

where as usual, the p_i are the integers nearest $q\theta_i$. Lagarias proved the following result: [La]

Theorem 6.1 *Each Hermite denominator q_j of θ is also a Euclidean-norm best denominator of θ. Furthermore,*

(1) $C(\theta) = \overline{\lim}_{j \to \infty} q_j^{2/(n-1)} \sum_{i=1}^{n-1} (p_i - q_j\theta_i)^2$.
(2) $q_{j+2^n} \geq 3q_j$.
(3) *If $\alpha = Q^{-2-2/(n-1)}$, then the Hermite denominator q corresponding to α satisfies two inequalities:*

  (a) $1 \leq q \leq \sqrt{n}Q$
  (b) $\sum_1^{n-1}(p_i - q\theta_i)^2 \leq nQ^{-2/(n-1)}$.

The next issue is how to compute Hermite denominators. By definition, a Hermite denominator is a positive integer q which appears as the final entry in a nonzero vector of minimal norm in $\Lambda_\alpha[\theta]$.

A basis $(\mathbf{b}_1, \ldots, \mathbf{b}_n)$ for an n-dimensional lattice Λ is *Minkowski-reduced* if:

(1) $\mathbf{b}_1$ is a nonzero lattice vector of minimal norm.
(2) For $2 \leq k \leq n$, $\|\mathbf{b}_k\|$ is minimal among all lattice vectors $\mathbf{v}$ such that there exists a basis of Λ that starts with $(\mathbf{b}_1, \ldots, \mathbf{b}_{k-1}, \mathbf{v})$.

The definition here is not quite what we might have expected. Why not instead specify that $\mathbf{b}_k$ be that vector $\mathbf{v}$ of minimal length, among those not in the span of $\mathbf{b}_1$ through $\mathbf{b}_{k-1}$? The reason is that for large n, there are examples of lattices in which the shortest such $\mathbf{v}$ cannot be appended to $\mathbf{b}_1$ through $\mathbf{b}_{k-1}$ and then augmented to arrive at a lattice basis.

This wrinkle on the definition gives us a clue to how we might convert the definition, which does not on its face tell us how we might compute a reduced basis, into an algorithm for Minkowski reduction.

If B is a (matrix) lattice basis of Λ, and $P \in GL(n, \mathbb{Z})$, then PB is a lattice basis of Λ, and vice-versa. What are the possibilities for the first row of such a P? Clearly, the gcd of the entries of this first row must be one, for otherwise, $|\det P|$ could not be one. On the other hand, if $\gcd(a_1, a_2, \ldots, a_n) = 1$, then there exists a unimodular integer matrix P with first row equal to $\mathbf{a}$. This is because we can transform the row vector $\mathbf{a}$, by a sequence of post-multiplications by simple integer matrices of determinant ± 1, into successive rows of the form $(0, \gcd(a_1, a_2), a_3, \ldots, a_n)$, then $(0, 0, \gcd(a_1, a_2, a_3), a_4, \ldots, a_n)$, and finally into $(0, \ldots, 0, 1)$.

If $\mathbf{b}_1$ is a nonzero vector of minimal norm in Λ, then the set of all P which leave $\mathbf{b}_1$ intact as the first row of a matrix basis, is exactly the set of all P which have $(1, 0, \ldots, 0)$ as a first row. The second row of PB is determined by the second row of P, so we are led to think about what the possibilities are for that second row. This time, it is necessary and sufficient that the gcd of $(a_2, \ldots, a_n)$ be 1. This line of reasoning leads ultimately to an effective criterion for Minkowski reduction: A basis $B = (\mathbf{b}_1, \ldots, \mathbf{b}_n)$ of a lattice Λ is Minkowski-reduced, if for all $1 \leq d \leq n$, and all $a \in \mathbb{Z}^n$ for which $\gcd(a_d, \ldots, a_n) = 1$, we have

$$\|\sum_1^n a_k \mathbf{b}_k\| \geq \|\mathbf{b}_d\|.$$

When $\alpha = 1$, the lattice $\Lambda_\alpha[\theta]$ is Minkowski-reduced. As α decreases, it drifts out of reduction, and we must conduct some kind of search and restore reduction, by premultiplying the basis we are currently using for $\Lambda_\alpha[\theta]$ by some matrix in $GL(n, \mathbb{Z})$. The computational task, then, is to determine the correct 'update matrix' by which to multiply the current basis.

Suppose we have a lattice basis $B = \{b_1, b_2, \ldots b_n\}$ and we wish to determine whether or not it is Minkowski-reduced, and if not, determine a unimodular, integral 'update matrix' M so that MB is in some sense closer to being Minkowski-reduced. We know from the general theory of Minkowski reduction [La] that a matrix B with rows $\mathbf{b}_1, \mathbf{b}_2, \ldots \mathbf{b}_n$ is Minkowski-reduced if and only if the rows of B are presented in increasing order of length, and if for every integer vector $\mathbf{s} \in \mathbb{Z}^n$, $|\sum_1^n s_j \mathbf{b}_j| \leq |\mathbf{b}_d|$ where $d = d[\mathbf{s}]$ is the largest integer so that $\gcd[s_d, s_{d+1}, \ldots s_n] = 1$.

There is a simple reason why every Hermite denominator occurs as one of the q_j associated with the shortest nonzero vector in $\Lambda_\alpha[\theta]$. If not, then for all α in the interval (s, t) on which q meets the conditions to be a Hermite denominator, there would have to be some positive integer $q_j \neq q$

such that, with an appropriate choice of $\mathbf{p}_j \in \mathbb{Z}^{n-1}$,

$$\|\mathbf{p}_j 0 - q_j(\theta)1\|_\alpha = \|\mathbf{p}0 - q\theta 1\|_\alpha.$$

Since the interval (s,t) is uncountable, there would then have to be two distinct numbers $\alpha_1 < \alpha_2$ in (s,t) so that this equality held for both α_1 and α_2. But then $q = q_j$.

The converse is a bit tricky. If q is the last entry in the first row of a Minkowski reduced basis of $\Lambda_\alpha[\theta]$, then that first row is a nonzero vector of minimal length in the lattice. This alone almost qualifies it as a Hermite denominator, but a bit more is needed: there must exist an interval in which this same vector, but with respect to a different norm, is still minimal. We cannot be sure of this; there can be values of α, and Minkowski-reduced bases of $\Lambda_\alpha[\theta]$, so that the shortest nonzero vector in this basis is not shortest, for any other value of α.

It is for just this reason that we need a stricter notion of reduction than is provided by Minkowski reduction. Lagarias chose to work with lexicographically Minkowski reduced bases. A lexicographically reduced basis is a basis so that not only is $\mathbf{b}_1$ a shortest nonzero vector in the lattice, but so that, among all such vectors, the eventual choice of $\mathbf{b}_2$ is shortest, and to further break ties if needed, the eventual choice of $\mathbf{b}_3$ is shortest, and so on.

This definition has the disadvantage that it requires more or less full knowledge of the structure of the lattice, to determine of a given basis whether or not it is lexicographically reduced. Thus, we choose instead to work with another variant upon the notion of Minkowski reduction. Our new notion, that of a 'forward reduced' basis, is contrived so that a forward-reduced basis will have three properties:

(1) If B is a forward-reduced basis of $\Lambda_\alpha[\theta]$, then there exists a $\beta < \alpha$ so that for all $\gamma \in (\beta, \alpha]$, B is a forward-reduced basis of $\Lambda_\gamma[\theta]$.
(2) If q is the absolute value of the last entry of the first row of a forward reduced basis, then q is a Hermite denominator of θ.
(3) Computing a forward-reduced basis is not much more difficult than computing a Minkowski reduced basis for $\Lambda_\alpha[\theta]$.

A basis B of a lattice $\Lambda_\alpha[\theta]$ is *forward reduced* if it is Minkowski reduced, and if, moreover, for each integer $1 \leq d \leq n$, among all $\mathbf{s} \in \mathbb{Z}^n$ with $\gcd(s_d, \ldots, s_n) = 1$, either $\|\sum_1^n s_j b_j\|_\alpha > \|b_d\|_\alpha$, or $\|\sum_1^n s_j b_j\|_\alpha = \|b_d\|_\alpha$ and $|\sum_1^n s_j b_j[n]| \leq |b_d[n]|$.

(Another way to say this: a basis is not forward reduced, if there is any way to replace one of the vectors in the basis with a vector that is no longer, and for which the last entry has larger absolute value.)

We now prove the first claim: If B is a forward reduced basis of $\Lambda_\alpha[\theta]$ for $\alpha = \alpha_0$, then it is also a forward reduced basis on some interval $(\beta, \alpha_0]$.

Suppose B is a forward reduced matrix of $\Lambda_\alpha[\theta]$. Suppose also that $1 \leq d \leq n$, that $s \in \mathbb{Z}^n$, and that $\gcd(s_d, \ldots, s_n) = 1$. Then either $\|\sum s_j b_j\|_\alpha > \|b_d\|_\alpha$, which will continue to hold with β in place of α, on some interval below α, or $\|\sum s_j b_j\|_\alpha = \|b_d\|_\alpha$ and $|\sum s_j b_j[n]| \leq |b_d[n]|$, which will continue to hold with any $\beta < \alpha$ in place of α.

We now prove the second claim. If q is the final entry of the first row of B, and positive, and if B is a forward reduced basis of $\Lambda_\alpha[\theta]$, then there exists $\beta < \alpha$ so that B is also a forward reduced basis of $\Lambda_\gamma[\theta]$, for all $\gamma \in (\beta, \alpha]$. Thus, the first row of B serves as a shortest nonzero lattice vector of $\Lambda_\gamma[\theta]$, for $\beta < \gamma < \alpha$, so that q, as the last entry of that first row, serves as a Hermite denominator.

The third claim was never articulated exactly. But the point is that we can decide upon one vector, and then the next and the next, and thereby construct a forward reduced basis, without having to as it were think all the way through to the end, the chain of cascading consequences.

It is unfortunately neither trivial, nor, in general, fast, to find the shortest nonzero vector in a lattice, or to find all the short vectors. Nevertheless, if we are to arrive at an executable version of the Lagarias multidimensional continued fraction algorithm, we shall have to do just this.

Consider the lattice Λ_B with basis the rows of B, and a positive number r. The set of all s so that

$$\|\sum_1^n s_j \mathbf{b}_j\| \leq r$$

is just the set of coefficients, with respect to the basis B, of lattice points in Λ_B and within a ball of radius r about the origin. There is an algorithm (though unfortunately this task is NP-hard, so it may be slow if the dimension n is large) for this; the Fincke-Pohst algorithm [Co]. pages 102-104. It is based on the observation that there is a hyperplane $\langle \mathbf{b}_2, \ldots, \mathbf{b}_n \rangle$, the span of the $\mathbf{b}_j$'s apart from the first, such that every lattice vector can be written as the sum of an integer multiple of the orthogonal complement $\mathbf{b}_1^*$ of $\mathbf{b}_1$ with respect to this hyperplane, and a vector in the hyperplane. Thus, the only integers which can appear as coefficients of $\mathbf{b}_1$ are those integers s_1 so that $|s_1| \leq r/\|\mathbf{b}_1^*\|$.

So, given a lattice basis $B = (\mathbf{b}_1, \ldots, \mathbf{b}_n)$ of $\Lambda_\alpha(\theta)$, with its inner product $\|\cdot\|_\alpha$, we first find the shortest nonzero vector in the lattice, by taking $r = \|\mathbf{b}_1\|_\alpha$ and executing the Fincke-Pohst algorithm on that input. (In general, we might want to run the LLL algorithm first on our lattice basis, but in our current circumstances, we are working always with a matrix which is already close to Minkowski reduced, so this is not needed.)

The corresponding list s of integers will not be zero, and its entries will have no common divisor greater than one, or we should divide through by that divisor and arrive at a shorter nonzero vector after all. Thus, there is a unimodular matrix for which the first row is s.

It may happen that, even apart from the inevitable choice of $\pm$, there are multiple vectors of minimal norm. Since we want our basis to be forward reduced, we choose, among those s which lead to minimal norm $\sum s_j \mathbf{b}_j$, the one for which $|\sum_1^n s_j \mathbf{b}_j[n]|$ is maximal. If there are still choices (a rare situation), we break ties by choosing the lexicographically least s.

Now that we know how to get a forward reduced basis for $\Lambda_{\alpha_j}(\theta)$, we still need a way to find the next critical point α_{j+1}; that is, the largest real number α for which $\Lambda_\beta(\theta)$ is not forward-reduced for any $\beta \leq \alpha$. Here, the idea is to make an exploratory reduction of α, setting $\beta = \alpha/2$. This may, or may not, result in a situation in which some vector of $\Lambda_\beta(\theta)$ is shorter, with respect to $\|\cdot\|_\beta$, than one of the basis vectors. We calculate all $\mathbf{s} \in \mathbb{Z}^n$ so that $\|\sum_1^n s_j \mathbf{b}_j\|_\beta < \|\mathbf{b}_n\|_\alpha$, thus casting a wide net. We inspect these to determine whether any of them allow for a change of basis at some value between $\alpha/2$ and α, and we select that one which triggers a change, at the greatest β. If there are competing alternatives at this point, we select the one for which the index d of the least $\mathbf{b}_d$ affected, is minimal. And if there are still competing alternatives, we select that one for which the final entry of the resulting new $\mathbf{b}_d$ is greatest in absolute value. If no such changes occur, we halve α again. This will yield a sequence of decreasing critical values α_j, tending to zero, so that the corresponding bases B_j are forward reduced on $(\alpha_{j+1}, \alpha_j]$. If any of the θ_k are irrational, the sequence is infinite. But if all are rational, then we must eventually arrive at a situation in which the final entry of the first row of B_j is equal to the least common multiple of the denominators of the θ_k. At this point, the algorithm can go no further, so it terminates.

Thus, modulo filling in some details which belong more properly to the realm of writing code, than explaining mathematics, we have described how to implement the Lagarias multidimensional continued fraction algorithm on an ideal computer which does exact arithmetic with arbitrary real numbers.

In practice, such computers are not available. On the face of it, we cannot be sure our algorithm can be executed, except when θ has rational entries. If there should be 'ties', in the norms of various competing sums, and if we were working with real numbers known only by the availability of arbitrarily good rational approximations, we might not be able to break the ties and proceed. What is more, we do not yet know anything about 'continuity' of the Lagarias continued fraction algorithm output. We are, in effect, trying to map lists of n real numbers, to sequences of unimodular matrices, so as to arrive at the sequence of Hermite denominators. There must be discontinuities. Thus, one needs information about the geometry of the domains in which a given expansion prevails.

6.3 Convexity of Expansion Domains in the Lagarias Algorithm

Suppose we execute the algorithm to a finite depth r and return only the first r such matrices. We deem $\theta_1 \sim \theta_2$ if and only if the first r matrices generated by the algorithm, on the two inputs, are equal.

The induced partition of $(-1/2, 1/2)^{n-1}$ is thus refined, as r increases. But what sort of things, geometrically, are the pieces of any one such partition?

They are, we claim, convex sets. This is convenient, because if we have some irrational θ and we want to obtain M_1 through M_r for θ, we could enclose θ in a simplex of n rational θ_i, and execute our algorithm on all of them. If the results agreed to depth r, we would then know that this was indeed the result we would have got, from an ideal computer, working with θ itself.

Theorem 6.2 *For $M_1, \ldots, M_r$ in $GL(n, \mathbb{Z})$, let $Z[M_1, \ldots, M_r]$ be the set of all $\theta \subset (-1/2, 1/2)^{n-1}$ such, for $1 \leq j \leq r$, $M_j B_0[\theta]$ is the forward reduced lattice basis of $\Lambda_\alpha[\theta]$ in our algorithm, for some $\alpha > 0$ (the α can depend on θ as well as on j). Then either $Z[M_1, \ldots, M_r]$ is empty, or it is a convex subset of $\mathbb{R}^n$.*

Remark 6.1 *It would be very nice if we knew that not only are these pieces of the induced partition convex, but relatively round. There is room for further research on this topic.*

Proof. The algorithm works by passing from one pair (M, α) to another, where in each case, $MB_0[\theta]$ is forward-reduced on some interval $(\alpha - \epsilon, \alpha]$, but is not (even) Minkowski reduced for any $\beta > \alpha$. The new M, call it M', is a product PM, where P itself is a product $P_m P_{m-1} \ldots P_1$ of matrices of a certain form, generated by an inner loop of the algorithm.

At the outset of one of these inner loops, we have a lattice basis B which was forward reduced for $\beta > \alpha$, but which is not reduced for $\beta = \alpha$, because one or more of the inequalities which define forward reduction is at the cusp of flipping from valid, to not valid, as β decreases through α.

The inner loop identifies the least d so that there exists $\mathbf{s} \in \mathbb{Z}^n$ so that $\gcd(s_d, \ldots, s_n) = 1$, for which $\|\sum_1^n s_j \mathbf{b}_j\|_\alpha = \|\mathbf{b}_d\|_\alpha$, and $|\sum_1^n s_j \mathbf{b}_j[n]| > |\mathbf{b}_d[n]|$. This $\mathbf{s}$ is used to construct a matrix $P_1 \in GL(n, \mathbb{Z})$ of the form

$$P = \begin{pmatrix} 1 & 0 & 0 & \cdots & 0 \\ 0 & 1 & 0 & \cdots & 0 \\ \vdots & 0 & \ddots & & \vdots \\ s_1 & \cdots & s_d & \cdots & s_n \\ 0 & \cdots & t_{d+1,d} & \cdots & t_{d+1,n} \\ \vdots & & \vdots & & \vdots \\ 0 & \cdots & t_{n,d} & \cdots & t_{n,n} \end{pmatrix}.$$

This matrix has for its first $d-1$ rows, the elementary unit vectors $\mathbf{e}_{i,n}$, $1 \leq i \leq d-1$. The dth row is $\mathbf{s}$. The remaining rows have zeros in the first $d-1$ entries, all entries are integers, and the determinant is one.

This matrix P_1, when multiplied from the left by B, leaves invariant the vectors $\mathbf{b}_i$, $1 \leq i \leq d-s$, it replaces $\mathbf{b}_d$ with $\sum s_j \mathbf{b}_j$, and the remaining rows of $P_1 B$ are elements of $\langle \mathbf{b}_d, \ldots, \mathbf{b}_n \rangle$. Insofar as $\mathbf{b}_i$, $1 \leq i \leq d-1$ were concerned, the current B was already forward reduced at α, and no improvements were possible. An improvement has been made in $\mathbf{b}_d$, and the new $\mathbf{b}'_d$ will be shorter than the old $\mathbf{b}_d$, for all $\beta < \alpha$. In the wake of this improvement, there is chaos. True, the new list of vectors $\mathbf{b}'_k$, $d+1 \leq k \leq n$, appended to $\mathbf{b}_1, \ldots, \mathbf{b}_{d-1}, \mathbf{b}'_d$, constitutes a lattice basis for $\Lambda_\alpha[\theta]$. But we have lost all control over reduction upstream of d.

On the other hand, the good news is that no further work on $\mathbf{b}'_d$ is needed or possible, at this value of $\beta = \alpha$. This claim requires proof.

Suppose that, to the contrary, there were now a new $\mathbf{s}'$, satisfying

$$\gcd(s'_d, s'_{d+1}, \ldots, s'_n) = 1,$$

with
$$\|\sum s'_j \mathbf{b}'_j\|_\alpha = \|\mathbf{b}'_d\|_\alpha = \|\mathbf{b}_d\|_\alpha$$
and with
$$\left|\sum s'_j \mathbf{b}'_j[n]\right| > |\mathbf{b}'_d|.$$

There would then exist a matrix $P_{\mathbf{s}'}$, constructed in the same way as $P_{\mathbf{s}}$. The product $P_{\mathbf{s}'}P_{\mathbf{s}}$ would once again have, for its first $d-1$ rows, the first rows of the identity matrix. It would have integer entries in the dth row. It would have, for each of its rows from $d+1$ to n, integer linear combinations of $\mathbf{b}_d$ through $\mathbf{b}_n$. And finally, it would have determinant one. Thus, the gcd of the entries d through n, in row d of $P_{\mathbf{s}'}P_{\mathbf{s}}$, would be 1. By assumption, the final entry of the dth row of $P_{\mathbf{s}'}P_{\mathbf{s}}B_0[\theta]$ has absolute value greater than that of $\mathbf{b}'_d$. But this means that the dth row of $P_{\mathbf{s}'}P_{\mathbf{s}}$ would have been a superior alternative to the $\mathbf{s}$ which we assumed to been chosen, and that contradicts our rules for how $\mathbf{s}$ is to be chosen. Thus, once $\mathbf{b}_d$ has been modified, the new value is permanent for the duration of the inner loop.

This means that the inner loop visits each $\mathbf{b}_d$ at most once, and the inner loop executes at most n times, updating some subset of rows 1 through n of B, one after the other.

The product of the P's for each such update, constitutes the overall update implemented by the outer loop. The *trigger time* for $(\mathbf{s}, d, B)$ is that positive real number β, if it exists, such that
$$\left\|\sum_1^n s_j \mathbf{b}_j\right\|_\gamma < \|\mathbf{b}_d\|_\gamma$$
if and only if $\gamma < \beta$. Equivalently, $(\mathbf{s}, d)$ has trigger time β if and only if $|\sum_1^n s_j \mathbf{b}_j[n]| > |\mathbf{b}_d|$ and $\|\sum_1^n s_j \mathbf{b}_j\|_\beta = \|\mathbf{b}_d\|_\beta$. Let $T[\mathbf{s}, d, M, \theta]$ denote the trigger time of the move $(\mathbf{s}, d)$ when the basis B is $MB_0[\theta]$.

We are now in a position to directly address the issue of why it is that the zones in which the Lagarias algorithm (as modified here) has common execution history to depth r, are convex. For purposes of induction, we may assume that up to some value r, it is already established that the set of all θ so that successive matrices M_1 through M_r are held in common, is a convex polyhedron. Some or all of the faces may be present in this polyhedron, and some part of the faces may belong to an adjoining zone. No matter: it is a convex polyhedron. Suppose now that θ and ϕ have this same execution history to depth r, and that ψ lies on the interior of the line segment joining them. All three inputs eagerly await the result of the next inner loop. What $P = P_m \ldots P_1$ will be chosen? For θ and for ϕ, the same P is output by the inner loop, albeit perhaps associated with different values of α. But for ψ, a different inner loop update matrix P' is selected! This, we claim, cannot happen.

If it did, there would be a number $1 \leq d \leq n$, and an $\mathbf{s} \in \mathbb{Z}^n$, so that for $\mu = \theta$, and for $\mu = \phi$, but *not* for $\mu = \psi$,

(1)
$$\left\|\sum_1^n s_j \mathbf{b}_j[\mu]\right\|_{\alpha[\mu]} \leq \|\mathbf{b}_d[\mu]\|_{\alpha[\mu]}$$

(2)
$$\left|\sum_1^n s_j \mathbf{b}_j[n]\right| > |b_d[n]|$$

(3) The move associated with $(\mathbf{s}, d)$ has priority, according to our inner loop tiebreakers, and with respect to the basis $B[\mu]$, over any other moves triggered at the same time.

There are a number of ways such a situation might occur; none of them are, as we shall see, viable.

Suppose we have an $\mathbf{s}$, an integer d with $1 \leq d \leq n$, and a lattice basis $B = (\mathbf{b}_1, \ldots, \mathbf{b}_n)$.

First, it might be the case that the move associated with $\mathbf{s}$ is never triggered, at any $\beta < \alpha[\psi]$. That is, for all $\beta < \alpha[\psi]$, it might be that $|\sum_1^n \mathbf{b}_j[n]| \leq |\mathbf{b}_d[n]|$, or that for all $\beta < \alpha[\psi]$, $\|\sum_1^n s_j \mathbf{b}_j[\psi]\|_\beta > \|\mathbf{b}_d[\mu]\|_\beta$.

Second, it might be that there was a different pair $(\mathbf{s}', d')$ so that the trigger time $\beta(\mathbf{s}', d')$ was greater than that for $\mathbf{s}$.

Third, it might be that there was a different $\mathbf{s}'$, and a $d' < d$, so that the trigger time for $(\mathbf{s}', d')$ was equal to that for $(\mathbf{s}, d)$. The move associated with $\mathbf{s}'$ would then have taken priority.

Finally, it might be that there as a different $\mathbf{s}'$, with the same d, and triggered at the same time, but that $\mathbf{s}'$ had priority, in the inner loop, on tiebreakers.

We eliminate these hypothetical failure modes for the claim, in order. If $(\mathbf{s}, d)$ is never triggered at ψ, then for all $\alpha[\psi] > \epsilon > 0$, $\|\sum_1^n s_j \mathbf{b}_j[\psi]\|_\epsilon > \|\mathbf{b}_d[\psi]\|_\epsilon$.

Now since s is presumed to be triggered for θ and ϕ, $|\sum_1^n s_j \mathbf{b}_j[n]| > |\mathbf{b}_d[n]|$, (this will give $q > v$) but for ϵ sufficiently small, $\|\sum s_j \mathbf{b}_j[\theta]\| < \|\mathbf{b}_d[\theta]\|$ and $\|\sum s_j \mathbf{b}_j[\phi]\| < \|\mathbf{b}_d[\phi]\|$. Thus, in particular,

$$\left\|\sum s_j \mathbf{b}_j[\theta]\right\|_0 < \|\mathbf{b}_d[\theta]\|_0$$

and

$$\left\|\sum s_j \mathbf{b}_j[\phi]\right\|_0 < \|\mathbf{b}_d[\phi]\|_0$$

while

$$\|\sum_1^n s_j \mathbf{b}_j[\psi]\|_0 \geq \|\mathbf{b}_d[\psi]\|_0.$$

Now there are integers p_j, $1 \leq j \leq n-1$, and q, and u_j, $1 \leq j \leq n-1$, and v, with $q > v > 0$, so that for all $\mu \in \mathbb{R}^{n-1}$,

$$\sum s_j \mathbf{b}_j[\mu] = (p_1 - q\mu_1, \ldots, p_{n-1} - q\mu_{n-1}, q)$$

and

$$\mathbf{b}_d[\mu] = (u_1 - v\mu_1, \ldots, u_{n-1} - v\mu_{n-1}, v).$$

With this notation, the previous batch of inequalities now reads

$$\sum_1^{n-1}(p_j - q\theta_j)^2 < \sum_1^{n-1}(u_j - v\theta_j)^2$$

and

$$\sum_1^{n-1}(p_j - q\phi_j)^2 < \sum_1^{n-1}(u_j - v\phi_j)^2$$

and

$$\sum_1^{n-1}(p_j - q\psi_j)^2 \geq \sum_1^{n-1}(u_j - v\psi_j)^2.$$

Now this would mean that

$$(\mathbf{p} - q\theta) \cdot (\mathbf{p} - q\theta) < (\mathbf{u} - v\theta) \cdot (\mathbf{u} - v\theta)$$

and

$$(\mathbf{p} - q\phi) \cdot (\mathbf{p} - q\phi) < (\mathbf{u} - v\phi) \cdot (\mathbf{u} - v\phi)$$

and $\psi = (\sigma)\theta + (1-\sigma)\phi$, with $0 < \sigma < 1$, and yet
$$(\mathbf{p} - q\psi) \cdot (\mathbf{p} - q\psi) \geq (\mathbf{u} - v\psi) \cdot (\mathbf{u} - v\psi).$$
With a bit of algebra, this becomes
$$\mathbf{p} \cdot \mathbf{p} - 2q\mathbf{p} \cdot \psi - \mathbf{u} \cdot \mathbf{u} + 2v\mathbf{u} \cdot \psi < (v^2 - q^2)(\sigma\theta \cdot \theta + (1-\sigma)\phi \cdot \phi)$$
and
$$\mathbf{p} \cdot \mathbf{p} - 2q\mathbf{p} \cdot \psi - \mathbf{u} \cdot \mathbf{u} + 2v\mathbf{u} \cdot \psi \geq (v^2 - q^2)\psi \cdot \psi.$$
But since $q > v$, from this it follows that
$$(q^2 - v^2)\psi \cdot \psi \geq X \geq (q^2 - v^2)(\sigma\theta \cdot \theta + (1-\sigma)\phi \cdot \phi)$$
where $X = -(\mathbf{p} \cdot \mathbf{p} - 2q\mathbf{p} \cdot \psi - \mathbf{u} \cdot \mathbf{u} + 2v\mathbf{u} \cdot \psi)$. Dividing by $(q^2 - v^2)$ which is positive, and combining inequalities, we are faced with
$$\psi \cdot \psi > (\sigma\theta \cdot \theta + (1-\sigma)\phi \cdot \phi).$$
Putting $\psi = \sigma\theta + (1-\sigma)\phi$ and expanding, gives
$$\sigma(\sigma - 1)\theta \cdot \theta + 2\sigma(1-\sigma)\theta \cdot \phi + (1-\sigma)(-\sigma)\phi \cdot \phi > 0.$$
Dividing now by $\sigma(1-\sigma)$, we would have
$$-\theta \cdot \theta + 2\theta \cdot \phi - \phi \cdot \phi > 0$$
which is impossible. This shows that the first possible failure mode of the claim cannot, after all, occur.

In the second hypothetical failure mode for convexity, the trigger time for a competing move $(\mathbf{s}', d')$ was greater than that for $(\mathbf{s}, d)$ for $\mu = \psi$, although not for θ or ϕ. We have a current value for M, common to θ, ϕ, and ψ, so that the current basis in each case is $MB_0[\theta]$, $MB_0[\phi]$, and $MB_0[\psi]$ respectively. Thus,
$$T[\mathbf{s}', d', M, \psi] > T[\mathbf{s}, d, M, \psi]$$
$$T[\mathbf{s}', d', M, \theta] \leq T[\mathbf{s}, d, M, \theta]$$
$$T[\mathbf{s}', d', M, \phi] \leq T[\mathbf{s}, d, M, \phi].$$
Now $T[\mathbf{t}, \delta, M, \mu]$ is that real number γ so that $(t_1, \ldots, t_n)MB_0[\mu]$ satisfies
$$\|t \cdot M \cdot B_0[\theta]\|_\gamma = \|\mathbf{e}_d[n] \cdot M \cdot B_0[\theta]\|_\gamma$$

or equivalently,

$$\sum_{j=1}^{n-1}\left(\sum_{k=1}^{n} t_k(m_{kj} - \mu_j m_{kn})\right)^2 + \gamma \left(\sum_{k=1}^{n} t_k m_{kn}\right)^2$$
$$= \sum_{j=1}^{n} (m_{dj} - \mu_j m_{dn})^2 + \gamma m_{dn}^2.$$

Now we need some more notation. Let

$$A[\mathbf{s}, d, M] = \sum_{k=1}^{n-1}\left(\left(\sum_{j=1}^{n} s_j m_{jk}\right)^2 - m_{dk}^2\right),$$

$$\mathbf{B}[\mathbf{s}, M] = \left(\sum_{j=1}^{n} s_j m_{jn}\right)\left(\sum_{l=1}^{n} s_l m_{lk}, 1 \le k \le n-1\right), \text{(so } \mathbf{B}[\mathbf{s}, M] \in \mathbb{R}^{n-1})$$

$$C[\mathbf{s}, d, M] = \left(\sum_{j=1}^{n} s_j m_{jn}\right)^2 - m_{dn}^2,$$

$$D[\mathbf{s}, d, M] = -\frac{A[\mathbf{s}, d, M]}{C[\mathbf{s}, d, M]},$$

$$\mathbf{E}[\mathbf{s}, d, M] = \frac{2}{C[\mathbf{s}, d, M]} \mathbf{B}[\mathbf{s}, M].$$

Note that $C[\mathbf{s}, d, M] > 0$ because $\mathbf{s}$ is assumed to be a valid move, so that the new vector has a final entry greater, in absolute value, than the vector $\mathbf{b}_d$ that it replaces. With our new notation,

$$T[\mathbf{s}, d, M, \mu] = D[\mathbf{s}, d, M] + \mathbf{E}[\mathbf{s}, d, M] \cdot \mu - \|\mu\|^2.$$

The danger to convexity was that $T[\mathbf{s}', d', M, \psi] > T[\mathbf{s}, d, M, \psi]$ while at θ and at ϕ it was the other way around.

That is,

$$D[\mathbf{s}', d', M] + \mathbf{E}[\mathbf{s}', d', M] \cdot \psi - \|\psi\|^2 > D[\mathbf{s}, d, M] + \mathbf{E}[\mathbf{s}, d, M] \cdot \psi - \|\psi\|^2$$
$$D[\mathbf{s}', d', M] + \mathbf{E}[\mathbf{s}', d', M] \cdot \theta - \|\theta\|^2 \le D[\mathbf{s}, d, M] + \mathbf{E}[\mathbf{s}, d, M] \cdot \theta - \|\theta\|^2$$
$$D[\mathbf{s}', d', M] + \mathbf{E}[\mathbf{s}', d', M] \cdot \phi - \|\phi\|^2 \le D[\mathbf{s}, d, M] + \mathbf{E}[\mathbf{s}, d, M] \cdot \phi - \|\phi\|^2$$

With the shorthand D_1 for $D[\mathbf{s}', d', M]$, $\mathbf{E}_1$ for $\mathbf{E}[\mathbf{s}', d', M]$, and D_2 and E_2

when **s** and d take the place of $\mathbf{s}'$ and d', this boils down to

$$D_1 + \mathbf{E}_1 \cdot \psi > D_2 + \mathbf{E}_2 \cdot \psi,$$
$$D_1 + \mathbf{E}_1 \cdot \theta \le D_2 + \mathbf{E}_2 \cdot \theta,$$
$$D_1 + \mathbf{E}_1 \cdot \phi > D_2 + \mathbf{E}_2 \cdot \phi.$$

But this is impossible because there is a positive linear combination of the bottom two inequalities that flatly contradicts the top inequality.

Remark 6.2 *This is the heart of the matter. This shows that the interiors of the regions in question are convex. The rest of the calculations go to establishing that convexity holds also with respect to the zones themselves, and not just their interiors.*

In the third failure mode, we have the possibility that $(\mathbf{s}, d)$ generates the next move, for θ and for ϕ, but that there is a move $(\mathbf{s}', d')$ with $d' < d$ which triggers at the same time as $(\mathbf{s}, d)$ at ϕ.

Since $d' < d$, and since it is given that $(\mathbf{s}, d)$ was the move chosen by the algorithm at both θ and ϕ, we must have $T[\mathbf{s}, d, M, \theta] < T[\mathbf{s}', d', M, \theta]$ and $T[\mathbf{s}, d, M, \phi] < T[\mathbf{s}', d', M, \phi]$. But then

$$D[\mathbf{s}, d, M] + \mathbf{E}[\mathbf{s}, d, M] \cdot \theta - \|\theta\|^2 < D[\mathbf{s}', d', M] + \mathbf{E}[\mathbf{s}', d', M] \cdot \theta - \|\theta\|^2$$

and

$$D[\mathbf{s}, d, M] + \mathbf{E}[\mathbf{s}, d, M] \cdot \phi - \|\phi\|^2 < D[\mathbf{s}', d', M] + \mathbf{E}[\mathbf{s}', d', M] \cdot \phi - \|\phi\|^2.$$

From these two inequalities, though, it follows that

$$D[\mathbf{s}, d, M] + \mathbf{E}[\mathbf{s}, d, M] \cdot \psi - \|\psi\|^2 < D[\mathbf{s}', d', M] + \mathbf{E}[\mathbf{s}', d', M] \cdot \psi - \|\psi\|^2,$$

which contradicts the assumption that at ψ, the algorithm takes a different step based on $(\mathbf{s}', d')$ than what it took at the two points bracketing ψ.

The last possibility for failure of convexity was that there might have been a different $\mathbf{s}'$, but with the same d, and triggered at the same time for input ψ as was **s**, which took priority on tiebreakers.

But if it had come to tiebreakers at either of the endpoints of our line segment, this $\mathbf{s}'$ would have been selected there too. Tiebreakers do not depend in any way on the value of θ, ψ, or ϕ, but only on M and **s**. Thus, as it is given that $(\mathbf{s}, d)$ was the chosen move for inputs θ and ϕ, this move can only have been chosen because it was triggered prior to the trigger time for $(\mathbf{s}', d)$. The same calculation shows that in this case, too, $(\mathbf{s}, d)$ must have been triggered before $(\mathbf{s}', d)$ at ψ. □

Having completed the proof of the convexity result, we pause to take stock: For each input θ, there is a sequence, finite if all the entries of θ are rational, and infinite otherwise, of matrices $M_1, M_2, \ldots$ which is output by the Lagarias algorithm, as given here. The set of all $\theta \in (-1/2, 1/2)^{n-1}$ so that the algorithm output, up to depth r, agrees with a certain list of $M's$, is the region $Z[M_1, M_2, \ldots, M_r]$, and it is a convex set. Its interior is either a convex polytope, or empty. (The author has no example of an instance in which a region has empty interior, but has not found a proof that this cannot happen.)

There are two factors that work against speed in this algorithm. One is intrinsic: finding the shortest vector in a lattice is in general an NP-hard problem, and this algorithm requires finding that shortest vector along with other things, and finding it repeatedly as α decreases and the geometry of the lattice changes.

The other obstruction, which can make the algorithm slow even when the dimension is modest, is less forbidding. Even in the case of ordinary continued fraction expansion of a real number, it can [and in general will] happen that several consecutive steps are all of the form $b_2 \leftarrow b_1 + b_2$, or all of the form $b_2 \leftarrow -b_1 + b_2$. The regular continued fraction algorithm captures all these steps in a single step $b_2 \leftarrow b_2 \pm ab_1$.

In higher dimensional cases, it can happen from time to time that the vector to be approximated lies very nearly parallel to a single lattice vector, or very nearly parallel to a lattice subspace of some intermediate dimension. The algorithm in its simplest form fails to take advantage of these 'lucky breaks', and plods along, despite the inherent advantage in simplicity of dealing with lower dimensions for a time. It should be possible to refine the procedure so that these situations are recognized and exploited. If this hope were realized, the resulting algorithm could well crank out the successive q_r in time that while exponential in n was polynomial in the number of binary digits in q_r. That is, one can hope that for any fixed dimension n, finding the q_r for an arbitrary θ can be done rapidly.

There are special cases of particular interest in which these hazards seem in practice not to materialize. If the target vector θ for approximation has algebraic entries which constitute a basis for a certain algebraic extension of $\mathbb{Q}$, then the resulting vector is badly approximable. But from our perspective, the nonexistence of extraordinarily good Hermite denominators means that we are assured a steady supply $(q_r, r \geq 1)$ of new Hermite denominators, all just barely as good as the theoretical minimum quality. On general principles, we can be sure that when we have a vector $\theta \in \mathbb{R}^{n-1}$

associated with an algebraic extension of $\mathbb{Q}$ of degree n, the approximation errors $\mathbf{p}_r - q_r \theta$ will be comparable, in magnitude, to $q_r^{-1/(n-1)}$. Thus, if we scale these errors by multiplying by $q_r^{1/(n-1)}$, we shall find all the points somewhere inside a hollow ball about the origin, with finite inner and outer radius.

It so happens that in the case in which the target vector is, say, the list $(\xi, \xi^2, \ldots, \xi^{n-1})$ of successive powers of a single algebraic *integer* ξ satisfying the polynomial $\sum_0^n a_k X^k = 0$, these scaled approximation errors all fall, to within increasingly narrow tolerances as $r \to \infty$, on a Russian-doll set of nested copies of a single surface; a hyperboloid of sorts. Moreover, the ratios q_{r+1}/q_r, while of course themselves rational, are approximately equal to one or another of a finite list of algebraic integers in $\mathbb{Q}[\xi]$. This phenomenon is the topic of our next chapter.

Chapter 7

Powers of an Algebraic Integer

7.1 Introduction

Let $\alpha \in \mathbb{R}$ be an algebraic integer of degree n. For $\sigma > 0$ and for integer $q > 0$, we call the triple (σ, q, α) *good* if there exist integers $p_1, p_2, \ldots p_{n-1}$ so that for $1 \leq j \leq n-1$,

$$|p_j - q\alpha^j| \leq \sigma q^{-1/(n-1)}.$$

It is an elementary exercise in the pigeonhole principle that for any integer $Q \geq 1$ there exist q, $1 \leq q \leq Q^{n-1}$, and integers $p_1, p_2, \ldots p_{n-1}$, such that $|p_j - q\alpha^j| \leq 1/Q$. Thus, for any $\sigma \geq 1$, there is an infinite sequence of positive integers q so that (σ, q, α) is good. On the other hand, none of the resulting approximations are all *that* good: Drmota and Tichy [DT] have shown that for any algebraic number α of degree n, there is a $\sigma > 0$ so that for no q is (σ, q, α) good. Such vectors are termed 'badly approximable'. The scaled error associated with q is

$$q^{1/(n-1)}(p_1 - q\alpha, p_2 - q\alpha^2, \ldots p_{n-1} - q\alpha^{n-1}),$$

and it is bounded between two balls about the origin.

If we were to calculate, for some $\sigma \geq 1$, the first several good q and plot the associated scaled error vectors, we should see some sort of scatterplot of points, all belonging to this hollow ball. Until recently, the calculation of a robust set of good q, for a goodly sample of algebraic α, has been difficult. The LLL algorithm gives passable industrial grade good q. The Jacobi-Perron multidimensional continued fraction algorithm converges rapidly, but the q it gives are not necessarily even *good* q as we have defined them. The literature on the Jacobi Perron algorithm deems the sequence of approximations associated with a sequence of q's to be *strongly convergent* to the approximation target $(\theta_1, \theta_2, \ldots \theta_d)$ provided $|(p_1, p_2, \ldots p_d - q(\theta_1, \theta_2, \ldots \theta_d)| \to 0$. The q we want must give approximations involving an error that is roughly $O(q^{-1/d})$ times that of the acceptable error in strong convergence.

In 1994, Lagarias [La] gave a description and analysis of his geodesic multidimensional continued fraction algorithm. This algorithm takes as input a vector $\theta \in \mathbb{R}^n$, and returns a sequence of unimodular matrices, the rows of which are increasingly near to parallel to the original θ. The first row is always, in a certain technical sense (due to Hermite), an optimal approximation, taking into account an appropriate tradeoff between quality of approximation and the size of the q involved.

(An integer $q > 0$ is Hermite optimal for a target $\theta \in \mathbb{R}^n$ if there exist $0 < u < v$ and $\mathbf{p} \in \mathbb{Z}^n$ so that $|\mathbf{p} - q\theta|^2 + t^2 q^2$ is less, for all $u < t < v$, than any other $|\mathbf{p}' - q'\theta|^2 + t^2 q'^2$. Here, $|\mathbf{x}|^2 = \mathbf{x} \cdot \mathbf{x} = \sum_1^n x_j^2$.)

The Lagarias algorithm is of exponential complexity in the dimension, but it is tolerably fast for small dimension, and with current technology, feasible up to dimension ten or so. The author implemented this algorithm in Mathematica and eventually got around to looking at the scatterplots described above, for a variety of cubic algebraic integers, expecting to see a sort of cloud of points, with perhaps some clues to a density function. A scatterplot of the errors associated with the first several Hermite q for $(\log 2, \log 3)$ is representative of what one might have expected to see, except that in this case, there is no evidence of the 'hole in the middle' that must occur in the case of badly approximable pairs.

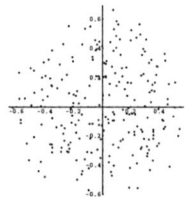

Fig. 7.1   Scaled error vectors for a pair of transcendentals.

But in every case, the sequence of 'Hermite optimal' q generated by the algorithm gave rise to a scatterplot in which all the points appeared to lie on one or several concentric, similar ellipses or hyperbolas about the origin; with ellipses when α had a pair of complex conjugates, and hyperbolas when α was totally real. The figures at the end of this section show typical results, for $x^3 - 7x - 11$ and for $x^3 - 7x - 3$, respectively.

Numerical work confirmed that the curves really were (very nearly) ellipses and hyperbolas, and the hunt was on for an explanation. In this work we explain why this was observed, and extend the analysis to the general case of α an algebraic integer of degree $n > 1$. There is a second result concerning the good approximations. For σ sufficiently large, if $q[k, \sigma, \alpha]$ denotes the kth successive $q \geq 1$ so that (σ, q, α) is good, then there is a finite set of numbers so that $q[k+1, \sigma, \alpha]/q[k, \sigma, \alpha]$ lies within $O(q^{-n/(n-1)})$ of some element of that set, as $q \to \infty$. So, even as the scaled errors fall nearly on a finite set of surfaces, the ratios of the underlying q's fall nearly on a finite set of numbers.

When β is an algebraic *number*, but not an algebraic integer, of degree n, then there is a positive integer $N = N[\beta]$ so that $N\beta$ is an algebraic integer. Any particularly good approximation $(p_1/q, p_2/q, \ldots p_{n-1}/q)$ to $(\beta, \beta^2, \ldots \beta^{n-1})$ would yield a comparably good approximation, namely $(Np_1/q, N^2p_2/q, \ldots N^{n-1}p_{n-1}/q)$, to $(N\beta, (N\beta)^2, \ldots, (N\beta)^{n-1})$.

While our analysis and results apply to the case $n = 2$, they give nothing new for that case. The best approximations come from the continued fraction expansion of α. Everything we prove, as well as much that has no evident analogue in higher dimensions, is quite well understood in the case of quadratic irrationals. Still, the analysis here does represent an extension to higher dimensions of many of the properties of the continued fraction expansion of a quadratic irrational.

We illustrate, and foreshadow, with the example $\alpha = \sqrt{3} - 1$, which has continued fraction expansion

$$\alpha = \cfrac{1}{1 + \cfrac{1}{2 + \cfrac{1}{1 + \cfrac{1}{2 + \ldots}}}}$$

and for which $\mu = 2 + \sqrt{3}$ is a fundamental unit of $\mathbb{Q}(\alpha)$. The minimal polynomial for μ is $x^2 - 4x + 1$, and the associated matrix M for which that polynomial is the minimal polynomial is $M = \begin{pmatrix} 0 & 1 \\ -1 & 4 \end{pmatrix}$. The matrix $P = \begin{pmatrix} 1 & -1 \\ 2 & 1 \end{pmatrix}$ has the property that for $k \geq 1$, $M^k P$ has rows nearly parallel to $(1, \alpha)$. Moreover,

(1) The rows of $M^k P$ provide good approximations to the desired (target) direction $(1, \alpha)$.
(2) All 'good' approximations (q, p) to $(q, q\alpha)$ can be written as simple integer linear combinations of the rows of some $(1/3)M^k P$. For in-

stance, $(56, 41)$, associated with $[0, 1, 2, 1, 2, 1, 2, 1]$, can be written as $(-1/3)(26, 19) + (2/3)(97, 71)$, and thus as a combination of the rows of $M^3 P$, while $(7953, 5822)$, another continued fraction convergent of $\sqrt{3} - 1$, is $(1/3)$ times the sum of the rows of $M^7 P$.

(3) The approximation errors $e(q) = q(1, \alpha) - (q, p)$ tend to zero and are always comparable to $1/q$. Thus, to properly assess which q do the best job, it is necessary to rescale the errors by multiplying by q. This done, the scaled errors converge to a number of one-dimensional 'circles' about the origin of a fixed radii. The errors 'rotate' their way around these 'circles', flipping alternately from negative to positive and back, as well as hopping from level to level.

So, what happens in higher dimensions? Calculation of a few hundred consecutive Hermite optimal denominators for some typical cubic algebraic integers, followed by plotting of the scaled errors, gives a clue. Here we show two such plots. These show scaled errors for Hermite optimal denominators associated with (α, α^2) when α is the real root of $x^3 - 7x - 11$, or the largest of the three real roots of $x^3 - 7x - 3$, respectively:

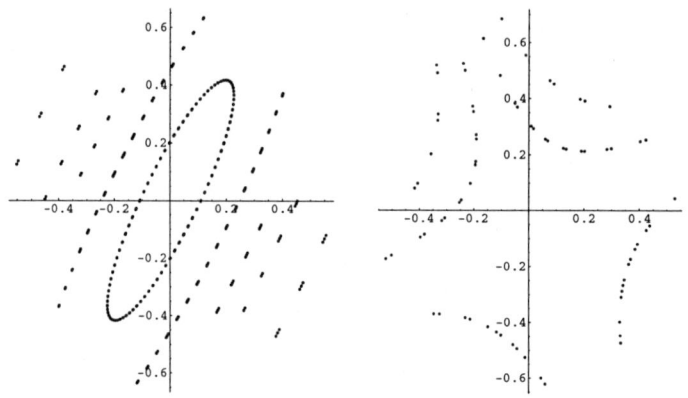

Fig. 7.2 Scaled error vectors for two algebraic integers.

7.2 Outline and Plan of Proof

The full details of the results require additional notation and are thus deferred to the appropriate sections. But we can now indicate the major stepping stones along the path to these results.

(1) Given an algebraic integer α of degree $n > 1$, there is a unit $\mu > 1$ in $\mathbb{Q}(\alpha)$ so that $\mathbb{Q}(\mu) = \mathbb{Q}(\alpha)$. For this unit μ, there is a positive integer $d = d[\mu]$ so that every algebraic integer $\beta \in \mathbb{Q}(\alpha)$ has the form

$$\beta = \frac{1}{d}\sum_{k=0}^{n-1} b_k \mu^k.$$

(This is not to say that all such sums give an algebraic integer. A simple example would be $\mathbb{Q}(\sqrt{5})$, where the algebraic integers have the form $\frac{1}{2}(a + b(\sqrt{5}-1))$ with a even.)

(2) Let

$$V_\mu := \begin{pmatrix} 1 & 1 & \cdots & 1 \\ \mu_1 & \mu_2 & \cdots & \mu_n \\ \mu_1^2 & \mu_2^2 & \cdots & \mu_n^2 \\ \vdots & \vdots & \vdots & \vdots \\ \mu_1^{n-1} & \mu_2^{n-1} & \cdots & \mu_n^{n-1} \end{pmatrix}$$

be the Vandermonde matrix associated with $\mu = \mu_1$, where the remaining conjugates of $\mu_2, \ldots, \mu_n$ of μ are numbered so that real conjugates come first, and so that if there are r_1 real conjugates and r_2 complex conjugate pairs, then $\mu_{r_1+r_2+k} = \overline{\mu}_{r_1+k}$ for $1 \le k \le r_2$. Then the $n \times n$ matrix $P_\mu = V_\mu V_\alpha^t$ is nonsingular and has rational integer entries.

(3) There is a sequence $(\nu[k])$, $(k \ge 1)$ of units, each greater than one, in $\mathbb{Q}(\alpha)$, and a positive constant C, so that for all $k \ge 1$, if $\nu[k]_j$ denotes the jth conjugate of $\nu[k]$, numbered consistent with the conjugates of μ, then for $2 \le i, j \le n$,

$$|\nu[k]_i/\nu[k]_j| \le C \text{ and } 2 \le \nu[k+1]/\nu[k] \le C.$$

As the other conjugates of $\nu[k]$ are comparable and have product with absolute value $1/\nu[k]$, each must be $O(\nu[k]^{-1/(n-1)})$.
Since $\sum_{j=1}^{n} \nu[k]_j = q[k]$ is an integer, and since the other conjugates are small, $q[k] = \nu[k] + O(q[k]^{-1/(n-1)})$.

Remark 7.1 *For k sufficiently large, these units are a special case of PV numbers.*

(4) For any real algebraic integer α of degree $n > 1$, there is a $\sigma_0 > 0$ and a positive integer m so that for every $\sigma > \sigma_0$ there is a finite set

$A(\sigma,\alpha) \subseteq \mathbb{Z}^n$ so that for all q with (σ,q,α) good, there is an $l \geq 1$ and an $\mathbf{a} \in A(\sigma,\alpha) \subset \mathbb{Z}^n$ so that

$$q = \frac{1}{m} \sum_{k=1}^{n} a_k \sum_{j=1}^{n} \mu_j^{k-1} \nu[l]_j.$$

Furthermore, the corresponding $\mathbf{p} \in \mathbb{Z}^{n-1}$ has, for $1 \leq i \leq n-1$,

$$p_i = \frac{1}{m} \sum_{k=1}^{n} a_k \sum_{j=1}^{n} \mu_j^{k-1} \nu[l]_j \alpha_j^{i-1}.$$

This formula will lead to the announced results about the scaled errors associated with good q.

7.3 Proof of the Existence of a Unit $\mu \in \mathbb{Q}(\alpha)$ of Degree n

The set U_α of units in $\mathbb{Q}(\alpha)$ is a finitely generated multiplicative group with structure $\{\pm 1\} \times \mathbb{Z}^r$ with $r = r_1 + r_2 - 1$ where (r_1, r_2) is the signature of $\mathbb{Q}(\alpha)$, that is, where α has r_1 real conjugates and $2r_2$ that are not real. [Co]

Given a fundamental set of units $\{\mu_1, \mu_2, \ldots, \mu_r)\}$ for $\mathbb{Q}(\alpha)$, every unit $\gamma \in \mathbb{Q}(\alpha)$ has the form

$$\gamma = \pm \prod_{j=1}^{r} \mu_j^{a_j}, \quad (a_1, a_2, \ldots a_r) \in \mathbb{Z}^r.$$

Let $\lambda(\gamma) = (a_1, a_2, \ldots a_r)$ be the list of exponents in this characterization, so that $\lambda : U_\alpha \to \mathbb{Z}^r$. The mapping is two to one, and the inverse image of 0 is ± 1. For any subfield K of $\mathbb{Q}(\alpha)$, let $\Lambda(K)$ be the lattice of all $a \in \mathbb{Z}^r$ so that $\lambda^{-1}(a) \subseteq K$. If, contrary to our claim, there were to be a real algebraic integer α so that no one unit μ of $\mathbb{Q}(\alpha)$ generated $\mathbb{Q}(\alpha)$, then for each unit $\mu \in U_\alpha$, the subfield $K(\mu)$ generated by μ would be a proper subfield of $\mathbb{Q}(\alpha)$. From Galois theory, the set $\mathcal{K}$ of proper subfields of $\mathbb{Q}(\alpha)$ is finite. This would lead to a representation of $\mathbb{Z}^r$ as the union of a finite set of lattices $\Lambda(K)$.

In one dimension we are at once done since no finite set of proper sublattices of $\mathbb{Z}$ has union $\mathbb{Z}$; ± 1 would be excluded nor could the density of the union be 1. There are, after all, infinitely many primes. But even in two dimensions, one may readily construct a set of three proper lattices of $\mathbb{Z}^2$, each of determinant 2, so that their union is all of $\mathbb{Z}^2$.

This does not much impede our argument, because, we claim, the sublattices $\Lambda(K) \subset \mathbb{Z}^r$ corresponding to proper subfields K have dimension less than r, and thus density zero in $\mathbb{Z}^r$. From this, the existence of a generating μ is immediate.

To prove the claim, we note that the degree of K over $\mathbb{Q}(\alpha)$ is n/d for some $d > 1$. Thus if K has signature (r_1', r_2') then $r_1' + 2r_2' = n/d \leq n/2$ while $r = r_1 + r_2 - 1 \geq (n-1)/2$. Thus $r_1' + r_2' - 1 < r$, and so the number of generators of U_μ in K is less than r, and the lattice $\Lambda(K)$ has dimension less than r. From this, it is apparent that not only does there exist a unit $\mu \in \mathbb{Q}(\alpha)$ so that $\mathbb{Q}(\mu) = \mathbb{Q}(\alpha)$, but that almost all units are like that. We choose for our fixed μ associated with $\mathbb{Q}(\alpha)$ any such unit that is greater than 1.

In the outline, we mentioned that the matrix $P_\mu = V_\mu V_\alpha^t$ had rational integer entries and was nonsingular. The proof of this claim is simple now that we know that μ has n distinct conjugates: We now know that we can indeed construct V_μ, and the entry (j, k) of P_μ is $P_\mu[j, k] = \sum_{i=1}^n \mu_i^{j-1} \alpha_i^{k-1}$ and is invariant under automorphisms of $\mathbb{Q}(\alpha)$. It is thus both an algebraic integer, and a rational number. Since the conjugates of μ are distinct, V_μ is nonsingular and of course for the same reason so is V_α^t so P_μ is nonsingular.

7.4 The Sequence $\nu[k]$ of Units with Comparable Conjugates

For a positive unit $\nu \in \mathbb{Q}(\alpha)$, we call (C, ν) good if, for all conjugates of ν other than ν itself, $|\nu_j| \leq C\nu^{-1/(n-1)}$. If the signature of α is (r_1, r_2), we put $r = r_1 + r_2 - 1$. This is the number of units, apart from -1, needed as generators for the multiplicative group of units in $\mathbb{Q}(\alpha)$. We prove that there exists $C = C(\alpha) > 1$ and a sequence $(\nu[k]), (k \geq 1)$ of units of $\mathbb{Q}(\alpha)$, each of degree n, so that $2 \leq \nu[k+1]/\nu[k] \leq C$ and $(C, \nu[k])$ is good.

We associate to any positive unit $\nu \in \mathbb{Q}(\alpha)$ the real vector

$$\mathbf{x}(\nu) = (x_1, x_2 \ldots x_{r+1}) \in \mathbb{R}^{r+1}$$

given by

$$\mathbf{x}(\nu) = (\log(\nu_1), \log|\nu_2|, \ldots \log|\nu_{r_1}|, 2\log|\nu_{1+r_1}|, \ldots 2\log|\nu_{r_1+r_2}|),$$

which lies in the hyperplane $H = \{x \in \mathbb{R}^{r+1} : \sum_1^{r+1} x_k = 0\}$.

Now consider the lattice $\Lambda(\alpha) \subset H$ consisting of all $\mathbf{x}(\nu)$ so that ν is a unit of $\mathbb{Q}(\alpha)$. We claim that there is a set of r such units, each of degree

n, which generates the group of units of $\mathbb{Q}(\alpha)$ apart from the torsional component ± 1. Only the claim that we may take our units to have degree n requires proof. The lattice points **a** in the lattice of the previous section, corresponding to units of degree less than n, have zero density in $\mathbb{Z}^r$ and lie in a finite set of sub-dimensional hyperplanes. Thus if we select, as we may, a lattice basis of $\mathbb{Z}^r$ consisting of vectors all nearly parallel to some vector not in any of the hyperplanes, the units corresponding to that lattice basis will all have degree n. This proves the claim.

Thus in the representation of units as $\nu = \pm \prod_1^r \mu[j]^{a_j}$, we may as well, and we do, assume that all the $\mu[j]$ have degree n, and each is greater than 1. Each of these units has a list of n conjugates, $\mu[j] = \mu[j]_1, \mu[j]_2, \ldots \mu[j]_n$, which we number consistently with the numbering of the conjugates of α. We take our basic μ, used in V_μ and so on, to be $\mu[1]$, and we define $\delta[j] \in \mathbb{R}^{r+1}$ by

$$\delta[j] = (\log|\mu[j]_1|, \ldots, \log|\mu[j]_{r_1}|, 2\log|\mu[j]_{1+r_1}|, \ldots 2\log|\mu[j]_{r_1+r_2}|).$$

The set $\Delta = \{(\delta[1], \delta[2], \ldots, \delta[r]\}$ is then a basis for H. The question of the existence of units ν fitting our conditions, can now be rephrased as a question about the lattice $L \subset H$ generated by Δ: Does there exist a sequence $\phi[k](k \geq 1)$ of elements in L, confined to a cylinder about the ray $\{t(r, -1, -1, \ldots - 1) : t > 0\}$ of radius small enough that any ν corresponding to a point inside the cylinder must have (C, ν) good, and so that the sequence of first components of $\phi[k]$ is increasing to infinity with $\log 2 \leq \phi[k+1]_1 - \phi[k]_1 \leq \log C$?

Put this way, a simple construction shows that the answer is yes when C is large enough. Let $\Delta = \Delta(L) := \{\sum_1^n s_j \delta[j] : -1/2 < s_j \leq 1/2$ for $1 \leq j \leq n\}$. Let $\theta > 1$ be large enough that the first coordinate of any element of $\theta(r, -1, \ldots, -1) + \Delta$ is greater than the first coordinate of any element of Δ. Any translate of Δ contains exactly one element of L. We take $\mathbf{x} \in L$ to be an element sufficiently distant from the cylinder axis that no element of $\Delta + \mathbf{x}$ lies on the cylinder. (This displacement of the cell from the axis forecloses the possibility that the sequence of lattice elements would lie in some fixed subspace of dimension less than that of H.) We then take $\nu[k]$ to be the unit corresponding to the element $\phi[k]$ of L in $\Delta + x + k\theta(r, -1, \ldots, -1)+$. The units then satisfy $\nu[k] = \exp(kr\theta + O(1))$, $2 \leq \nu[k+1]/\nu[k] \leq C$, and $|\nu[k]_j/\nu[k]_i| < C$ for $2 \leq i, j \leq n$, as claimed. Furthermore, the set of ratios $\nu[k+1]/\nu[k]$ is finite because the set of differences $\phi[k+1] - \phi[k]$ is a subset of the set of all lattice points in

$\theta(r, -1, \ldots -1) + 2\Delta$ and there are just 2^n such lattice points. Finally, we purge the sequence of any elements of degree less than n; as there are only finitely many subfields of $\mathbb{Q}(\alpha)$ to avoid, and as each corresponding subspace of H can touch only finitely many of the boxes from which we originally took our sequence of lattice points, this leaves an infinite sequence of good units, each of degree n, and with all the asymptotic properties of the original sequence.

7.5 Good Units and Good Denominators

Let $\nu = \nu_k$ be one of our sequence of good units. For each such ν, there is a polynomial f_ν with integer coefficients and degree less than n, so that $\nu = \frac{1}{d} f_\nu(\mu)$.

From this point on, we dispense with keeping track of the details of the constants in good units or good denominators, and use 'Big Oh' terminology. When we say that some object depending on k is good, what we mean is that there exist constants so that for all sufficiently large k, that object is good with respect to those constants, and 'O' will also be with respect to k.

Now for any $\beta \in \mathbb{Q}(\alpha)$, let D_β be the diagonal matrix with diagonal entries $D_\beta[i, i] = \beta_i$, the ith conjugate of β. Let

$$M_\nu = V_\mu D_\nu V_\mu^{-1} P_\mu = V_\mu D_\nu V_\alpha^t.$$

Now the (k, j) entry of M_ν may be written as $M_\nu[j, k] = \sum_{i=1}^{n} \mu_i^{j-1} \nu_i \alpha_i^{k-1}$ which is a sum of products of integers of $\mathbb{Q}(\alpha)$ and thus an integer of $\mathbb{Q}(\alpha)$, and since the sum is a symmetric polynomial in $\alpha_1, \ldots, \alpha_n$, it is a rational number. Thus, the entries of M_ν are integers. Since both V_μ and V_ν have determinants of absolute value 1, $|\det M_\nu| = |\det P_\mu| = p$ say, and $p \geq 1$.

On the other hand, the rows $M\nu[j]$ of M_ν satisfy the estimate

$$M_\nu[j] = \mu^{j-1}\nu(1, \alpha, \alpha^2, \ldots, \alpha^{n-1}) + O(\nu^{-1/(n-1)}).$$

In particular, the $(1, 1)$ entry q_ν, which is also (for large ν) the integer nearest the PV number ν, is a good denominator for simultaneous diophantine approximation of the powers of α.

Now suppose $\mathbf{c} = (c_1, c_2, \ldots c_n) \neq 0 \in \mathbb{Z}^n$. Then $\sum_1^n c_k \mu^{k-1} \neq 0$. Thus for any such **c** and for any $\nu = \nu_k$ from our sequence of good units,

$$\mathbf{c} \cdot M_\nu = \nu \sum_{k=1}^{n} c_k \mu^{k-1}(1, \alpha, \alpha^2, \ldots, \alpha^{n-1}) + O(\nu^{-1/(n-1)})$$

so that the first entry of this vector is a good denominator for α. More important is that a kind of converse is also true.

Theorem 7.1 *Let α be a real algebraic number of degree $n > 1$. Let $\mu > 1$ be a unit of $\mathbb{Q}(\alpha)$ of degree n. Let $p = |\det P_\mu|$. Suppose (ν) is a sequence of units of degree n in $\mathbb{Q}(\alpha)$, with the property that if $\nu_j[k]$ are the conjugates of ν_j, ordered so that $\nu_j = \nu_j[1]$, then $\nu_j[1] > 1$, $2 < \nu_{k+1}/\nu_k < C$ for all k, and $|\nu_j[k]| \in [1/C, C]\nu_j^{-1/(n-1)}$ for all conjugates $\nu_j[k]$ of ν_j, $2 \leq j \leq n$. Let q_k be the sum of ν_k and all its conjugates. Then for all σ large enough that (σ, q_k, α) is good for all k, there exists $N \geq 1$ so that if $q \geq 1$ and (σ, q, α) is good, there exist integers $(c_1, c_2, \ldots, c_n)$ with $|c_i| \leq N$ for $1 \leq i \leq n$, not all zero, and an integer $m \geq 1$, so that $\nu_m < q < \nu_{m+1}$ and, writing $\nu_m = \nu$,*

$$q = \frac{1}{p}(\mathbf{c} \cdot M_\nu)_1.$$

Remark 7.2 *Informally, this says that all good denominators for α come from simple rational linear combinations of the rows of some M_ν, for some good unit ν in our sequence. Stripped to a mnemonic, good denominators come from good units.*

Proof. Suppose (σ, q, α) is good. Then the integers nearest $q\alpha_j$ respectively, call them $(p_1, \ldots, p_{n-1})$, satisfy $|p_j - q\alpha^j| \leq \sigma q^{-1/(n-1)}$ for $1 \leq j \leq n-1$. Choose m so that $\nu = \nu_m < q < \nu_{m+1}$. Note that M_ν has integer entries, but on the other hand, we can express the kth row of M_ν in the form

$$M_\nu[k] = (1 + O(q^{-1/(n-1)}))q_m \mu^{k-1}(1, \alpha, \alpha^2, \ldots \alpha^{n-1}) + \mathbf{e}$$

where $\mathbf{e} \cdot (1, \alpha, \alpha^2, \ldots \alpha^{n-1}) = 0$ and where $|\mathbf{e}| = O(q^{-1/(n-1)})$.

Now consider the $n \times n$ matrix T_q determined by the twin requirements that $(1, \alpha, \alpha^2, \ldots \alpha^{n-1}) \cdot T_q = q^{-1}(1, \alpha, \alpha^2, \ldots \alpha^{n-1})$ and that for any vector $\mathbf{x} \in \mathbb{R}^n$ with $\mathbf{x} \cdot (1, \alpha, \alpha^2, \ldots \alpha^{n-1}) = 0$, $\mathbf{x} \cdot T_q = q^{1/(n-1)}\mathbf{x}$. From the foregoing observations about M_ν, it follows that the rows of $M_\nu \cdot T_q$ have length $O(1)$. Since $\det(T_q) = 1$, it follows that the rows of $M_\nu \cdot T_q$ are in fact comparable to 1 in length, and from this, it follows that $(M_\nu \cdot T_q)^{-1}$ has bounded entries, independent of m and q. Now let

$\mathbf{r} = \mathbf{r}[\nu] = (r_1, r_2, \ldots, r_n) = (q, p_1, p_2, \ldots, p_{n-1}) \cdot M_\nu^{-1}$. Then pr_j is an integer for all j. Also, $(q, p_1, p_2, \ldots, p_{n-1}) \cdot T_q = O(1)$ and $\|(M_\nu \cdot T_q)^{-1}\| \ll 1$, so $|(q, p_1, p_2, \ldots, p_{n-1}) \cdot T_q \cdot T_q^{-1} M_\nu^{-1}| \ll 1$. That is, $|r_j|$ is bounded independent of m and q. From this it follows that $(q, p_1, p_2, \ldots, p_{n-1}) = p^{-1} \mathbf{c} \cdot M_\nu$ with $c_j = Pr_j \in \mathbb{Z}$, and the $|c_j|$ are bounded, independent of m and q. The bound does depend on α, μ, and σ. This completes the proof of the theorem. In a nutshell, every good q is the first entry in an integer vector of the form $p^{-1} \mathbf{c} \cdot M_\nu$ for some $\nu = \nu[m]$ in our sequence of good units, and $\mathbf{c}$ with $\mathbf{c} \in \mathbb{Z}^n$ bounded, independent of m. □

7.6 Ratios of Consecutive Good q

We are now in position to prove our second theorem, to the effect that the ratios of consecutive good q are asymptotically close to one or another of a finite set of elements of $\mathbb{Q}(\alpha)$.

Theorem 7.2 *Let α be a real algebraic integer of degree $n > 1$. Then for all sufficiently large σ, there exists a finite set $B(\sigma, \alpha) \subset \mathbb{Q}(\alpha) \cap (1, \infty)$, and a constant $C = C(\sigma, \alpha) > 1$, so that if (σ, q, α) and (σ, q', α) are consecutive good triples for α, then there exists $\beta \in B(\sigma, \alpha)$ so that*

$$|\beta - q'/q| \le Cq^{-n/(n-1)}.$$

Proof. Choose σ so that for all q_k associated with ν_k, (σ, q_k, α) is good. Choose N so that, for all q such that (σ, q, α) is good, there exists a $\mathbf{c} \in [-N, N]^n$ and a $k \ge 1$ such that q is the first entry of $p^{-1} \mathbf{c} \cdot M_{\nu_k}$. Choose m so that $q_m \le q < q_{m+1}$. Then $q' \le q_{m+1}$. Let $\nu = \nu_m$. Then as in Theorem 7.1, there also exists $\mathbf{c}' \in [-N, N]$ such that

$$q' = \frac{1}{p} (\mathbf{c}' \cdot M_\nu)[1].$$

Furthermore,

$$(\mathbf{c} \cdot M_\nu)[1] = \nu \sum_{j=1}^n c_j \mu^{j-1} + O(\nu^{-1/(n-1)}),$$

and similarly for $\mathbf{c}'$. Thus, since $q_m = \nu + O(\nu^{-1/(n-1)})$,

$$(\mathbf{c} \cdot M_\nu)[1] = q_m \sum_{j=1}^n c_j \mu^{j-1} + O(q^{-1/(n-1)}),$$

and similarly for $\mathbf{c}'$ and q'. Thus

$$\frac{q'}{q} = \frac{\sum_{j=1}^n c'_j \mu^{j-1} + O(q^{-1/(n-1)})}{\sum_{j=1}^n c_j \mu^{j-1} + O(q^{-1/(n-1)})}.$$

Now the entries of $\mathbf{c}$ and $\mathbf{c}'$ are bounded by N, so the quantity $|\sum_{j=1}^n c_j \mu^{j-1}|$ is bounded and bounded away from zero, independently of m, as is the corresponding quantity for $\mathbf{c}'$. Thus

$$\frac{q'}{q} = \frac{\sum_{j=1}^n c'_j \mu^{j-1}}{\sum_{j=1}^n c_j \mu^{j-1}} + O(q^{-n/(n-1)}).$$

The set $B(\sigma, \alpha)$ of ratios of this type is finite because $\mathbf{c}$ and $\mathbf{c}'$ come from a finite set. □

7.7 The Surfaces Associated With the Scaled Errors

In this section we explain the experimentally observed ellipses and hyperbolas mentioned in the introduction, and prove a theorem generalizing what was observed.

Any hyperbola centered at the origin has a parametric representation as the image, under a nonsingular linear transformation of $\mathbb{R}^2$, of $\{(t_1, t_2) : t_1 t_2 = 1\}$, and any ellipse is a like transformation of the unit circle. For $(r_1, r_2) \in \mathbb{Z}^2$ with $r_1 + 2r_2 = n$ and $r_1 > 0, r_2 \geq 0$, let

$$U[r_1, r_2] = \{(z_2, z_3, \ldots z_n) \in \mathbb{C}^{n-1} : z_j \in \mathbb{R}, 2 \leq j \leq r_1,$$

$$z_{r_1+r_2+k} = \overline{z}_{r_1+k}, 1 \leq k \leq r_2, \text{ and } \prod_{j=2}^n |z_j| = 1\}.$$

Also, if M is an $(n-1) \times (n-1)$ invertible matrix with complex entries, and if in each column of M, the first $r_1 - 1$ entries are real, and the last r_2 entries are the respective conjugates of the middle r_2 entries, we say that M is of *type* (r_1, r_2).

If M has type (r_1, r_2), then $U[r_1, r_2]M$ is a surface in $\mathbb{R}^{n-1}$ of dimension $n - 2$, and $0 \notin U[r_1, r_2]M$. In the special case $r_1 = 3, r_2 = 0$, it is a pair of hyperbolas, a linear transformation of $\{(t_1, t_2) : t_1 t_2 = \pm 1\}$, whereas if $r_1 = r_2 = 1$, it is an ellipse.

Theorem 7.3 *Let α be an algebraic integer of degree $n > 1$ and signature (r_1, r_2). Then for all $\sigma > 1$ there is a finite set $R(\sigma, \alpha)$ of positive elements* of $\mathbb{Q}(\alpha)$, and a nonsingular matrix W_α of type (r_1, r_2), so that for each $q \geq 1$ with (σ, q, α) good, there is a $\gamma \in R(\sigma, \alpha)$ such that the scaled error

$$q^{1/(n-1)} \left[(p_1, \ldots p_{n-1}) - q(\alpha, \ldots \alpha^{n-1}) \right]$$

lies, to within $O(q^{-n/(n-1)})$, on the surface $\gamma U[r_1, r_2] W_\alpha$. The (k, l) entry of W_α is $W_\alpha[k, l] = \alpha_{k+1}^l - \alpha_1^l$.

Proof. Without loss of generality we may take σ arbitrarily large, because if (σ_1, q, α) is good and $\sigma_2 > \sigma_1$, then (σ_2, q, α) is good. We may also take q arbitrarily large. We recall that $p = |\det P_\mu|$. Now by Theorem 7.2, there is a finite set $B(\sigma, \alpha)$, and a positive integer N, so that for all large q with (σ, q, α) good,

$$(q, p_1, \ldots p_{n-1}) = \frac{1}{p}(\mathbf{c} \cdot M_\nu)$$

for some $\nu = \nu[m]$ with $m \geq 1$ and some integer-entried $\mathbf{c} \in [-N, N]^n$. Now $M_\nu = V_\mu D_\nu V_\alpha^t$. A routine calculation then yields

$$(\mathbf{c} \cdot M_\nu)_j = \sum_{i=1}^{n} c_i \sum_{k=1}^{n} \mu_k^{i-1} \nu_k \alpha_k^{j-1}.$$

Thus

$$q = \frac{1}{p} \sum_{i=1}^{n} \sum_{k=1}^{n} c_i \mu_k^{i-1} \nu_k$$

so that the jth entry of $q(\alpha, \alpha^2, \ldots \alpha^{n-1})$ is

$$q\alpha^j = \frac{1}{p} \sum_{i=1}^{n} \sum_{k=1}^{n} c_i \mu_k^{i-1} \nu_k \alpha^j, \quad (1 \leq j \leq n-1).$$

The resulting approximation error

$$\mathbf{e}(q) = (p_1, \ldots, p_{n-1}) - q(\alpha, \ldots, \alpha^{n-1})$$

has jth entry

$$e_j(q) = \frac{1}{p} \left(\sum_{i=1}^{n} \sum_{k=2}^{n} c_i \mu_k^{i-1} \nu_k (\alpha_k^j - \alpha_1^j) \right), \quad (1 \leq j \leq n-1).$$

This sum can be written as a sum of matrix products, writing $D(x_1, x_2, \ldots)$ for the diagonal matrix based on the given entries:

$$\mathbf{e}(q) = \frac{1}{p} \sum_{i=1}^{n} c_i(\nu_2, \ldots, \nu_n) \cdot D(\mu_2, \ldots, \mu_n)^{i-1} W_\alpha,$$

where $W_\alpha[k, j] = (\alpha_{k+1}^j - \alpha_1^j)$ for $1 \leq j, k \leq n-1$. We now show that W_α is nonsingular. For if $\mathbf{x} \cdot W_\alpha = 0$, consider $\mathbf{y} = (-\sum_1^{n-1} x_j, x_1, x_2, \ldots, x_{n-1}) \in \mathbb{R}^n$, and consider the $n \times n$ matrix $V_\alpha - V'$, where V' has n identical rows, each being $(1, \alpha, \ldots, \alpha^{n-1})$. Then $\mathbf{y} \cdot (V_\alpha - V') = 0$. But $\mathbf{y} \cdot V' = 0$, so $\mathbf{y} \cdot V_\alpha = 0$ so that $\mathbf{y} = \mathbf{0}$ and thus $\mathbf{x} = \mathbf{0}$.

For $\mathbf{c} \in \mathbb{Z}^n$, let $\xi(\mathbf{c}) = \left(\sum_1^n c_i \mu_1^{i-1}\right)^{1/(n-1)}$, and let

$$S_\alpha(\mathbf{c}) = \{p^{-1}\mathbf{c} \cdot V'_\mu D(s_2, s_3, \ldots s_n) : s \in U[r_1, r_2]\}$$

where V'_μ is the $n \times (n-1)$ matrix got by deleting the leftmost column from V_μ. Then

$$q^{1/(n-1)}\mathbf{e}(q) = (1 + O(q^{-n/(n-1)}))p^{-1/(n-1)}\nu^{1/(n-1)}\xi(\mathbf{c})\mathbf{e}(q).$$

Now

$$\nu^{1/(n-1)}\mathbf{e}(q)W_\alpha^{-1} = p^{-1}\sum_{i=1}^{n} c_i(\nu_2\nu^{1/(n-1)}, \ldots, \nu_n\nu^{1/(n-1)})D(\mu_2, \ldots, \mu_n)^{i-1},$$

and

$$(\nu_2\nu^{1/(n-1)}, \ldots, \nu_n\nu^{1/(n-1)}) \in U[r_1, r_2].$$

Equivalently, $\nu^{1/(n-1)}\xi(\mathbf{c})\mathbf{e}(q)W_\alpha^{-1} \in \xi(\mathbf{c})S_\alpha(\mathbf{c})$, and so the scaled errors lie, to within a factor of $(1 + O(q^{-n/(n-1)}))$, on $p^{-1/(n-1)}\xi(\mathbf{c})S_\alpha(\mathbf{c})W_\alpha$.

It remains only to show that for each nonzero $\mathbf{c} \in \mathbb{Z}^n$ there is a nonzero scaling coefficient $\rho(\mathbf{c})$ so that every point of $\xi(\mathbf{c})S_\alpha(\mathbf{c})$ belongs to $\rho(\mathbf{c})S_\alpha(1, 0, \ldots, 0)$. The scaling coefficient must be

$$\rho(\mathbf{c}) = \left(\prod_{j=1}^{n} |f_\mathbf{c}(\mu_j)|\right)^{1/(n-1)} = \left|\mathcal{N}\left(\sum_1^n c_j \mu_1^{j-1}\right)\right|^{1/(n-1)}$$

where $f_\mathbf{c}(x) = \sum_{k=1}^{n} c_k x^{k-1}$. Note that $\rho(\mathbf{c}) \neq 0$, because $f_\mathbf{c}(\mu) \neq 0$ since μ is algebraic of degree n, and because the other conjugates of μ are likewise of degree n.

To prove that these surfaces are scalar multiples of each other, it is sufficient to show that for any $s \in U[r_1, r_2]$ and any $\mathbf{c} \neq \mathbf{0}$, there exists $t \in U[r_1, r_2]$ so that

$$\mathbf{c} \cdot V'_\mu D(s_2, \ldots s_n) = (1, 0, , \ldots, 0) \cdot V'_\mu D(t_2, \ldots, t_n).$$

So suppose $\mathbf{c} \in \mathbb{Z}^n$, $\mathbf{c} \neq (0, 0, \ldots, 0)$ and suppose $s \in U[r_1, r_2]$. Then $\rho(\mathbf{c}) \neq 0$, and

$$\mathbf{c} \cdot V'_\mu D(s) = (f_\mathbf{c}(\mu_2), \ldots, f_\mathbf{c}(\mu_n)) D(s) = \rho(\mathbf{c})(s'_2 f_\mathbf{c}(\mu_2), \ldots, s'_n f_\mathbf{c}(\mu_n))$$

where $s'_j = s_j / \rho(\mathbf{c})$. But

$$(s'_2 f_\mathbf{c}(\mu_2), s'_3 f_\mathbf{c}(\mu_3), \ldots, s'_n f_\mathbf{c}(\mu_n)) \in U[r_1, r_2]$$

so $\mathbf{c} V'_\mu D(U[r_1, r_2]) \subseteq \rho(\mathbf{c}) U[r_1, r_2]$. □

A direct illustration of Theorem 7.3 in three or more dimensions is far more difficult than in two dimensions. The two dimensional paper does not readily reveal nested three dimensional surfaces. But what can be done without great trouble is to calculate the ρ associated with each q, that is, by what number must we multiply the standard surface so that the current scaled error falls on the surface? This was done for $x^4 - 4x + 2$, and the resulting plot is shown in the next figure.

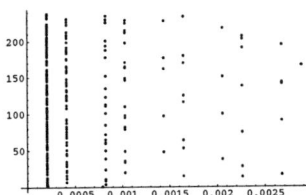

Fig. 7.3 Point frequencies for the various nested surfaces.

7.8 The General Case of Algebraic Numbers in $\mathbb{Q}(\alpha)$

Suppose now that α is as before an algebraic integer of degree n, and a real number, and it has signature (r_1, r_2). Suppose also that

$$\{1, \beta_1, \beta_2, \ldots, \beta_{n-1}\} \subset \mathbb{Q}(\alpha)$$

is linearly independent over $\mathbb{Q}$. Recall that (q, σ, β) is *good* if there exist integers $p_1, \ldots, p_{n-1}$ so that $|q\beta_j - p_j| \leq q^{-1/(n-1)}\sigma$ for $1 \leq j \leq n-1$. Then we have the same theorems for $\beta = (\beta_1, \beta_2, \ldots, \beta_{n-1})$ that we had for algebraic integers α.

We first note that there is an invertible matrix B with integer entries, and a positive integer N, such that

$$(1, \beta_1, \ldots, \beta_n) = N^{-1}(1, \alpha, \ldots, \alpha^{n-1})B.$$

Let $d = |\det B|$, so that $B^{-1} = d^{-1}B'$ with B' again an invertible matrix with integer entries. If (q, σ, β) is good, then $(Nq, N^{n/(n-1)}\sigma, \beta)$ is also good, so

$$(Nq, Nq\beta) = (Nq, Np_1, \ldots, Np_{n-1}) + (0, e_1, \ldots, e_{n-1})$$

where $|e_j| \leq N^{n/(n-1)}\sigma(Nq)^{-1/(n-1)}$ for $1 \leq j \leq n-1$. Thus with $b' = \max_{i,j}|B'_{i,j}|$, we have

$$(Nq, Nq\beta)B' = r + (0, e'_1, e'_2, \ldots, e'_{n-1})$$

where $r \in \mathbb{Z}^n$ and where $|e'_j| \leq (nN^{n/(n-1)}\sigma b')(Nq)^{-1/(n-1)}$. Now the first entry of r is dq, so we restate this as

$$|e'_j| \leq (nNd^{1/(n-1)}\sigma b')(dq)^{-1/(n-1)}.$$

Thus, $(dq, (nNd^{1/(n-1)}\sigma b'), \alpha)$ is good. Thus, Theorem 7.3 applies and $(dq)^{1/(n-1)}(e'_1, e'_2, \ldots, e'_{n-1})$ lies, to within $O(q^{-1/(n-1)})$, on a surface of the form $\rho U[r_1, r_2]W_\alpha$, where ρ is one of a finite set $R[\sigma', \alpha]$ of positive real numbers. Let B^* be what remains of B after deleting the top row and the left column. Then $|\det B^*| = d/N$, and B^* is an invertible matrix with integer entries. Now $(e_1, \ldots, e_{n-1}) = d^{-1}e'B^*$, so $e \in q^{-1/(n-1)}d^{-n/(n-1)}\rho U[r_1, r_2]W_\alpha B^* + O(q^{-2/(n-1)})$, from which we conclude that the scaled approximation errors for $(q\beta)$ lie nearly on one of a finite list of surfaces of the same sort as before.

Our theorem about ratios of consecutive good approximations also extends to this nominally more general case. There is a finite set $G(\sigma, \beta) \subset \mathbb{Q}(\alpha) \cap (1, \infty)$, and a constant $C = C(\sigma, \beta) > 1$, such that if q and q' are consecutive good numbers in the context of $(*, \sigma, \beta)$, then there exists $\gamma \in G(\sigma, \beta)$ such that $|q'/q - \gamma| \leq Cq^{-n/(n-1)}$.

To see this, we first note that as before, $(Nq, N^{n/(n-1)}\sigma, \beta)$ as well as $(Nq', N^{n/(n-1)}\sigma, \beta)$ are good, as are (dq, σ', α) and (dq', σ', α), where $\sigma' = nNd^{1/(n-1)}b'$

Now choose m_1 and m_2 so that $\nu[m_1] < q < \nu[1+m_1]$ and $\nu[m_2] < q' < \nu[1+m_2]$. Because β is badly approximable, there is a number $K > 1$ such that for all q, q' that are consecutive good (in context of σ) denominators, $q'/q < K$. Now

$$\frac{q'}{q} = \frac{q'}{q(\nu[m_2])} \cdot \frac{q(\nu[m_2])}{q(\nu[m_1])} \cdot \frac{q(\nu[m_1])}{q}.$$

But the the two ratios at either end of this product of ratios are covered by our earlier theorem on ratios of q's, while the central ratio is a ratio that, to our usual tolerances, coincides with $\nu[m_2]/\nu[m_1]$. But this ratio is an algebraic integer corresponding to a bounded lattice displacement, from one lattice point in our cylinder to another. The number of choices for it is therefore finite.

Chapter 8

Marshall Hall's Theorem

For $A, B \subseteq \mathbb{R}$, let $A + B = \{a + b \mid a \in A \text{ and } b \in B\}$, $A - B = \{a - b \mid a \in A \text{ and } b \in B\}$, and $AB = \{ab \mid a \in A \text{ and } b \in B\}$. Let E_N denote the continued fraction Cantor set

$$E_N = \{[a_1, \ldots a_n] \mid 1 \le a_k \le N \text{ for } k \ge 1\}, \text{ and}$$
$$F_N = E_N + \mathbb{Z}.$$

Marshall Hall [Hall] proved in 1947 that $F_4 + F_4 = \mathbb{R}$. It has since been shown that $F_4 + F_2 \ne \mathbb{R}$ and $F_3 + F_3 \ne \mathbb{R}$, but $F_3 - F_3 = \mathbb{R}$, $F_3 + F_4 = \mathbb{R}$ and $F_2 + F_5 = \mathbb{R}$. In this chapter, we give Astels' proof [As] of the positive result that $F_3 + F_4 = \mathbb{R}$ and discuss how, on the other hand, it may be verified by means of interval computations that $F_3 + F_3 \ne \mathbb{R}$. We close with a discussion of connections to the Lagrange and Markoff spectra.

For both the positive and the negative result, one key idea is a variant of a familiar theme: there are trees associated with continued fraction Cantor sets. But this time, we shall be using *binary* trees. The vertices of the tree T_N associated with E_N will be labelled in three different ways. The first labeling associates to each vertex a finite string $\mathbf{a} = (a_1, a_2, \ldots a_n)$ of integers, each in $\{1, 2, \ldots, N\}$, with the last entry less than N. The second associates to each vertex a closed real interval, termed a *bridge* $A[\mathbf{a}]$, and the third associates to the same vertex, an open interval $O[\mathbf{a}] \subset A[\mathbf{a}]$. These will be called the formal tree, the bridge tree, and the gap tree, respectively.

8.1 The Binary Trees of E_N

The root of the formal tree $T_{N,f}$ is the string (1). The children of a vertex $\mathbf{a} = (a_1, \ldots, a_n)$ are two strings. The first is $(a_1, \ldots a_n, 1)$. The second depends on a_n. If $a_n < N - 1$, then the second string is $(a_1, \ldots, 1 + a_n)$. If $a_n = N - 1$, then the second string is instead $(a_1, \ldots, a_{n-1}, N, 1)$. Thus the vertices of the formal tree are finite strings of positive integers, each in the closed interval $[1, N]$, with the last entry less than N.

The root of the interval tree $T_{N,i}$ is the closed interval $[\alpha_N, \omega_N]$, where α_N and ω_N are the least and greatest element of E_N. The interval $A[\mathbf{a}]$ associated with the string $(a_1, a_2, \ldots, a_n)$ is determined by its endpoints: in one order or the other depending on the parity of n, these are $[a_1, \ldots, a_{n-1}, a_n + \alpha]$ and $[a_1, \ldots, a_{n-1} + \alpha]$; if $n = 1$ then this latter endpoint is simply α.

Now let $\beta = \beta_N = 1 - \omega_n$. The root of the gap tree $T_{N,g}$ is the open interval about $1/2$ extending from $1/(2 + \alpha)$ to $1/(2 - \beta)$. The endpoints (not included) of the open gap interval $O[\mathbf{a}]$ corresponding to a string $(a_1, \ldots, a_n)$ are $[a_1, \ldots, a_{n-1}, 1 + a_n - \alpha]$ and $[a_1, \ldots, 1 + a_n - \beta]$; the rational number $g[\mathbf{a}]$ with smallest denominator in the gap is $[a_1, \ldots, a_{n-1}, 1 + a_n]$.

Lemma 8.1 *The following hold for all integers $N \geq 2$:*

(1) *Each of the trees $T_{N,*}$ is a binary tree. Each vertex has exactly two children.*
(2) *Every bridge $A[\mathbf{a}]$ in $T_{N,i}$ is the disjoint union of its two children and of the gap $O[\mathbf{a}]$.*
(3) *The endpoints of each $A[\mathbf{a}]$, and of each $O[\mathbf{a}]$, are elements of E_N, while the gaps themselves are disjoint from E_N.*
(4) *The longest interval of $A[\mathbf{a}] \backslash E_N$ is the gap $G[\mathbf{a}]$.*
(5) *The union of the gaps of $T_{N,g}$ is $[\alpha, \omega] \backslash E_N$.*
(6) *If $A = A[(a_1, a_2, \ldots, a_{n-1}, b)] = A[(\mathbf{a}b)]$, the length of $A[(\mathbf{a}b)]$ is $|A[(\mathbf{a}b)]| = |\mathbf{a}|^{-2} \left(\frac{1}{b + \{\mathbf{a}\} + \alpha} - \frac{1}{N + \{\mathbf{a}\} + \alpha} \right)$.*
(7) *The length of $O[(\mathbf{a}b)]$ is $|\mathbf{a}|^{-2} \left(\frac{1}{b+1-\beta+\{\mathbf{a}\}} - \frac{1}{b+1+\alpha+\{\mathbf{a}\}} \right)$.*
(8) *Equivalently, $|O[\mathbf{a}]| = (\alpha + \beta)|\mathbf{a}|^{-2}(1 + \alpha\{\mathbf{a}\})^{-1}(1 - \beta\{\mathbf{a}\})^{-1}$. (This second formulation helps in seeing that $O[\mathbf{a}]$ is a larger gap than any of its descendants.)*
(9) *Given a bridge $A = A[(a_1, a_2, \ldots, a_{n-1}, b)]$,*
let A_0 be the bridge $A[(a_1, a_2, \ldots, a_{n-1}, b, 1)]$
and let A_1 be the other descendant. Then

(a)

$$|A_0| = |\mathbf{a}|^{-2} \left(\frac{1}{b + \alpha + \{\mathbf{a}\}} - \frac{1}{b + \omega + \{\mathbf{a}\}} \right)$$

(b)
$$|A_1| = |\mathbf{a}|^{-2}\left(\frac{1}{b+1+\alpha+\{\mathbf{a}\}} - \frac{1}{N+\omega+\{\mathbf{a}\}}\right)$$

(c)
$$\frac{|A_0|}{|O[ab]|} = \frac{\omega-\alpha}{\alpha+\beta}\frac{b+1+\{\mathbf{a}\}+\alpha}{b+\{\mathbf{a}\}+\alpha}$$

(d)
$$\frac{|A_1|}{|O[ab]|} = \frac{N-b-\alpha-\beta}{\alpha+\beta}\frac{b+\{\mathbf{a}\}+\omega}{N+\{\mathbf{a}\}+\omega}.$$

The statements about the lengths of the intervals and gaps are a consequence of a more general fact. If $x = [a_1, \ldots, a_n, s]$ and $y = [a_1, \ldots, a_n, t]$, then

$$|x-y| = |\mathbf{a}|^{-2}\frac{|s-t|}{(s+\{\mathbf{a}\})(t+\{\mathbf{a}\})}.$$

The proof is inductive. For the null list, the default values of $|\mathbf{a}|$ and $\{\mathbf{a}\}$ are 1 and 0 respectively, so the identity collapses to $|1/s - 1/t| = |s-t|/(st)$. For a list $\mathbf{a} = (a_1)$, $\{\mathbf{a}\} = 1/a_1$ and $|\mathbf{a}| = a_1$, so the identity simplifies to the algebraic fact that

$$\left|\frac{1}{(a_1+1/s)} - \frac{1}{(a_1+1/t)}\right| = \frac{|s-t|}{a_1^2(1+s/a_1)(a_1+t/a_1)}.$$

Assuming the identity to hold for all lists $\mathbf{a}$ of length n, we consider the case of a list $(a_1, a_2, \ldots, a_n, b)$ of length $n+1$. By the inductive assumption,

$$|[a_1, \ldots, a_n, b+1/s] - [a_1, \ldots, a_n, b+1/t]| = \frac{|\mathbf{a}|^{-2}|(b+1/s) - (b+1/t)|}{(b+\{\mathbf{a}\}+1/s)(b+\{\mathbf{a}\}+1/t)}.$$

In view of the fact that $|ab| = (b + \{\mathbf{a}\})|\mathbf{a}|$, and that $\{ab\} = |\mathbf{a}|/|ab|$, this simplifies to the required expression $|ab|^{-2}|s-t|/((s+\{ab\})(t+\{ab\}))$.

There are various characterizations of E_N. One is that

$$E_N = \bigcup_1^N \frac{1}{k+E_N}.$$

Another is that if one sets $E_N^0 = [\alpha, \omega]$, and $E_N^k = \bigcup_1^N 1/(k + E_N)$, then

$$E_N = \bigcap_1^\infty E_N^k.$$

The length of any gap associated with a list $\mathbf{a} = (a_1, a_2, \ldots, a_n)$ is at most comparable to $|(1, 1, 1, \ldots, 1)|^{-2}$ and this tends to zero exponentially. Thus, there is a one to one correspondence between E_N and infinite binary strings designating paths to infinity down the tree. The complement $[\alpha, \omega] \backslash E_N$ is, we claim, the union of the gaps $O[\mathbf{a}]$, taken over all lists $\mathbf{a} = (a_1, \ldots, a_n)$ with $n \geq 1$, $1 \leq a_k \leq N$ for $k < n$, and with $a_n < N$.

Consider $x \in [\alpha, \omega] \cap O[N]$. This x has a continued fraction expansion $x = [a_1, a_2, \ldots]$ in which not all the a_j belong to $[1, 2, \ldots, N]$. There must, therefore, be a least k so that $x = [a_1, a_2, \ldots, a_k + y]$ with $y \notin [\alpha, \omega]$. If $a_k = 1$ then $y > \omega$, since otherwise we should have $x = [a_1, a_2, \ldots, a_{k-1} + 1/(1+y)]$ with $1/(1+y) > \omega$ contrary to the definition of k. For $2 \leq a_k \leq N$, we could have either $0 \leq y < \alpha$ or $y > \omega$. If $y < \alpha$ then $x \in O[(a_1, a_2, a_{k-1}, a_k - 1)]$, while if $y > \omega$, then $x \in O[(a_1, a_2, \ldots a_k)]$.

In the other direction, if x belongs to $O[\mathbf{a}]$ for some $\mathbf{a}$, then the continued fraction expansion of x has the form $[a_1, a_2, \ldots a_k + y]$ where $y \notin [\alpha, \omega]$. But then, either y is rational so that x is rational and not in E_N, or y has a continued fraction expansion which eventually involves some digit greater than N.

We now prove point (4) of the lemma: The gap $O[\mathbf{a}]$ is the longest interval of $O[N, \infty] \cap A[\mathbf{a}]$. It is sufficient to prove that each gap is longer than the gaps associated with the immediate descendants. That is,

$$|O[(a_1, \ldots, a_n)]| > |O[(a_1, \ldots, a_n), 1]|,$$
$$|O[(a_1, \ldots, a_n)]| > |O[(a_1, \ldots, 1 + a_n)]| \text{ if } a_n < N - 1,$$
$$|O[(a_1, \ldots, a_n)]| > |O[(a_1, \ldots, 1 + a_n, 1)]| \text{ if } a_n = N - 1.$$

To prove this, we first note that $1/(N+1) < \alpha_N < 1/N$, $1/(N+2) < \beta_N < 1/(N+1)$, and $\alpha - \beta < 1/N^2$. For $\mathbf{a} = (a_1, a_2, \ldots, a_n)$, let $\mathbf{a'} = (a_1, a_2, \ldots, 1 + a_n)$ and $\mathbf{a''} = (a_1, \ldots, 2 + a_n)$. Let $G[\mathbf{a}] = 1/((\alpha + \beta)|O[\mathbf{a}]|)$ and note that

$$G[\mathbf{a}] = |\mathbf{a'}|^2 \cdot (1 + \alpha\{\mathbf{a'}\})(1 - \beta\{\mathbf{a'}\}).$$

Now

$$G[\mathbf{a}1] = |\mathbf{a}2|^2(1 + \alpha/(2 + \{\mathbf{a}\}))(1 - \beta/(2 + \{\mathbf{a}\}))$$

and
$$G[\mathbf{a}'] = |\mathbf{a}''|^2(1+\alpha\{\mathbf{a}''\})(1-\beta\{\mathbf{a}''\}).$$

Let $\epsilon = \{\mathbf{a}'\} \leq 1/2$, so that $\{\mathbf{a}''\} = \epsilon/(1+\epsilon)$ and $\{\mathbf{a}2\} = (1-\epsilon)/(2-\epsilon)$. Since $\alpha < 1/N$ and $\beta < 1/(N+1)$,

$$\frac{(1+\alpha\epsilon)(1-\beta\epsilon)}{\left(1+\frac{\alpha\epsilon}{1+\epsilon}\right)\left(1-\frac{\beta\epsilon}{1+\epsilon}\right)} > \frac{1}{2}.$$

Also,
$$\frac{|\mathbf{a}2|^2}{|\mathbf{a}'|^2} = (2-\epsilon)^2 > 6$$

so that
$$|G[\mathbf{a}1]|/|G[\mathbf{a}']| > 1$$

and with room to spare.

Matters are a little more delicate in the case of the comparison of $G[\mathbf{a}'']$ and $G[\mathbf{a}']$. We have

$$\frac{G[\mathbf{a}'']}{G[\mathbf{a}']} = \frac{(1+\epsilon)^2\left(1+\frac{\alpha\epsilon}{1+\epsilon}\right)\left(1-\frac{\beta\epsilon}{1+\epsilon}\right)}{(1+\alpha\epsilon)(1-\beta\epsilon)} = \frac{(1+\epsilon+\alpha\epsilon)(1+\epsilon-\beta\epsilon)}{(1+\alpha\epsilon)(1-\beta\epsilon)} > 1.$$

When $a_n = N-1$, the calculation about $G[\mathbf{a}2]/G[\mathbf{a}]$ is unaffected, but the other case is affected. The claim to be established is that $G[(a_1, a_2, \ldots, a_n+1, 1)] > G[(a_1, a_2, \ldots, a_n)]$. We just proved, though, that $G[\mathbf{a}'] > G[\mathbf{a}]$ and that $G[\mathbf{a}1] > G[\mathbf{a}']$, and concatenating these inequalities gives us the desired conclusion. This calculation establishes, in the terminology of Astels' work, that the system of bridges and gaps used here is what he terms an *ordered derivation*.

The next important point is that the children intervals A_0 and A_1 of $A[\mathbf{a}]$ have lengths comparable to $O[\mathbf{a}]$, and that the minimum value of $|A_j[\mathbf{a}]|/|O[\mathbf{a}]|$, taken over all intervals A, and over both children, is readily calculated.

Lemma 8.2 *For all finite lists* $\mathbf{a} = (a_1, a_2, \ldots, a_n)$ *of positive integers satisfying* $1 \leq a_k \leq N$ *for* $1 \leq k < n$, *and* $1 \leq a_n < N$,

$$\min[\{|A_0[\mathbf{a}]|/|O[\mathbf{a}]|, |A_1[\mathbf{a}]|/|O[\mathbf{a}]|\}] \geq \frac{\omega_N - \alpha_N}{\alpha_N + \beta_N} \cdot \frac{N - \beta_N}{N + \omega_N}.$$

This lower bound is achieved by the choice $\mathbf{a} = (N-1)$, using the sub-bridge $A_1 = [1/(N+1/(1+\alpha_N)), 1/(N+\alpha_N)]$.

Proof. From (9) of Lemma 8.1, if $(\mathbf{a}b) = (a_1, a_2, \ldots, a_n, b)$, then $O[\mathbf{a}b]$, $A_0[\mathbf{a}b]$ and $A_1[\mathbf{a}b]$ satisfy

$$\frac{|A_0[\mathbf{a}b]|}{|O[\mathbf{a}b]|} = \frac{\omega - \alpha}{\alpha + \beta} \cdot \frac{b + 1 + \{\mathbf{a}\} + \alpha}{b + \{\mathbf{a}\} + \alpha}$$

$$\frac{|A_1[\mathbf{a}b]|}{|O[\mathbf{a}b]|} = \frac{N - b - \alpha - \beta}{\alpha + \beta} \cdot \frac{b + \{\mathbf{a}\} + \omega}{N + \{\mathbf{a}\} + \omega}.$$

The first of these expressions is greater than $(\omega - \alpha)/(\beta - \alpha)$, while the second is at least

$$\frac{\omega - \alpha}{\alpha + \beta} \cdot \frac{N - 1 + \omega}{N + \omega},$$

with the minimum being achieved when $\mathbf{a}$ is the null list, so that $\{\mathbf{a}\} = 0$, and with $b = N - 1$. □

For $N = 2, 3, 4$, and 5, this lower bound evaluates to $\tau_2 = (\sqrt{3} - 1)/2$, $\tau_3 = (4\sqrt{21} - 6)/15$, $\tau_4 = (15\sqrt{2} - 3)/14$, and $\tau_5 = 4\sqrt{5}/5$.

8.2 Sums of Bridges Covering $[\alpha_N, \omega_N]$

Let I and J be the root bridges of E_3 and E_4, respectively. Then $I_3 + I_4 = [\alpha_3, \omega_3] + [\alpha_4, \omega_4]$ and this is an interval of length greater than 1. The idea is that given $\gamma \in I + J$, we can construct a sequence of bridges A_n in the tree of E_3, and B_n in the tree of E_4, so that (A_n) descends on a path to infinity in its tree, (B_n) does likewise in its tree, and for each n, $\gamma \in A_n + B_n$. This not only will prove that $\gamma \in E_3 + E_4$, it will provide an algorithm which computes an appropriate pair of elements from E_3 and E_4 which sum to γ.

The descent is determined inductively. We set $A_0 = I = A^3[(1)]$, and in the tree for E_4, we set $B_0 = J = A^4[(1)]$. The gap corresponding to A_0 we call $OA_0 = A[(1)]$; it is the open interval

$$OA_0 = O^3[(1)] = \left(\frac{1}{2+\alpha_3}, \frac{1}{2-\beta_3}\right).$$

The gap associated with B_0 is $OB_0 = A^4[(1)]$ and is given similarly. It may readily be verified that A_0 is longer than OB_0 and that B_0 is longer than OA_0.

Now assume that A_n and B_n are determined, with $|A_n| > |OB_n|$ and $|B_n| > |OA_n|$, and $\gamma \in A_n + B_n$. From our results about the ratio of bridge length to gap length, for each of the two descendants A_n^i of A_n, and each of the descendants B_n^i of B_n,

$$\frac{|A_n^i|}{|OA_n|} \geq \tau_3, \quad \frac{|B_n^i|}{|OB_n|} \geq \tau_4 > 1.$$

Thus $|A_n^i||B_n^j| > |OA_n||OB_n|$, and it follows that in particular if we choose i and j so that A_n^i and B_n^j are each the shorter sibling, then $|A_n^i||B_n^j| > |OA_n||OB_n|$. There are now two cases. If $|A_n^i| > |OB_n|$, then $A_n + B_n = (A_n^0 \cup A_n^1) + B_n$, because B_n is longer than the gap between the two pieces of A_n. Since the gap OA_n is longer than the gaps OA_n^0 and OA_n^1 of the children of A, and since B_n was longer than OA_n, B_n is longer than the gap of either A-child. By the case hypothesis, either child of A_n is longer than OB_n. We choose $B_{n+1} = B_n$.

We choose A_{n+1} to be whichever of the two intervals A_n^0, A_n^1 happens to have γ contained in $A_n^i \cup B_n$; if γ belongs to both, we choose the one in which the list $(a_1, a_2, \ldots, a_n)$ is replaced by $(a_1, a_2, \ldots, a_n, 1)$. Our inductive premises are preserved:

$$|A_{n+1}| > |OB_{n+1}| \text{ and } |B_{n+1}| > |OA_{n+1}|.$$

In the other case, we do the same thing but split B_n instead. There cannot be arbitrarily long runs of splitting A and leaving B unchanged, since the interval A_n would eventually shrink to a length less than $|OB|$ which is forbidden, nor can there be arbitrarily long runs of splitting B only. Thus, A_n and B_n each descend to infinity in their respective trees. Since every element of $[\alpha_3 + \alpha_4, \omega_3 + \omega_4]$ belongs to $E_3 + E_4$, this set contains an interval of length greater than one, so that $F_3 + F_4 = \mathbb{R}$ as claimed.

The same proof works to show that $E_2 + E_5$ contains an interval of length greater than 1 so that $F_2 + F_5 = \mathbb{R}$.

Remark 8.1 *Astels investigates the more general situation in which a finite or infinite set B of positive integers is designated. If B is finite, then F_B is defined to be the set of all real numbers of the form $a + [b_1, b_2, \ldots]$ where $a \in \mathbb{Z}$ and where the infinitely many b_i must all come from B. If B is infinite, the definition of F_B is modified to include all numbers of the form $a + [b_1, b_2, \ldots, b_r]$ where $r \geq 0$ and all b_i, if any, belong to B.*

In the other direction, B. Diviš showed in [Di] that $F_3 + F_3 \neq \mathbb{R}$. We now present a reason that $F_3 + F_3 \neq \mathbb{R}$. It is forthrightly computational.

We compute the intervals corresponding to all the $A[\mathbf{a}]$ to depth 4 in the tree for E_3. From this we may compute the set of sums of these intervals, and we do the same for $(E_3 + 1) + E_3$. Sorting these by their left endpoint, we use a computer to check that there is a gap. This does *not* suffice to establish that $E_3 + E_3$ contains no interval.

The integer lists corresponding to this depth 4 penetration of the tree for E_3 are $(1,1,1,1,1)$, $(1,1,1,2)$, $(1,1,2,1)$, $(1,1,3,1)$, $(1,2,1,1)$, $(1,2,2)$, $(1,3,1,1)$, $(1,3,2)$, $(2,1,1,1)$, $(2,1,2)$, $(2,2,1)$, $(2,3,1)$, $(3,1,1,1)$, $(3,1,2)$, $(3,2,1)$, and $(3,3,1)$.

One gets two gaps out of this: $(1.6041682996, 1.6048884362933)$, and $(1.6127896650, 1.62202018532)$. Pressing ahead to depth eleven gives nineteen gaps. These gaps are actual gaps in $F_3 + F_3$. Using exact arithmetic, the union E_3^{11} of the bridges of depth 11 in the tree for E_3 was calculated. The intersection of $F_3 + F_3$ with the closed interval $[1, 2]$ is a subset of $F_3^{11} = (E_3^{11} + E_3^{11} + \{0, 1\}) \cap [1, 2]$. This set is the union of a finite number of closed intervals of the form $[a, b]$ where a and b are in turn numbers of the form $r_1 + r_2\sqrt{21}$, r_j rational. Consolidating overlapping intervals leads to a representation of F_3^{11} as a disjoint union of twenty finite intervals. Of these, the thirteenth and fourteenth are

$$\left[\frac{99404761}{61918457} + \frac{39241\sqrt{21}}{1300287597}, \frac{12235699}{7615457} - \frac{37819\sqrt{21}}{159924597} \right]$$

and

$$\left[\frac{980382}{610601} + \frac{\sqrt{21}}{610601}, \frac{500}{311} - \frac{\sqrt{21}}{2177} \right].$$

On the basis of these calculations, it would appear that $F_3 + F_3$ contains a long central interval, but that there are infinitely many intervals in both $(F_3 + F_3) \cap [1, 2]$ and its complement in $[1, 2]$, and that the boundary region is a maze of a few large gaps and many smaller ones. This is indeed the case, as has been shown by Freiman [Frei]. In fact, $F(3) + F(3)$ contains the interval $[a, b]$, where

$$a = [0; \overline{3, 1}] + [0; 2, \overline{1, 3}] \approx 0.62202, \; b = [0; \overline{1, 3}] + [0; 1, 2, \overline{1, 3}] \approx 1.52753.$$

8.3 The Lagrange and Markoff Spectra

There is a connection between Hall's theorem and the subsets of $\mathbb{R}^+$ known as the *Lagrange* and *Markoff* spectra. Both may be defined in terms of continued fractions associated with a doubly infinite sequence $A: \mathbb{Z} \to \mathbb{Z}^+$. If A is such a sequence and if $i \in \mathbb{Z}$, let

$$\lambda_i(A) = [a_i; a_{i+1}, a_{i+2} \ldots] + [0; a_{i-1}, a_{i-2}, \ldots].$$

Let $L(A) = \varlimsup_{i \in \mathbb{Z}} \lambda_i(A)$ and $M(A) = \sup_{i \in \mathbb{Z}} \lambda_i(A)$. Let $\mathbf{L} = \{L(A) \mid A : \mathbb{Z} \to \mathbb{Z}^+\}$ and $\mathbf{M} = \{M(A) \mid A : \mathbb{Z} \to \mathbb{Z}^+\}$. It is known that $[6, \infty) \subset \mathbf{L} \subset \mathbf{M} \subset [\sqrt{5}, \infty)$, and that $\mathbf{L}$ and $\mathbf{M}$ are closed. The first inclusion is a straightforward corollary of Hall's theorem. Recall that F_4 is the set of all real numbers of the form $[a_0; a_1, a_2, \ldots]$ so that for $k \geq 1$, $1 \leq a_k \leq 4$, while $E_4 = F_4 \cap (0, 1)$. Now since $F_4 + F_4 = F_4 + E_4 = \mathbb{R}$, if $x > 6$ then there must be an integer $a \geq 5$, and numbers $e_1, e_2 \in E_4$, such that $x = (a + e_1) + e_2 = [a; b_1, \ldots] + [0; c_1, \ldots]$ say. Let $A : \mathbb{Z} \to \mathbb{Z}^+$ be the sequence with right half $(a_0, a_1, \ldots)$ given by $(a, c_1, b_1, a, c_1, c_2, b_2, b_1, a, c_1, c_2, c_3, b_3, b_2, b_1, \ldots)$, and with $a_i = 1$ for $i < 0$. Since $a \geq 5$, the only i for which $\lambda_i(A)$ is relevant to the determination of $L(A)$ are those for which $a_i = 5$, and thus, $L(A) = a + e_1 + e_2$.

Another of the inclusions, that $\mathbf{M} \subset [\sqrt{5}, \infty)$, while unconnected to Hall's theorem, nonetheless has a simple proof using continued fractions, which we give here.

If $a_i \geq 3$ for any i, then clearly $M(A) \geq 3$. Thus we may restrict attention to the case that all $A_i \leq 2$. If any A_i is 2, though, then $\lambda_i(A) = 2 + e_1 + e_2$ where $e_1, e_2 \in E_2$. But then $\min[E_2] = \overline{[2,1]} = (\sqrt{3} - 1)/2$ so $\lambda_i(A) \geq \sqrt{3} + 1 > \sqrt{5}$. Taking all $a_i = 1$ yields $M(A) = \sqrt{5}$.

There is a deep connection between $\mathbf{M}$, the theory of *quadratic forms*, and solutions in positive integers (m, m_1, m_2) to the equation

$$m^2 + m_1^2 + m_2^2 = 3mm_1m_2.$$

It is known that [CuFla]

$$\mathbf{M} \cap [0, 3] = \mathbf{L} \cap [0, 3] =$$
$$\{\sqrt{9m^2 - 4}/m \mid m \in \mathbb{Z}^+, \exists m_1, m_2 \in \mathbb{Z}^+, m_1 \leq m, m_2 \leq m, \text{ and }$$
$$m^2 + m_1^2 + m_2^2 = 3mm_1m_2\}.$$

The set of ordered triples $(m_1 \leq m_2 \leq m) \in (\mathbb{Z}^+)^3$ satisfying the Markoff condition has a *tree* structure. The root is $(1,1,1)$, the trunk is $(1,1,2)$, the crotch is $(1,2,5)$, and from each such triple beyond $(1,1,2)$ there spring two further triples, $(m_2, m, 3mm_2 - m_1)$ and $(m_1, m, 3mm_1 - m_2)$. Going down the tree, the vertex below (m_1, m_2, m) is $(3m_1m_2 - m, m_1, m_2)$. It is conjectured that for each m associated with a Markoff triple, there is but one such triple in which m is the largest entry. At any rate, if there is a counterexample, it involves an m greater than 10^{105} [Bo].

Chapter 9

Functional-Analytic Techniques

Pick a random number $X = X_0$ in the unit interval. Generate its continued fraction expansion $X = [v_1, v_2, v_3, \ldots]$ by calculating $v_n := \lfloor 1/X_{n-1} \rfloor$ and $X_n := T^n X$ where $TX = 1/X - \lfloor 1/X \rfloor$. What can be said about the asymptotic distribution of X_n? (If we know this, then the corresponding probability that $v_{n+1} = k$ is simply $\text{prob}[1/(k+1) < X_n \leq 1/k]$.)

In a letter to Laplace, Gauss [Gauss] stated that the event $T^n X < a$ has asymptotic probability $\log_2(1+a)$ for $0 \leq a \leq 1$. It may be that Gauss was aware of facts which we would now interpret as saying that the probability density function $g(t) := 1/((1+t)\log 2)$ is invariant under T: If a random variable X has density g, then so does TX.

The reason for this invariance is that for $0 \leq a < a+h \leq 1$, TX lies between a and $a+h$ if and only if there exists $k \geq 1$, so that X lies between $1/(k+a+h)$ and $1/(k+a)$. Thus

$$\text{prob}[a \leq TX \leq a+h] = \sum_{k=1}^{\infty} \text{prob}\left[\frac{1}{k+a+h} \leq X \leq \frac{1}{k+a}\right].$$

Taking limits as $h \to 0$ gives, for an arbitrary probability density function f for X, the corresponding density Gf for TX:

$$G[f](x) = \sum_{k=1}^{\infty} (k+x)^{-2} f\left(\frac{1}{k+x}\right) \qquad (*)$$

Clearly $Gg = g$.

From another perspective, the operator T is an ergodic operator on the unit interval (see Chapter 4), g is the density of the invariant measure, and G is the so-called *transfer operator* for T.

Returning to the case at hand, if X starts with density g, $T^n X$ also has that density. The difficulty is that, in general, it can happen that a dynamic process with a fixed point, or even with a unique fixed point, fails for some initial points to converge to that fixed point. Here, our 'points' are probability density functions f_n. The initial density f_0 is not g, but the constant function 1. A proof must demonstrate convergence of this initial density to g. That is, it must be shown that $f_n := G^n 1 \to g$. The first published proof is due to Kuz'min in 1928 [K].

Subsequent work has focused on the question of which initial probability distributions converge, and at what rate, to a distribution with density g, and on what happens when we restrict our continued fractions to those in which all v_k come from a restricted subset of the positive integers.

The modern reader, seeing equation (*), will want to bring in the theory of linear operators. This requires some care in specifying the linear space and the norm. Linear operators in general are not necessarily mere 'big matrices' and we might expect to run into the full range of complexities these present. We have a little good luck immediately in that G is *positive*, that is, if f is positive on $[0,1]$ then so is $G[f]$. Furthermore, as we might expect given its connection to probability densities,

$$\int_0^1 f(x)\,dx = \int_0^1 (Gf)(x)\,dx.$$

Among the best early treatments of the Gauss-Kuz'min theorem is that of Eduard Wirsing. [W] Instead of dealing directly with G, he uses an operator T, given by

$$T[f](x) := \sum_{k=1}^\infty \frac{1+x}{(k+x)(k+x+1)} f(1/(k+x)),$$

which we will consider to be acting on the space of functions which are continuously differentiable on $[0,1]$, equipped with the sup norm. If further $R[f](x) := (1+x)f(x)$, then

$$RG = TR \text{ and } T[1] = 1.$$

The integral invariance property of G translates to the following property for T:

$$\int_0^1 \frac{f(x)\,dx}{1+x} = \int_0^1 \frac{(Tf)(x)\,dx}{1+x}.$$

Thus convergence of $G^r 1$ to g is equivalent to convergence of $T^r[1+x]$ to 1, or to the convergence of $(T^r[1+x])'$ to zero. The leading eigenvalue of G and T is of course 1. The next eigenvalue, call it λ_2, (now known as the Wirsing constant) will determine the rate of convergence in the Gauss-Kuz'min theorem. One way to study interior eigenvalues and eigenfunctions of an operator is to invent a related operator with the property that the interior eigenvalue of the original operator is the leading eigenvalue of the new operator. Wirsing does this.

He defines an operator U acting on the set of all continuous functions h on $[0,1]$ so that $(T[f](x))' = -U[f'](x)$, so that

$$\frac{d}{dx}T^r[f](x) = (-1)^r U^r[f'](x).$$

This U is given explicitly by

$$Uh = \sum_{k=1}^{\infty}\left\{\frac{1+x}{(k+x)^3(k+1+x)}h\left(\frac{1}{k+x}\right) + \frac{k}{(k+1+x)^2}\int_{1/(k+1+x)}^{1/(k+x)}h\right\},$$

from which it follows that U is also a positive operator. Our expectation that $G^r[1]$ tends to g translates into the expectation that $U^r[1]$ tends to zero and this is the path Wirsing takes.

Wirsing's approach depends on the fact, special to the circumstances at hand, that if one has a (tractable) $\phi^*(x)$ and if $-\lambda_2^- \phi^* < U\phi^* < -\lambda_2^+ \phi^*$ for $0 \leq x \leq 1$ then $-\lambda_2^- \leq -\lambda_2 \leq -\lambda_2^+$.

Wirsing defines

$$\phi_a(x) := \frac{1-a}{(1+ax)^2} + \frac{1+a}{(1+(1+a)x)^2} \quad \text{so that } U\phi_a = (1+a+x)^{-2}.$$

He then takes $a_1 = 0.6247$ and $a_2 = 0.7$ and sets $\phi^* := 8\phi_{a_1} - 7\phi_{a_2}$, and calculates that $0.3020\phi^*(x) < U\phi^*(x) < 0.3043\phi^*(x)$ for $0 \leq x \leq 1$. This has the advantage that $U\phi_a$ can be calculated in closed form. But no systematic method for constructing further such functions is offered, and so we have recourse to the series methods developed later in this Chapter. For the moment, the reader may think of our ϕ as a rabbit drawn from a hat.

Let Φ be the leading eigenfunction of U, normalized so that $\Phi(0) = 1$. Equivalently, $\Phi(x) = ((1+x)g_2(x))'$, where g_2 is the second eigenfunction of G. Then $U\Phi(x) = -\lambda_2\Phi(x)$. The first eigenfunction of T is the constant function 1 by construction; the second eigenfunction of T, which we call $\Psi(x)$, is simply $(1+x)g_2(x)$.

We have a further purpose with these kind of calculations. We need some explicit bounds for the rate of convergence of G^r acting on functions

$f_{a,b} : [0,1] \to \mathbb{R}$ given by

$$f_{a,b}(x) := \begin{cases} \frac{(a-b)}{\log(1+a)-\log(1+b)}(1+ax)^{-1}(1+bx)^{-1}, & \text{if } a \neq b; \\ f_{a,a}(x) = \frac{1+a}{(1+ax)^2} & \text{if } a = b. \end{cases}$$

When $a, b > 0$, these functions are probability density functions associated with the conditional density of $T^{n+k}X$ given a fixed value for $(a_1, \ldots a_n)[X]$, and taking the initial density to be the Gaussian density g, and the strongest 'mixing' results for T require sharp estimates for the rate at which this conditional density settles back to the Gauss density g as we consider events depending only on $A_m[X], m \geq n+k$.

Theorem 9.1 *For $0 \leq s \leq 1$, $0 \leq x \leq 1$, and $r \geq 1$,*

$$\left| G^r \left[f_{a,b}(x) - \frac{1}{\log 2(1+x)} \right] \right| \leq \frac{|\lambda_2|^r}{(1+x)\log 2}.$$

Proof. We shall need some auxiliary results. Wirsing shows that if ϕ_0 is a positive, decreasing function on $[0,1]$, if s_0, t_0 are positive numbers so that $s_0 \phi_0 \leq U\phi_0 \leq t_0 \phi_0$, if

$$q_0 \leq \frac{1}{t_0 \phi_0(0)} \int_0^{1/3} \frac{y(1-y)\phi_0(y)\,dy}{(1+y)^2} + \frac{1}{9} \int_{1/3}^{1/2} \phi_0(y)\,dy,$$

if $\langle d_n \rangle$ is the sequence determined by $d_0 = t_0 - s_0$, $d_n := d_0(1 - q_0 + d_0/t_0)^n$, and if $\gamma = \prod_1^\infty (1 + d_n/s_0)$, then there is a multiple ϕ_0 of the leading eigenfunction $\Phi(x)$ for U such that for $0 \leq x \leq 1$,

$$\phi_0(x) \leq \Phi(x) \leq \gamma \phi_0(x).$$

We take for $\phi_0(x)$ a computer-generated polynomial of degree 256 which arises out of calculations using PARI of a matrix representation for U. This gives us the following values, which bracket the Wirsing constant:

$$s_0 = 0.30366300289873265859744812190155574000$$

$$t_0 = 0.30366300289873265859744812190155674000$$

and $\phi_0(x) := \sum_{n=0}^{256} a_n (x-1)^n$ where the coefficient list is (here we give a sharply truncated version of a long list of very high precision numbers:)

$$(a_n) = \{2.45921, -2.1303, 1.35604, -0.75315, 0.3856, -0.1866, 0.08655,$$
$$-0.03881, 0.016924, -0.007206, 0.00300435, -0.001229\}$$

This function is positive and strictly decreasing, with

$$\phi(0) \doteq 7.42540262862209183698500783286418114553 6366119$$

and

$$\phi(1) \doteq 2.45920694984189824692360970015518372582480723466.$$

The matrix calculations involve first finding a similar polynomial approximation in powers of $(x-1)$ to the second eigenfunction of G, which we call $g2$; the first few coefficients to $g2$ are

$$(1., 0.7296, -0.8974, 0.674695, -0.43149, 0.254304, -0.14270, 0.0775).$$

The first few coefficients to $\Psi(x)$ as a series in powers of $(x-1)$, while we're at it, are

$$(2., 2.45921, -1.06515, 0.452014, -0.18829, 0.077115, -0.031098, 0.012364).$$

For $F[\phi]$ we do a numerical integration and get 0.250409 which we round down to .25 to be safe, while $\|U\phi_0\| \le t_0\|\phi_0\| \doteq 2.25482$ which means that we can take $q_0 := 0.11105$. We start also with $d_0 = 10^{-33}$, and

$$\gamma = 1.00000000000000000000000000000000026361303970466689 2807$$

so that $\Phi(x)$, the exact (positive, leading) eigenfunction for U corresponding to $-\lambda_2$, satisfies $\phi_0(x) \le \Phi(x) \le \gamma\phi_0(x)$ for $0 \le x \le 1$.

Now consider iteration of G on an arbitrary continuously differentiable function f. Naturally, the second eigenfunction g_2 for G is involved. Its counterpart for T is $(1+x)g_2(x) = \Psi(x)$, following Wirsing's notation. $\Psi'(x) = \Phi(x)$ and $\int_0^1 \Psi(x)\,dx/(1+x) = 0$.

Next, recall that $T[f] = (1+x)G[f(x)/(1+x)]$ and that $U[f] = -\frac{d}{dx}[T[\int[f]]]$. Thus

$$\frac{d}{dx}T^r[f][x] = (-1)^r U^r[f'][x],$$

and in view of the integral invariance property for T mentioned earlier, $T^r[f]$ tends to the constant $Q[f] := \int_0^1 \frac{f(x)\,dx}{(1+x)\log 2}$. If the original 'error' $f - Q[f]$ has derivative f', then the next error has derivative $-U[f']$, and so on. Thus if $|f'| \le C_f\Phi(x)$ for $0 \le x \le 1$, then $|T^r[f]'| \le |\lambda_2|^r C_f\Phi(x)$ for $0 \le x \le 1$. For $f = f_{a,b}(x)$ the maximal C_f occurs with $a = b = 0$ and is $1/\Phi(1) \doteq .40663515531472035361552127$, while $\int_0^1 (1+x)^{-1}f_{a,b}(x)\,dx = 1$.

The second of these assertions is easily verified by formal integration. The first requires some justification. We begin with a simple estimate for $\phi_0(x)$:

$$\phi_0(x)/\phi_0(1) \le (1 + (5/6)(1-x)).$$

Since $\phi_0(x)$ is a polynomial of degree 256, we may write

$$\phi_0(x)/\phi_0(1) - (1 + (5/6)(1-x)) = \sum_{k=1}^{256} b_k(1-x)^k.$$

We then set $c_k := \sum_{j=2}^{k} b_j$. If $c_k > 0$ for $2 \le k \le 256$ then the polynomial $\phi(x)/\phi(1) - (1 + (5/6)(1-x))$ takes positive values for all $0 \le x < 1$ and the claim is established; the short version of $\phi_0(x)$ given above makes this assertion about $\phi_0(x)$ plausible, and *PARI* code is provided in the electronic appendix which will enable the reader with access to a computer to reconstruct ϕ_0 and check this calculation.

Without loss of generality we may assume that $a \ge b$ in $f_{a,b}(x)$. If $a > b$ then $(a-b)/(\log(1+a) - \log(1+b)) < 1+a$ because $(a-b) - (1+a)(\log(1+a) - \log(1-b)) < 0$, because this last expression tends to zero as $b \to a$ and because its derivative with respect to b is $-1 + (1+a)/(a-b) > 0$. Thus to establish our claim that $f'_{a,b}(x) \le \phi_0(x)/\phi_0(1)$ it will suffice to show that for $0 \le b \le a \le 1$ and $0 \le x \le 1$,

$$(1 + (5/6)(1-x))(1+ax)^2(1+bx)^2 \ge (1+a)(1-a-b-2abx-abx^2).$$

Expanding both sides and subtracting, this is equivalent to the claim that for all a, b and x in the range in question,

$$\frac{5}{6} + a^2 + b + ab$$
$$+ \left(-\left(\frac{5}{6}\right) + \frac{11\,a}{3} + \frac{11\,b}{3} + 2\,a\,b + 2\,a^2\,b \right) x$$
$$+ \left(\frac{-5\,a}{3} + \frac{11\,a^2}{6} - \frac{5\,b}{3} + \frac{25\,a\,b}{3} + a^2\,b + \frac{11\,b^2}{6} \right) x^2$$
$$+ \left(\frac{-5\,a^2}{6} - \frac{10\,a\,b}{3} + \frac{11\,a^2\,b}{3} - \frac{5\,b^2}{6} + \frac{11\,a\,b^2}{3} \right) x^3$$
$$+ \left(\frac{-5\,a^2\,b}{3} - \frac{5\,a\,b^2}{3} + \frac{11\,a^2\,b^2}{6} \right) x^4$$
$$- \frac{5\,a^2\,b^2\,x^5}{6} \ge 0.$$

Let $c_{a,b,j}$ be the coefficient of x^j above. It is sufficient for the claimed inequality that $\sum_{j=0}^{k} c_{a,b,j} \geq 0$ for all relevant a and b. On forming the required running totals, it turns out that every term, at every k, is positive.

Given that $|h'(x)| \leq c\phi(x)$ and that

$$\int_0^1 \frac{h(x)\,dx}{1+x} = 0,$$

it follows that $\max_{0 \leq x \leq 1} |h(x)| \leq c|\psi(0)|$, where $\psi(x)$ satisfies $\psi'(x) = \phi(x)$ and $\int_0^1 \psi(x)\,dx/(1+x) = 0$. Here, we have $\psi(0) \doteq -2.2931242543678787835$ so $|(T^r f_s)(x) - 1/\log 2| \leq 0.932464937331|\lambda_2|^r$. But then

$$\left| G^r f_{a,b}(x) - \frac{1}{\log 2(1+x)} \right| \leq 0.932464937331|\lambda_2|^r/(1+x) \leq (2/3)|\lambda_2|^r g(x),$$

which improves slightly on the claimed result. □

9.1 Continued Fraction Cantor Sets

Until now, we have restricted ourselves to a discussion of the classical continued fraction expansion of a real number drawn 'at random' from the real interval $[0,1]$. But the same methods are as well suited to the study of the statistics of continued fraction expansions of numbers with restricted partial quotients.

The *continued fraction Cantor sets* E_M corresponding to a subset M of the positive integers with at least two, and perhaps infinitely many, elements. The classical middle-third Cantor set (the set of all numbers in $[0,1]$ so that all digits in the base 3 expansion are zero or two) has analogues for continued fractions. The continued fraction digits of an irrational $x \in (0,1)$ are a sequence of positive integers $(a_1, a_2, \ldots)$. The set $E_M \subseteq [0,1]$ is defined by

$$E_M := \{x \in (0,1) : \forall k \in \mathbb{Z}^+, a_k \in M\}.$$

The question is, how big is this set? Trivially E_M is uncountable, while unless $M = \mathbb{Z}^+$, E_M has Lebesgue measure zero. There are, however, a couple of nice intermediate yardstick by which to measure E_M, about which more will be said shortly.

One can ask another question in connection with M and continued fractions: How many pairs (p,q) of positive integers are there so that $p < q \leq x$ and so that all of the (finitely many) a_k in the expansion $p/q =$

$1/(a_1 + 1/(a_2 + 1/(\ldots a_{r-1} + 1/a_r)))$ belong to M? This will depend on x, and asymptotically, this quantity will turn out to grow like x^σ where $0 < \sigma < 2$. Here, and in all that follows, M^r denotes the Cartesian product of r copies of M.

The value of $\sigma = \sigma(M)$ may be defined without reference to linear operators. Let

$$\lambda_M(\sigma) := \sup\{\lambda > 0 : \limsup_{r \to \infty} \lambda^{-r} \sum_{\mathbf{v} \in M^r} |\mathbf{v}|^{-\sigma} < \infty\}.$$

Equivalently,

$$\lambda_M(\sigma) = \exp \lim_{r \to \infty} r^{-1} \log \left(\sum_{\mathbf{v} \in M^r} |\mathbf{v}|^{-\sigma} \right).$$

Now

$$\sigma(M) := \inf\{s : \lambda_M(s) < 1\}.$$

If $\sigma(M) < 1$, then there are too few such fractions to permit the existence of one p for each large q. A conjecture of Zaremba[Za] asserts that if $M = \{1, 2, 3, 4, 5\}$ then for all q there exists p so that all the partial quotients in the continued fraction expansion of p/q belong to M. In view of the result above, it seems that more is true:

Conjecture 9.1 For all finite M such that $\sigma(M) > 1$, there exists $C = C(M)$ so that for $q > C$, there exists p so that all partial quotients of p/q belong to M.

There is evidence in favor of this conjecture, and partial results in that direction, but there is also computational evidence that runs counter to certain heuristic justifications for the conjecture.

The bulk of these questions can be addressed in the linear operator setting, which we now discuss. The most general case we will treat involves a complex parameter s with positive real part, and a subset M of the set $\mathbb{Z}^+$ of positive integers. The operator $G_{s,M}$, acting on any space of functions defined on $(0, 1)$, is given by

$$G_{s,M}[f](t) = \sum_{k \in M} (k+t)^{-s} f(1/(k+t))$$

provided $\sum_{k \in M} k^{-\Re[s]} < \infty$. Whatever the space on which G is deemed to live, G at least has the property that it maps linear combinations of functions of the form $(1 + \theta t)^{-s}$, $0 \le \theta \le 1$ to other such functions. We shall have to specify the linear space or spaces on which these G act to be definite, but one main point that will emerge from this will be that for appropriate real σ, the spectral radius of $G_{M,\sigma}$ is $\lambda_M(\sigma)$.

When M is finite, this operator is also closely linked to the fractal structure of E_M. One useful yardstick for such sets is the Hausdorff dimension. Another is the Minkowski dimension.

9.1.1 Hausdorff dimension of continued fraction Cantor sets

The Hausdorff dimension of a bounded set A in a metric space (as for instance the real line) is defined as the infimum of all $\sigma > 0$ so that, for all $\epsilon > 0$, there exists an open cover $\mathcal{P}$ of A with the property that for all $P \in \mathcal{P}$, diam $P < \epsilon$, and so that $\sum_{P \in \mathcal{P}} (\text{diam } P)^\sigma < \epsilon$. Informally, if the 'price' of an interval is the σ power of the length of the interval, then for $\sigma > \dim A$, we can buy open covers of A arbitrarily cheaply using only small pieces, while if $\sigma < \dim A$, we cannot.

A good way to get a feel for this definition is to verify that the Hausdorff dimension of the Cantor middle third set is $\log 2 / \log 3$, while the Hausdorff dimension of a countable set is zero. The Hausdorff dimension of the ordinary unit cube in n dimensions, is n.

For the continued fraction Cantor set E_M, we begin with some notation. Recall that M^r denotes the Cartesian product of r copies of M. The basic identity is that

$$G^r_{\sigma,M}[f](t) = \sum_{\mathbf{v} \in M^r} |\mathbf{v} + t|^{-\sigma} f([\mathbf{v}+t]).$$

(It doesn't matter, yet, just which space we regard G as being defined on. The space of all functions from $[0,1]$ to $\mathbb{C}$ will serve.) Taking $f \equiv 1$ in the identity above, and then setting $t = 0$, gives

$$G^r_{\sigma,M}[1](0) = \sum_{\mathbf{v} \in M^r} |\mathbf{v}|^{-\sigma}.$$

The importance of this identity is that the length of the interval $\{[\mathbf{v}+s] : 0 \leq s \leq 1\}$ is exactly $1/(|\mathbf{v}||\mathbf{v}+1|)$, and to within a factor of 2, is $|\mathbf{v}|^{-2}$. Thus, the *canonical open cover* $\{\{[\mathbf{v}+t] : 0 < t < 1\} : \mathbf{v} \in M^r\}$ has price comparable to $G^r_{2\sigma,M}[1](0)$. Clearly, if this tends to zero, then $\sigma \geq \dim E_M$. The converse is not obvious, but it is true: if $G^r_{2\sigma,M}[1](0)$ does not tend to zero as $r \to \infty$, or if it is infinite, (which can only happen if M is infinite), then $\sigma \leq \dim E_M$.

The proof involves two cases. Either M is finite, or infinite. We first show how the infinite case follows from the finite case. Suppose, then, that we already know the claim holds for finite sets M'. If M is infinite, and if E_M has only 'expensive' open covers for some particular σ, then the canonical open covers for E_M in particular are all expensive. Replace σ with any smaller σ', and the price will tend to infinity as the upper bound on interval length tends to zero. Thus for all $\sigma' < \sigma$, there exists r so that $\sum_{\mathbf{v} \in M^r} |\mathbf{v}|^{-2\sigma'} > 10$. (The sum may even be infinite.) But then there exists a finite $M' \subset M$ so that $\sum_{\mathbf{v} \in V_{M'}(r)} |\mathbf{v}|^{-2\sigma'} > 9$. From this it follows that $G^n_{\sigma',M'}[1](t) \to \infty$, so that $\dim E_{M'} \geq \sigma'$. But then, the larger set E_M also has Hausdorff dimension greater than σ'.

For the case of finite M, we need a tool from the literature. (A quote would be sufficient; the original proof is due to Cusick [Cu]. But the approach here serves to motivate further the study of the operators which are the main topic of this Chapter.)

Let λ be the spectral radius of $G = G_{2\sigma,M}$. There exists a positive, decreasing function g on $[0,1]$ with the property that $G[g] = \lambda g$, $g(0) = 1$, and $g(1) > 2^{2\sigma} g(0)$[He7]. Assume that there exist 'cheap' finite open covers of E_M with all interval lengths small. Let $\mathcal{P}$ be one such open cover, with the property that $\sum_{P \in \mathcal{P}} |P|^\sigma < 1/10$. $\mathcal{P}$ need not be a canonical open cover. However, such an open cover can be *converted* into a canonical open cover without much increasing its 'price'. First, for any $P \in \mathcal{P}$, we determine the least r so that there exist two distinct elements of the canonical open cover of order $r+1$ which meet P. That is, there should exist $\mathbf{v} \in M^r$ and $n_1, n_2 \in N$, $t_1, t_2 \in [0,1]$ so that $[\mathbf{v}+1/(n_1+t_1)] \in P$ and $[\mathbf{v}+1/(n_2+t_2)] \in P$, but $P \subset \{[\mathbf{v}+t] : 0 \leq t \leq 1\}$. The length of P, then, is comparable to the length of the canonical interval $[\mathbf{v}+1/(n_2+t_2)] \in \mathcal{P}$, so replacing P with this interval will not significantly increase the 'price' of the open cover. Should there be more than one such P associated with the same $\mathbf{v}$, so much the better, for we may discard the rest. Just because the components of our 'working' open cover are now canonical intervals, though, does not mean that the cover itself is canonical, for we may have different values of r for different intervals.

Now consider the tree structure of $\cup_{r=0}^\infty M^r$: Each $\mathbf{v}$ is a vertex of the tree, with edges between $\mathbf{v}$ and $\mathbf{u}$ if there exists $k \in N$ so that $\mathbf{u} = \mathbf{v}k$ or vice-versa. Our 'working' $\mathcal{P}$ is an open cover of E_M, and this is equivalent to the statement that every path from the root to infinity, outward along the tree, must pass through some vertex $\mathbf{v}$ for which the associated canonical interval belongs to $\mathcal{P}$. Should there be any redundant intervals, corresponding to cases in which both **v** and some descendant of **v** in the tree structure are represented in $\mathcal{P}$, we may remove the descendants, and reduce the 'price' of the cover, to arrive at a working cover $\mathcal{P}$ so that each path from root outward to infinity meets exactly one vertex **v**.

We now propose to recount the 'price' of the cover, by weighting intervals not by the price length$^{2\sigma} = |\mathbf{v}|^{-\sigma}|\mathbf{v}+1|^{-\sigma}$, nor by the essentially equivalent $|\mathbf{v}|^{-2\sigma}$, but instead by $|\mathbf{v}|^{-2\sigma}g(\{\mathbf{v}\})$. This does not significantly affect the price of our open cover, because g is comparable to 1 throughout the interval $[0,1]$. Now the claim to be proved is that 'cheap open covers' imply $\lambda = \lambda(2\sigma, M) < 1$. So assume to the contrary that $\lambda \geq 1$.

There must exist some maximal level r in the finite set S of vertices, corresponding to intervals in our open cover, 'cutting' the path from the origin to infinity. For this maximal level, there must exist a $\mathbf{v} \in M^{r-1}$ so that for all $k \in M$, $(\mathbf{v}, k) \in S$. We replace these $(\mathbf{v}, k)$ with **v**. This replacement reduces our cost (or leaves it unchanged if $\lambda = 1$), in the new way of counting prices. The calculation is that

$$\sum_{k \in M} |\mathbf{v}, k|^{-2\sigma} g(\{\mathbf{v}, k\}) = |\mathbf{v}|^{-2\sigma} \sum_{k \in M} (1 + \{\mathbf{v}\}/k)^{-2\sigma} g(1/(k + \{\mathbf{v}\}))$$
$$= |\mathbf{v}|^{-2\sigma} \lambda(2\sigma, M) g(\{\mathbf{v}\}).$$

Further such replacements eventually bring us all the way back to the single element cover corresponding to the root of the tree. But this cannot have arbitrarily small 'price', a contradiction. The contradiction stemmed from our assumption that $\lambda(2\sigma, M) \geq 1$ even though cheap open covers of E_M existed. This proves the converse, and with it, the full conclusion: For all sets $M \subset \mathbb{Z}^+$ with two or more elements, the Hausdorff dimension of E_M is equal to half the infimum of the set of all σ so that $G^r_{\sigma,M}[1](t) \to 0$ as $r \to \infty$.

This suggests a strategy for computing this Hausdorff dimension: Find a linear space which contains the constant function 1, or at any rate, contains the image under $G = G_{\sigma,M}$ of that constant function, and study the spectrum of G. If the spectral radius ρ of $G_{\sigma,M}$ is less than one then $\sigma \geq 2\dim E_M$, and conversely. Loosely speaking (but this is exactly correct if M is finite), the leading eigenvalue of $G_{\sigma,M}$ is one if and only if $\sigma = 2\dim E_M$.

9.1.2 Minkowski dimension

The other approach to sizing up these continued fraction Cantor sets is to look at the Minkowski dimension of the set complement. We discuss this in the special case where $M = M_N = \{1, 2, \ldots N\}$. The complement in $(0,1)$ of $E_N := E_{M_N}$ is a *fractal string*, that is, a disjoint union $\mathcal{L}_N$ of open intervals.

Let $\alpha_N := [N, 1, N, 1, N \ldots]$ and $\omega_N := [1, N, 1, N, \ldots]$ be the smallest and largest elements of E_N, respectively. Then $(0, \alpha_N)$ and $(\omega_N, 1)$ are the two outermost intervals of $\mathcal{L}_N$. Given any element $\mathbf{v} \in M^r$ for some positive integer r, with $v_r > 1$, there is an interval $L_\mathbf{v}$ with endpoints $[v_1, v_2, \ldots v_{r-1}, v_r + \alpha]$ and $[v_1, v_2, \ldots v_{r-1}, v_r + \omega - 1]$ (in one order or the other depending on the parity of r) of $\mathcal{L}_N$ corresponding to $\mathbf{v}$ and $L_\mathbf{v}$ has length comparable to the square of the denominator $|\mathbf{v}|$ of $[v_1, v_2, \ldots v_r]$. If $d(x, E_N)$ denotes the distance from x to E_N, then we may consider the set $\mathcal{L}_N[\epsilon]$ of all points in $\mathcal{L}_N$ within ϵ of the boundary E_N, and the aggregate length (Lebesgue measure) $V_N[\epsilon]$ of this set. The *Minkowski dimension* of $\mathcal{L}_N$ is then $\inf\{\sigma : V_N[\epsilon] = O(\epsilon^{1-\sigma})$ as $\epsilon \to 0^+\}$. Lapidus and Frankenhuysen have a recent book on *fractal strings* in which they discuss the topic of Minkowski dimension and the associated *geometric zeta function* of $\mathcal{L}$, defined as the sum of the s power of the lengths of the intervals in the string. In our case, this geometric zeta function $\zeta_{\mathcal{L}_N}(s)$ is comparable to $\sum_{\mathbf{v} \in M^*} |\mathbf{v}|^{-2s}$ since the length of $I_\mathbf{v}$ is comparable to $|\mathbf{v}|^{-2s}$. As we have seen, the abscissa of convergence for this sum is the Hausdorff dimension of E_N. But it is known [LvF] that the Minkowski dimension of a fractal string is also the abscissa of convergence of its zeta function. Thus in the case of the continued fraction Cantor sets E_N, the Hausdorff dimension of E_N is equal to the Minkowski dimension of the fractal string $(0,1) \backslash E_N$. (These authors cite [BT].) The *complex dimensions* of $\mathcal{L}$ are the poles of $\zeta_\mathcal{L}$. This requires an analytic extension of $\zeta_\mathcal{L}$ into a wider domain than the one in which convergence, and analyticity, is on the face of the matter assured. For the case of E_N, this topic is not well understood. Going by computational experiments and hope, it seems probable that $\zeta_{\mathcal{L}_N}(s)$ can be extended at least into some domain $\Re(s) > 2\dim[E_n] - \epsilon$. This would require that the associated eigenvalues of $G_{s,N}$, or at least the leading eigenvalue $\lambda_N(s)$, (leading, that is, for real s greater than twice the Hausdorff dimension), have an analytic extension into the same domain. With enough assumptions along these lines, the poles of the string's zeta function are the places where $\lambda_N(s) = 1$. While $\lambda_N(s)$ is not known to exist much less be analytic in such a wide domain, computation and speculation can forge ahead; it seems that the first non-real s so that $\lambda_2(s) = 1$ occurs at 0.91 ± 13.92. This was computed using the approach of the next section.

9.2 Spaces and Operators

In this section we discuss the linear spaces and operators of continued fractions. We use three linear spaces.

9.2.1 A simple linear space and the fruits of considering it

The first is in some ways the simplest, but there is an interaction between all three and the full spectrum of results seems to require hopping from one to the other as circumstances require. Let

$$H_A := \{f : [0,1] \to \mathbb{R} : f, f' \text{ are continuous on } [0,1]\}$$
$$\text{with } \|f\|_A := \sup_{0 \leq t \leq 1} |f(t)| + \sup_{0 \leq t \leq 1} |f'(t)|.$$

This is a Banach space. We shall need some conditions on M and s to ensure that the operators and sums we work with are defined. These *standard conditions* will be assumed from now on whenever we mention such a sum or operator without explicitly naming conditions.

$$\sigma = \Re(s) > 0 \text{ and } \sum_{k \in M} k^{-\sigma} < \infty.$$

The operators $G_{M,s} : f \to \sum_{k \in M}(k+t)^{-s}f(1/(k+t))$ are then bounded linear operators on H_A. Since H_A is not compact, the Krasnoselskii generalization of the Perron-Frobenius theorem does not apply directly. (That is one reason we bring in the other two linear spaces.) The point of using this space is that one can show the existence of a spectral gap that is uniform in M and σ satisfying the standard conditions. From this, it will follow by standard perturbation methods that there is a complex analytic extension of $\lambda_M(\sigma)$ to a domain D_M (that contains the real segment) satisfying the standard conditions (which depend on M), and an analytic extension of $g_M(\sigma, t)$ to an analytic function of two complex variables, s and t, in a domain of the form $D_M \times (\Re[t] > -1)$.

A simple calculation shows that the associated sums $\sum_{\mathbf{v} \in M^r} |\mathbf{v}|^{-\sigma}$ are log-convex in σ, from which it follows that $\lambda_M(\sigma)$ itself is log-convex so that $\log \lambda_M(\sigma)$ has non-negative second derivative. More work is needed to show that this $\lambda_M(\sigma)$ has positive second derivative on the standard interval, but it does.

Let $E_M(r)$ denote $\{x \in (0,1) : x = [v_1, v_2, \ldots v_r + \theta]$ with $0 \le \theta < 1\}$. What can we say about the conditional distribution of the random variable $\log Q_r(X) = \log(|v_1, v_2, \ldots v_r|)$ given that $X \in E_M(r)$? The results announced above, about $\lambda_M(\sigma)$, are sufficient to establish that it is asymptotically Gaussian, with mean and variance proportional to constants that depend on M through the first and second derivatives of $\lambda_M(\sigma)$ evaluated at $\sigma = 2$, for which the standard conditions are satisfied whatever the value of M.

9.2.2 A Hilbert space of power series

The Hilbert spaces H_α can be defined for arbitrary $\alpha > 0$, but in the application, we shall require $\alpha > 1$. Let H_α denote the set of all complex analytic functions on the disk $|z-1| < \sqrt{\alpha}$ so that if $f(z) = \sum_{n=0}^{\infty} c_n(z-1)^n$ then $\sum_{n=0}^{\infty} |c_n|^2 \alpha^n < \infty$, equipped with the norm

$$\|f\|_\alpha^2 = \sum_{n=0}^{\infty} |c_n|^2 \alpha^n = \lim_{r \to \sqrt{\alpha}} \oint_{|z-1|=r} |f^2(z)| \, ds.$$

This is clearly isomorphic, with the natural map, to the space of all sequences $a = (a_0, a_1, a_2, \ldots)$ so that $\sum |a_n|^2 \alpha^n$ converges, with the norm $\|a\|_\alpha^2 = \sum |a_n|^2 \alpha^n$, and we shall identify this sequence space with H_α.

Lemma 9.1 *For $\alpha > 1$, $G_{M,s}$ maps H_α into $H_{9/4}$.*

Proof. Suppose $\alpha > 1$ and choose ϵ, $0 < \epsilon < \min[1/10, 1/2 - 1/(1+\sqrt{\alpha})]$. Let $b = 9/4 + \epsilon$. The disk $|z - 1| < (3/2) + \epsilon$ is mapped by $z \to 1/(k+z)$ onto the open disk of diameter $(1/(k + 5/2 + \epsilon), 1/(k - 1/2 - \epsilon))$, and the union of these, for $k \in M$, is a subset of the disk $|z - 1| < 1 + 5\epsilon$ which in turn is a subset of the disk $|z - 1| < 3/2$. Thus

$$G_{M,s}[f](z) = \sum_{k \in M} (k+z)^{-s} f(1/(k+z))$$

converges uniformly on the disk $|z - 1| \le (3 + \epsilon)/2$ to an element of H_b. □

Our next lemma says that there is an 'infinite matrix' representation of $G_{M,s}$ with respect to the 'basis' $\{(z-1)^n, n \ge 0\}$.

Lemma 9.2 If $f(z) = \sum_{n=0}^{\infty} a_n(z-1)^n \in H_\alpha$ then $G_{M,s}[f](z) = \sum_{m=0}^{\infty} b_m(z-1)^m$ where

$$b_m = \sum_{n=0}^{\infty} \gamma_{M,s}[m,n] a_n \text{ and}$$

$$\gamma_{M,s}[m,n] = \frac{1}{2\pi i} \oint_{|z-1|=3/2} (z-1)^{-m-1} \sum_{k \in M} (k+z)^{-s} \left(\frac{1}{k+z} - 1\right)^m dz.$$

Proof. We have already seen that $G_{M,s}[f] \in H_{(9/4)+\epsilon}$ for some $\epsilon > 0$, so that there exist b_0, b_1, ... for which $G_{M,s}[f](z) = \sum_{m=0}^{\infty} b_m(z-1)^m$. What remains to be shown is that the announced values for b_m are correct, and that the sums and integrals involved converge.

We begin by noting that

$$b_m = \frac{1}{m!} \frac{d^m}{dz^m}\Big|_{z=1} (G_{M,s}[f])(z) =$$
$$\frac{1}{2\pi i} \oint_C (z-1)^{-m-1} \sum_{k \in M} (k+z)^{-s} f\left(\frac{1}{k+z}\right) dz,$$

where C is the circle $|z-1| = 3/2$, and this integral expands to

$$\frac{1}{2\pi i} \oint_C (z-1)^{-m-1} \sum_{k \in M} (k+z)^{-s} \sum_{n=0}^{\infty} a_n \left(\frac{1}{k+z} - 1\right)^n dz.$$

The order of summation and integration is arbitrary here because the integral and sum are absolutely convergent by Cauchy's inequality:

$$\sum_{n=0}^{\infty} |a_n| \left|\frac{1}{k+z} - 1\right|^n \ll \left(\sum_{n=0}^{\infty} |a_n|^2 \alpha^n\right)^{1/2}$$

so that

$$b_m = \sum_{n=0}^{\infty} a_n \frac{1}{2\pi i} \oint_C \sum_{k \in M} (k+z)^{-s} (z-1)^{-m-1} \left(\frac{1}{k+z} - 1\right)^n dz$$

as claimed. □

Our next lemma says that $G_{M,s}$ is a compact operator. Definitions of this concept differ from author to author but the difference is in appearance only. One definition, the one we shall use here, is that there be a sequence of finite dimensional operators that converge in norm to the compact operator.

The alternate definition terms an operator L on a Hilbert space H compact, if there exist orthonormal sequences (f_n) and (g_n) and positive real numbers (ρ_n) with $\rho_n \to 0$ so that $L = \sum \rho_n \langle \cdot, f_n \rangle g_n$ [May]. In [GK], p 28, it is shown that a compact operator, in the first sense, has such an expansion, there termed the Schmidt expansion; the ρ_n are precisely the so-called s-numbers of L, and will be defined and used later in this section.

The proof that $G_{M,s}$ is compact provided $s = \sigma + i\tau$ and $\sum_{k \in M} k^{-\sigma} := \sum_{k \in M} k^{-\sigma} < \infty$, depends on estimates for $|\gamma_{M,s}[m,n]|$ which we give as a lemma.

Lemma 9.3 *If $M \subseteq \mathbb{Z}^+$ with $|M| \geq 2$ (including infinite M), and if $\sum_{k \in M} k^{-\sigma} < \infty$, then for all $m, n \geq 0$,*

$$|\gamma_{M,s}[m,n]| \leq 2^\sigma (3/2)^{-m} \sum_{k \in M} k^{-\sigma}.$$

Proof. Let $C = C[M,s] := 2^\sigma \sum_{k \in M} k^{-\sigma}$. (For the moment, M and s are fixed so we drop the subscripts.) We must show $|\gamma[m,n]| \leq C(3/2)^{-m}$. We calculate:

$$|\gamma[m,n]| \leq \frac{1}{2\pi} \max_{|z-1|=3/2} (3/2)^{-m} \sum_{k \in M} |k+z|^{-\sigma} |1 - 1/(k+z)|^n$$

$$\leq (3/2)^{-m} \sum_{k \in M} (k-\tfrac{1}{2})^{-\sigma} \max\left[\left|1 - \frac{1}{k+5/2}\right|^n, \left|1 - \frac{1}{k-1/2}\right|^n \right]$$

$$\leq C(3/2)^{-m}.$$

$\square$

Remark 9.1 *Note that $(3/2)$ is best possible in this proof, for with a larger radius, the factor involving powers of n could not be safely replaced with 1.*

Now for $f(z) = \sum_{n=0}^\infty a_n (z-1)^n$, we set

$$G_N[f] = G_{M,s,N}[f] = \sum_{m=0}^{N-1} (z-1)^m \sum_{m=0}^\infty \gamma[m,n] a_n.$$

For $\|f\|_\alpha = 1$ we then have

$$\|(G - G_n)f\| \leq \sum_{m=N}^{\infty} (3/2)^m \left|\sum_{n=0}^{\infty} \gamma[m,n]a_n\right|^2$$

$$\leq \sum_{m=N}^{\infty} (3/2)^m C^2 \left(\sum_{n=0}^{\infty} |a_n|^2 \alpha^n\right)^{1/2} \left(\sum_{n=N}^{\infty} \alpha^{-n}\right)^{1/2}$$

$$\leq 3(3/2)^N C^2 \sqrt{\frac{1}{1-\alpha^{-1}}}.$$

Since the range of G_N is contained in the set of polynomials of degree less than N, this shows that G is approximated in norm by a sequence of finite dimensional operators and so is compact.

Remark 9.2 *G is not itself a finite dimensional operator, because the polynomials z^n have linearly independent images under G: if in*

$$\sum_{n=0}^{N} c_n \sum_{k \in M} (k+z)^{-n-s}$$

not all c_n are zero, then this sum has a singularity at $z = -\min_{k \in M} k$ and is thus not identically zero.

The spectrum of a compact operator on an infinite dimensional Hilbert space consists of a countable [finite or infinite] set of isolated eigenvalues, each of finite algebraic multiplicity, and zero, which is in the continuous spectrum. Here $G_{M,s}: H_\alpha \to H_\alpha$ provided $1 < \alpha \leq 9/4$. In general, the spectrum of an operator may depend on the space on which the operator acts. Not this time.

Lemma 9.4 *The spectrum of $G_{M,s}$ is independent of α.*

Proof. Suppose $\lambda \neq 0$ belongs to the spectrum of $G_{M,s}$ seen as an operator on H_α, $1 < \alpha \leq 9/4$. Then there exists $f \in H_\alpha$ so that $(G_{M,s} - \lambda I)[f] = 0$. Thus $f = \lambda^{-1} G_{M,s} f \in H_{9/4}$ so that λ is also an eigenvalue of $G_{M,s}$ in the context of $H_{9/4}$. Conversely, if λ is an eigenvalue of $G_{M,s}$ in this context, then there exists a nonzero $f \in H_{9/4}$ so that $G_{M,s} f = \lambda f$. Since $H_{9/4} \subset H_\alpha$, this selfsame f serves as an eigenfunction for λ in H_α, and this completes the proof. □

This brings us to the question of estimating the eigenvalues of $G_{M,\sigma}$. Briefly, most of them are small, as would be the case if they declined exponentially. The main idea is that the compact operator $G_{M,s}$ crams any unit cube of dimension N into a Hilbert cube which is, in most directions, quite narrow, and this is inconsistent with the existence of N relatively large eigenvalues.

The machinery to make this idea work is developed in [GK]. In Chapter 2, the "s-numbers" of a compact operator A on a Hilbert space H are described.

If $A : H \to H$ is compact, then the s-numbers of A are the eigenvalues, in decreasing order and taking multiplicity into account, of the self-adjoint operator $S = (A^*A)^{1/2}$. We cite two results.

Lemma 9.5 *If P is an orthogonal projection of H onto a subspace of H of dimension n, then $s_{n+1}(A) \leq \sup_{\|x\|=1} \|Ax - PAx\|$*

This is a corollary to Theorem 2.2, p. 31 of [GK].

Lemma 9.6 *If $\lambda_1, \lambda_2, \ldots \lambda_n$ and $s_1, s_2, \ldots s_n$ are the first n eigenvalues and s-numbers of A respectively, then $|\lambda_1 \lambda_2 \ldots \lambda_n| \leq s_1 s_2 \ldots s_n$.*

This is Lemma 3.3, pages 35 and 36 of [GK], and is due originally to H. Weyl.

We take $\alpha = 1 + 1/n$ and apply these results to $A = G_{M,\sigma}$ acting on H_α with $P = P_n[f] = \sum_{k=0}^{n-1} a_k(z-1)^k$ when $f(z) = \sum_0^\infty a_k(z-1)^k$. For $\|f\|_\alpha = 1$, with $f = \sum_0^\infty a_k(z-1)^k$, we calculate

$$\|G[f] - P_n G[f]\|_\alpha = \|\sum_{k=n}^\infty b_k(z-1)^k\| \text{ where } b_k = \sum_{j=0}^\infty \gamma[k,j]a_j.$$

Now

$$|b_k| \leq C(2/3)^k \sum_{j=0}^\infty |a_j|$$

$$\leq C(2/3)^k \left(\sum_0^\infty |a_j|^2 \alpha^n j\right)^{1/2} \left(\sum_0^\infty \alpha^{-j}\right)^{1/2}$$

$$= C(2/3)^k \sqrt{n+1}.$$

Thus

$$\|G - P_n G\|_\alpha \leq C\sqrt{n+1} \left(\sum_{k=n}^\infty (2/3)^{2k} \alpha^k\right)^{1/2} =$$
$$C(2/3)^n (1+1/n)^{n/2} (1 - (4/9)(1+1/n))^{-1/2} \leq 6C(2/3)^n.$$

From this, the result we have been aiming at (and now again making explicit the dependence on M and s) follows:

Theorem 9.2

$$|\lambda_1[M,s]\lambda_2[M,s]\ldots\lambda_n[M,s]| \leq (6C[M,s])^n(2/3)^{n(n+1)/2}$$

and so

$$|\lambda_n[M,s]| \leq 6C[M,s](2/3)^{(n+1)/2}.$$

This representation of G and g is computationally accessible, most readily in the case of finite M. It turns out to be computationally more efficient to use series in powers of $(z-1/2)$ rather than in powers of $(z-1)$. We define $\mathcal{G}: \mathbb{R}^n \to \mathbb{R}^n$ by

$$G_{M,s}[\sum_0^n a_k(z-1/2)^k] = \sum_0^n b_k(z-1/2)^k + O((z-1/2)^{n+1}) \Leftrightarrow$$
$$\mathcal{G}_{M,s}[(a_0,a_1,\ldots a_n)] = (b_0,b_1,\ldots b_n).$$

We can also get eigenvectors and eigenvalues of the matrix. But here, caution is in order. Although the leading eigenvalue of a positive operator is relatively stable under perturbation of the operator, small perturbations of G loom larger in comparison to small eigenvalues. This topic has been developed further just recently in [FV2].

(The famous illustration of this has to do with the roots of the polynomial $\prod_1^{2n}(x-k) + \epsilon x^n$.) We content ourselves here with an illustration of the method and make heuristic judgments to reject outright, or tentatively accept, the output of our calculations as reflecting the eigenvalues and eigenfunctions of $G_{M,\sigma}$ itself.

The procedure here, called gamma, takes as input the exponent σ, the list M =mlist of elements of M, the size of the square matrix to be evaluated, and the accuracy to which intermediate arithmetic is to be carried. It pays to keep plenty of digits, for there are cancellation errors to worry about and the interior eigenvalues of a matrix can be thrown off by apparently small roundoff errors.

One may then calculate the matrix of $G_{M,\sigma}$ to arbitrary depth and precision, continue by calculating the eigenvectors of that matrix, and finally, examine the output (best done by some sort of visual representation.) Here, the fact that *only* the values of f on the set E_M have any bearing on the behavior of $G_{M,\sigma}^r f$ as r goes to infinity, is apparent. Our eigenfunctions look very reasonable when plotted on the interval $[a,b] = [\min[E_M], \max[E_M]]$ just covering the continued fraction Cantor set under investigation. How fitting, after all, that this set should re-emerge after having been shunted far into the background of our investigations.

```
risingfactorial[u_, n_Integer] :=
  risingfactorial[u, n] = Product[u + k, {k, 0, n - 1}];

  gamma[sigma_, mlist_, size_, accuracy_] :=
  Module[{r, kpwrs, p, u, v, w, x, y, z, mm, n, m, j, k},
   mm = Length[mlist];
    Do[Do[p[m, n] = 0, {m, 0, size - 1}], {n, 0, size - 1}];
    Print["p established"];
    kpwrs =
    Table[N[(mlist[[j]] + 1/2)^(-sigma), accuracy],
    {j, 1, mm}];
    Print["kpwrs established"];
    Do[v = kpwrs[[j]];
    Print["j and k and v=", {j, mlist[[j]], v}];
      k = mlist[[j]];(*k loop*)
      Do[w = (k + 1/2)^(-m)*(-1)^m/m!;(*m loop*)
        Do[u = (-1)^n/2^n;(*n loop*)
          x = u v w; y = 0;
          Do[z = risingfactorial[r + sigma, m]*
            (k + 1/2)^(-r)*(-2)^r*
            Binomial[n, r]; y += z, {r, 0, n}];
          y *= x;
          p[m, n] = p[m, n] + y, {n, 0, size - 1}],
        {m, 0, size - 1}],(*closing m loop*)
      {j, 1, mm}](*closing k loop*);
    Table[p[m, n], {m, 0, size - 1}, {n, 0, size - 1}]]
```

Executing the command

```
gg = gamma[
     1.06256101255441,{1, 2},256, 320];
```

but with the given value of σ padded with many zeros followed by a 1 so that the machine accepts the input σ as a high precision decimal, gives a large matrix. Calling the Mathematica built-in routine for eigenvectors and eigenvalues confirms that this value of σ and M yields $\lambda_M(\sigma)$ near 1, so that the Hausdorff dimension of $E_{\{1,2\}}$ is, if we trust the calculations, approximately 0.5312805062772. Current computing machinery allows for these calculations to be extended to more places without great difficulty, and the whole business can be made rigorous. [He7] Converting the eigenvectors of our approximation to $\mathcal{G}$ corresponding to the first four eigenvalues (here rounded)

$$\{1.00000000000000018, -0.3117672, 0.109605, -0.04019\}$$

back into the corresponding power series gives the first four eigenfunctions of $G_{\{1,2\},\sigma}$, shown below on the interval $[a = 153/418, b = 418/571]$ The values of a and b are approximately the left and right boundaries of $E_{\{1,2\}}$. The eigenfunctions have been scaled to take the value 1 at a. The figure on the right shows the coefficients of the series itself, for the fourth eigenfunction, as a list-plot. It is typical of the others. Matrix truncation and

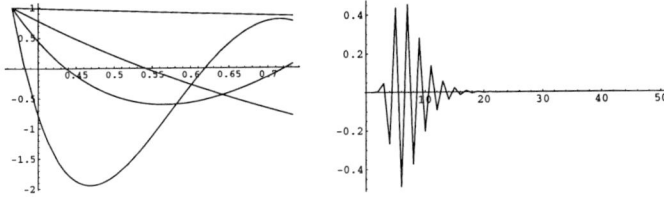

Fig. 9.1   Eigenfunctions.

numerical roundoff error can introduce spurious values and care is needed in interpreting numerical results. The sixth eigenvector reported by Mathematica, in a calculation taken to considerable depth, is in no way consistent with the overall pattern and must be judged spurious, for instance. More detailed calculations yield the following approximate value:

$$\dim E_2 \approx 0.53128050627720\allowbreak 51416244686473$$
$$6847178549305910901839877988839780392752953564\,.$$

This agrees, to all 54 places given there, with computations made by Oliver Jenkinson [Je], using the altogether different algorithm of [JP]. The source cited makes that calculation en passant; its main point is to partially prove a conjecture made independently by this author and by D. Mauldin and M. Urbański, to the effect that the set of Hausdorff dimensions associated to finite sets M is dense in $[0, 1]$. Jenkinson proves that it is, at any rate, dense in $[0, 1/2]$, and gives numerical and heuristic evidence in support of the rest of the conjecture. Just recently, a proof of the conjecture has been announced by Marc Kessenböhmer and Sanguo Zhu [KZ].

9.3 Positive Operators

In this section and in the next, we assume that $s = \sigma$ is real. The Perron-Frobenius theorem for finite $k \times k$ matrices says that if L is a matrix with non-negative entries and with the property that for some $n > 0$, L^n has all entries positive, then there is a unique eigenvector p of L so that all entries of p are positive, the corresponding eigenvalue λ is positive and has geometric multiplicity one (that is, the space of all vectors q so that for some $n > 0, (L - \lambda I)^n q = 0$ has dimension one and consists merely of the scalar multiples of p), and the rest of the spectrum of L lies inside a circle about the origin of radius less than λ. Thus uniformly on R^k,

$$(\lambda^{-1} L)^n v = c_v p + O((1 - \epsilon)^n \|v\|)$$

and the mapping $M : v \to c_v p$ is a projection. The restriction of the linear operator $v \to Lv$ to the null space of M has spectral radius less than λ.

Krasnoselskii has shown that the same sort of result holds for positive linear operators under suitable definitions and hypotheses. In a Banach space B, a *proper cone* P is a set P of vectors closed under addition and multiplication by non-negative scalars, and $P \cap -P = \emptyset$. If $B = P - P$, then the cone is called *reproducing*. Suppose the cone also has nonempty interior.

A compact operator L is *positive* if $LP \subseteq P$. It is u_0-positive if for all $p \neq 0$ in P there exist positive real numbers a and b and a positive integer n so that $bu_0 < L^n p < au_0$, where $u < v$ means that $v - u \in P$.

The cone P of elements $f \in H_\alpha$ so that $f(t) \geq 0$ for all $0 \leq t \leq 1$ is a proper, reproducing cone in the sense of Krasnoselskii. It has non-empty interior; the function $u_0(t) := 1$ lies in the interior of the cone. The operator $G_{M,s}$ is u_0-positive with respect to the cone P because for every $f \neq 0$ in P there exist real numbers $0 < a < b$, and a positive integer n, so that $b \leq G^n[f] \leq a$. For although f may have isolated zeros on $[0,1]$, the zeros of $G^n[f]$ are the set of all z so that for all $v \in M^n$, every number x of the form $x[v, z] = [v_1, v_2, \ldots v_{n-1}, v_n + z]$ is a zero of f itself. Thus if $G^r[f]$ had k zeros, f must have had at least $|M|^r k$ zeros. Finally, G is compact as we have seen.

With all this in place, Krasnoselskii's Perron-Frobenius theorem applies and we have that $G_{M,\sigma}$ has an isolated largest eigenvalue λ which is positive and simple, and a corresponding positive eigenfunction $g = g_{M,\sigma}$, and the rest of H_α tends to zero exponentially under iteration of $\lambda^{-1} G$. That is, the spectral radius of the $\lambda^{-1} G$ restricted to the set of $f \in H_M$ for which $(\lambda^{-1} G)^n f \to 0$ is less than 1. Clearly this largest eigenvalue λ is our old friend $\lambda_M(\sigma)$.

9.4 An Integral Representation of $g_{M,\sigma}$

From our earlier observation that $\lambda_{M,\sigma}^{-r} G_{M,\sigma}^r [1](t)$ is always a function of the form $\int_0^1 (1+\theta t)^{-\sigma} d\mu$ with μ a positive measure on $[0,1]$ of mass bounded between two positive constants, uniformly in r, and from the fact that the set of all such functions, with the Lévy metric, is compact, it follows that $g_M(\sigma, t)$ itself has that form:

$$g_M(\sigma, t) = \int_0^1 (1+\theta t)^{-\sigma} d\mu_{M,\sigma}(\theta).$$

Up to now, $g_M(\sigma, t)$ has been specified only up to scaling by a positive constant. We choose that constant so that the underlying measure is a probability measure. That is, $\mu_{M,\sigma}[0,1] = 1$, and equivalently, $g_M(\sigma, 0) = 1$.

The set P of all probability measures on the unit interval becomes a compact set when equipped with the Lévy metric. The mapping $L : \mu \to \int_0^1 (1+\theta t)^{-\sigma} d\mu(\theta)$ from P to $H_A \cap H_B$ is continuous, with the Lévy metric for the domain and either $\|\cdot\|_A$ or $\|\cdot\|_B$ for the domain. But it is not (yet) excluded that there could be two probability measures mapped by L to the same function.

This does not happen. The difference $L[\mu_1] - L[\mu_2]$ cannot be zero because it would be a function $f(t)$, analytic at least in the disk $|t - 1/2| < 1$, and it would have a power series expansion about zero. The derivatives $f^{(n)}(0)$ would be $\prod_{k=0}^{n-1} (-\sigma - k) \int_0^1 \theta^n d(\mu_1 - \mu_2)$. For these to all be zero would contradict the fact that if $\mu_1 \neq \mu_2$ then there would exist an interval $[a, b]$ so that $(\mu_1 - \mu_2)([0, \theta))$ has constant sign for $a < \theta < b$. Integrating by parts and constructing an appropriate polynomial, we arrive at a contradiction because we can force

$$\int_0^1 p(\theta)(\mu_1 - \mu_2)(\theta) \neq 0$$

yet we would be requiring
$$\int_0^1 \theta^n(\mu_1 - \mu_2)\,d\theta = 0.$$

So L is a one-to-one mapping, and thus the underlying measure $\mu_{M,\sigma}$ is unique.

We can say more about this measure. There is a formula for $\mu_{M,\sigma}$ in terms of $g_M(\sigma, t)$. This is not as circular in computational terms as it looks, because we can calculate g using truncations of the 'infinite matrix' representation $\mathcal{G}_{M,\sigma}$ of $G_{M,\sigma}$ acting on H_B.

Lemma 9.7 *For all M and σ satisfying the standard conditions, and for all x with $0 \le x \le 1$, we have*
$$\mu_{M,\sigma}[0,x] = \lim_{r\to\infty} \lambda_M(\sigma)^{-r} \sum_{\mathbf{v}\in M^r, [\mathbf{v}]\le x} |\mathbf{v}|^{-\sigma} g_{M,\sigma}(\{\mathbf{v}\}).$$

That is, $\mu_{M,\sigma}$ is the limiting case of discrete measures assigning mass
$$\lambda_M(\sigma)^{-r}|\mathbf{v}|^{-\sigma} g(\{\mathbf{v}\})$$
to the point $[\mathbf{v}]$, summed over all $\mathbf{v} \in M^r$. Incrementing r has the effect of splitting this atom of mass into an equal total mass
$$\lambda_M(\sigma)^{-r-1} \sum_{k\in M} |\mathbf{v}k|^{-\sigma} g(\{\mathbf{v}k\})$$
with point masses at the various $[\mathbf{v}k]$, $k \in M$.

The overall mass is invariant because
$$\lambda^{-1} \sum_{k\in M} |\mathbf{v}k|^{-\sigma} g(\{\mathbf{v}k\}) =$$
$$\lambda^{-1}|\mathbf{v}|^{-\sigma} \sum_{k\in M} (k+\{\mathbf{v}\})^{-\sigma} g(1/(k+\{\mathbf{v}\})) = |v|^{-\sigma} g(\{\mathbf{v}\})$$
from the definition of g. Backtracking this 'hair-splitting' argument to the beginning shows that the mass assigned to $[0,1]$ is $g(0) = 1$, for all r, and thus also in the limit.

Since the mass assigned at step r to a particular $\mathbf{v}$ is scattered a distance of at most $O(|\mathbf{v}|^{-2})$ during subsequent steps, ending up at various points $[\mathbf{v}+\theta]$ instead of at exactly $[\mathbf{v}]$, and since the function $(1+\theta t)^{-\sigma}$ is continuous in θ and in either of our norms (as an element of H_A or H_B), it follows that these sums converge both pointwise and as elements of H_A or H_B respectively. That is,

$$\lim_{r\to\infty} \lambda_M^{-r}(\sigma) \sum_{\mathbf{v}\in M^r} |\mathbf{v}|^{-\sigma} g_{M,\sigma}(\{\mathbf{v}\})(1+[\mathbf{v}]t)^{-\sigma} = \int_0^1 (1+\theta t)^{-\sigma} d\mu_{M,\sigma}(\theta).$$

It remains to show that this measure $\mu_{M,\sigma}$ gives rise to a function invariant under $\lambda_M^{-1}(\sigma) G_{M,\sigma}$. To show that μ gives rise to g we calculate

$$\lambda^{-1} G(\int_0^1 (1+\theta t)^{-\sigma} d\mu(\theta))$$
$$= \lim_{r\to\infty} \lambda^{-r-1} \sum_{k\in M} \sum_{\mathbf{v}\in M^r} |\mathbf{v}|^{-\sigma} (k+[\mathbf{v}]+t)^{-\sigma} g(\{\mathbf{v}\})$$
$$= \lim_{r\to\infty} \lambda^{-r-1} \sum_{k\in M} \sum_{\mathbf{v}\in M^r} |\mathbf{v}|^{-\sigma} g(\{\mathbf{v}\})(k+[\mathbf{v}])^{-\sigma} \left(1+\frac{t}{k+[\mathbf{v}]}\right)^{-\sigma}$$
$$= \lim_{r\to\infty} \lambda^{-r-1} \sum_{\mathbf{v}\in M^r} |k\mathbf{v}|^{-\sigma} g(\{k\mathbf{v}\})(1+t[k\mathbf{v}])^{-\sigma}$$
$$= \lim_{r\to\infty} \lambda^{-r-1} \sum_{\mathbf{w}\in M^{r+1}} |\mathbf{w}|^{-\sigma} g(\{\mathbf{w}\})(1+[\mathbf{w}]t)^{-\sigma}$$
$$= \int_0^1 (1+\theta t)^{-\sigma} d\mu_{M,\sigma}(\theta).$$

The step in which $\{\mathbf{v}\}$ is replaced with $\{k\mathbf{v}\}$ is valid because

$$\{k\mathbf{v}\} = (1 + O(((1+\sqrt{5})/2)^{-r}))\{\mathbf{v}\}.$$

From this calculation we see that the function defined by this integral is invariant under $\lambda^{-1}G$, and clearly it is positive and when $t=0$ it evaluates to 1. Thus it is g as claimed.

Now consider the linear functional $\Lambda_{M,\sigma} : H_B \to \mathbb{C}$ defined by

$$\Lambda_{M,\sigma} f := \lim_{r\to\infty} \lambda_M^{-r}(\upsilon) G_{M,\sigma}^r[f](0).$$

Let $\mathcal{N}_{M,\sigma}$ denote the null space of this linear functional. Then every $f \in H_B$ has a unique representation as

$$f = f_\mathcal{N} + \Lambda_{M,\sigma}[f] g,$$

and from the Krasnoselskii results, $(\lambda^{-1}G)^r f_\mathcal{N}$ tends to zero exponentially, while the second term is invariant under G. There is an explicit characterization of $\Lambda_{M,\sigma}$ as an integral, as we should expect from the Riesz representation theorem. The key fact is that

$$\int_0^1 \lambda_M(\sigma)^{-1} G_{M,\sigma}[f](x) \, d\mu_{M,\sigma}(x) = \int_0^1 f(x) \, d\mu_{M,\sigma}(x).$$

Proof. Abbreviating $G = G_{M,\sigma}$ and $g = g_{M,\sigma}$, we have $\int_0^1 \lambda^{-1} G f \, d\mu = \int_0^1 f \, d\mu$ because the second integral is

$$\lim_{r \to \infty} \lambda^{-r} \sum_{\mathbf{v} \in M^r} |\mathbf{v}|^{-\sigma} f([\mathbf{v}]) g(\{\mathbf{v}\})$$

while the first is

$$\sum_{k \in M} \lim_{r \to \infty} \sum_{\mathbf{v} \in M^r} \lambda^{-r-1} |\mathbf{v}|^{-\sigma} (k + [\mathbf{v}])^{-\sigma} f(1/(k + [\mathbf{v}])) g(\{\mathbf{v}\})$$
$$= \lim_{r \to \infty} \sum_{\mathbf{w} \in M^{r+1}} \lambda^{-r-1} |\mathbf{w}|^{-\sigma} f([\mathbf{w}]) g(\{\mathbf{w}\})$$

which simplifies to $\int_0^1 f \, d\mu$. Since $(\lambda^{-1} G)^r f$ tends to some multiple of g, the correct multiple must be the one with $\int_0^1 f \, d\mu = c \int_0^1 g \, d\mu$ and this is equivalent to our last claim. □

Lemma 9.8

$$\Lambda_{M,\sigma}[f] = C_{M,\sigma} \int_0^1 f(t) \, d\mu_{M,\sigma}(t)$$

where

$$C_{M,\sigma} := 1 / \int_0^1 g_{M,\sigma}(t) \, d\mu_{M,\sigma}(t).$$

Proof. Fix f, and write $\mu = \mu_{M,\sigma}$, $g = g_{M,\sigma}$ and so on. Let $c = C_{M,\sigma} \int_0^1 f \, d\mu$, and let $h = f - cg$. Then $\int_0^1 h \, d\mu = 0$. We claim that $(\lambda^{-1} G)^r h \to 0$. For if not, we should have $(\lambda^{-1} G)^r h \to \alpha g$ for some nonzero α, which would mean that $\int_0^1 h \, d\mu = C_{M,\sigma} \alpha$ rather than its assumed value of zero.

Observe that the operator $P_{M,\sigma} : f \to (\Lambda_{M,\sigma}[f]) g_{M,\sigma}$ is a projection, and that $\int_0^1 f \, d\mu_{M,\sigma}$ is invariant under $\lambda_{M,\sigma}^{-1} G_{M,\sigma}$.

The next best thing to an orthogonal projection on a Hilbert space H is a projection P with the property that the null space of the projection makes a positive angle with the range of the projection, or equivalently, that there is a positive constant c so that if $Q = I - P$, then for all $X \in H$,

$\|PX\| \geq c\|X\|$ for $X \in H$ and $\|QX\| \geq c\|X\|$. We have this property in the case at hand. Consider the sequence

$$c_n := (2/3)^n \int_0^1 (z-1)^n \, d\mu_{M,\sigma}(z).$$

If

$$f = \sum_{n=0}^{\infty} a_n(z-1)^n \in H_B$$

with $\|f\|_B = 1$ then the condition that $\int_0^1 f \, d\mu = 0$ is equivalent to requiring that

$$\sum_{n=0}^{\infty} (3/2)^n a_n c_n = 0$$

while if

$$g_{M,\sigma}(z) = \sum_{n=0}^{\infty} b_n(z-1)^n$$

then

$$2^{-\sigma} < \sum_{n=0}^{\infty} (3/2)^n b_n c_n = \int_0^1 g \, d\mu = C_{M,\sigma}^{-1} < 1.$$

In $\mathcal{H}_B$, then, (a_n) is orthogonal to (c_n) while (b_n) makes a certain positive inner product. All that remains to do is to show that indeed $(c_n) \in \mathcal{H}_B$, that is, that $\sum_{n=0}^{\infty} (3/2)^n c_n^2 < \infty$. But $c_n^2 \ll (2/3)^{2n}$ which completes the proof. □

9.5 A Hilbert Space Structure for G when $s = \sigma$ is Real

Our second space H_M depends on M. The idea of connecting G to a setting like the one presented here goes back to Babenko, who worked this kind of structure out for the case $M = \mathbb{Z}^+$, and to Mayer [May]. The existence of a Hilbert space structure, and the consequent reality of the eigenvalues of G acting on a suitably chosen space, for $G_{M,\sigma}$ and for the more general case in which the set M from which the a_r must be taken varies periodically in r, has been established by Vallée in [V2]. What is new here is the 'matrix expansion' of the associated operator $K_{M,\sigma}$ on this 'hidden Hilbert space' in terms of Laguerre polynomials. This expansion affords a second proof that the eigenvalues of $G_{M,\sigma}$ decay at least as fast as $(2/3)^n$. We take

$$H_M := \left\{ \phi : [0,\infty) \to \mathbb{C} : \phi \text{ is measurable and} \right.$$
$$\left. \int_0^\infty |\phi^2(x)| \sum_{k \in M} e^{-kx}\, dx < \infty \right\}.$$

Equipped with the norm

$$\|\phi\|_M^2 := \int_0^\infty |\phi^2(x)| \sum_{k \in M} e^{-kx}\, dx,$$

H_M is a Hilbert space. The point of introducing H_M is that there is a symmetric operator $K = K_{M,\sigma}$ on H_M that replicates the spectrum and the action of $G_{M,\sigma}$ on H_α. This time the proofs work only for $\alpha < 9/4$, so we assume from now on that $1 < \alpha < 9/4$. We also assume that we have a fixed M and real σ in mind, and the operators K and S to be defined will depend, but only implicitly, on these two parameters.

Let $x! := \Gamma(x+1)$, which, in the case of non-negative integers, agrees with the usual definition of $x!$. We also extend the notation of binomial coefficients; $\binom{x}{y} := \frac{x!}{y!(x-y)!}$. Recall that $\zeta_M(\sigma) = \sum_{k \in M} k^{-\sigma}$ and let $e_M(x) := \sum_{k \in M} e^{-kx}$. In all this, we lean heavily on the approach taken by Mayer [May], but using the spaces H_α in place of a certain space of analytic functions he used, which was tied closely to the choice $M = \mathbb{Z}^+$. (We must pay somehow for generality in M.)

Let $S = S_{M,\sigma} : H_M \to H_\alpha$ be the operator given by

$$S : \phi \to \int_0^\infty e^{-xz} \phi(x) x^{(\sigma-1)/2} \sum_{k \in M} e^{-kx}\, dx.$$

Lemma 9.9 *S maps H_M into H_α and is a bounded linear operator.*

Proof. By the Cauchy-Schwarz inequality, if $z = u + iv$ with $u > -1/2$ then

$$|S_{M,\sigma}[\phi](z)| \leq \left(\int_0^\infty x^{\sigma-1} e^{-2ux} \sum_{k \in M} e^{-kx}\, dx \right)^{1/2} \|\phi\|_M <$$
$$(\sigma-1)!^{1/2} ((1+2u)^{-\sigma} + 1)^{1/2} (\zeta_M(\sigma))^{1/2} \|\phi\|_M,$$

which is finite. To bound $\|S_{M,\sigma}[\phi]\|_B$ we need bounds for the various derivatives, evaluated at 1, of $S_{M,\sigma}[\phi]$. Using the same inequality, if

$$S_{M,\sigma}[\phi] = f = \sum_{n=0}^{\infty} a_n(z-1)^n$$

then using the facts that $\Gamma(\sigma+2j) \ll (1+\sigma^{-1})(1+2j)^\sigma \Gamma(2j+1)$ and that $\binom{2j}{j} \ll 2^{2j}$, we have

$$|a_j| \leq \frac{1}{j!} \left(\int_0^\infty x^{s-1+2j} e^{-2x} \sum_{k \in M} e^{-kx} \, dx \right)^{1/2} \|\phi\|_M$$

$$= \frac{\Gamma(\sigma+2j)^{1/2}}{j!} \left(\sum_{k \in M} (k+2)^{-\sigma-2j} \right)^{1/2} \|\phi\|_M$$

$$\leq j^{O(1)} (2/3)^j \left(\sum_{k \in M} k^{-\sigma} \right)^{1/2} \|\phi\|_M.$$

Thus

$$\|S[\phi]\|_\alpha / \|\phi\|_M \ll_{M,\sigma} \left(\sum_{j=0}^\infty j^{O(1)} (2/3)^{2j} \alpha^j \right)^{1/2},$$

which proves that S takes H_M into H_α and is bounded as claimed. □

Remark 9.3 *The bound tends to infinity as $\sigma \to 0$, as $\alpha \to 9/4$, or as $\sum_{k \in M} k^{-\sigma} \to \infty$; happily these cases have not arisen in applications.*

We now proceed with a string of lemmas.

Lemma 9.10 *$S = S_{M,\sigma}$ is nonsingular, that is, if $\phi \in H_M \neq 0$ then $S[\phi] \in H_\alpha \neq 0$.*

Proof. $S[\phi]$ is the classical laplace transform of $x^{(\sigma-1)/2} e_M(x) \phi$. If ϕ is not the zero function in H_M (that is, it is not the case that ϕ is zero almost everywhere), then this other function is likewise not zero, so that its laplace transform is not identically zero. □

Our next lemma says that if f is a linear combination of functions of the form $(1+\theta z)^{-\sigma}$ then $G[f] \in \text{Range}[S]$. The classical Gauss-Kuz'min question has to do with the convergence of $G^r_{\mathbb{Z}^+,2}[1]$ and the constant function

1 is covered by this lemma. Furthermore, it so happens that the eigenfunction $g_{M,\sigma}$ corresponding to $\lambda_M(\sigma)$ from Section 9.3 above has this form, so $g_{M,\sigma}$ is in the range of S.

Lemma 9.11 *If $f(z) = \int_0^\infty (1+\theta z)^{-\sigma}\,d\mu(\theta)$ where μ is a finite signed measure on $[0,\infty)$ then $G[f]$ has the same form, and*

$$S\left(\frac{1}{\Gamma(\sigma)}\int_0^\infty x^{(\sigma-1)/2}e^{-\theta x}\,d\mu(\theta)\right) = G_{M,\sigma}[f].$$

Proof. It is sufficient to establish the result for the case that μ is a probability measure on $[0,\infty)$. So suppose $f(z) = \int_0^\infty (1+\theta z)^{-\sigma}\,d\mu(\theta)$. Then

$$G[f](z) = \int_0^\infty \sum_{k\in M}(k+\theta)^{-\sigma}(1+z/(k+\theta))^{-\sigma}\,d\mu(\theta)$$

$$= \int_0^\infty (1+\theta z)^{-\sigma}\,d\nu(\theta)$$

where ν is the measure given by

$$\nu[0,\theta] = \sum_{k\in M}\int_{\max\,[1/\theta-k,0]}^\infty (k+\phi)^{-\sigma}\,d\mu(\phi).$$

On the other hand,

$$S\left[\frac{1}{\Gamma(\sigma)}\int_0^\infty x^{(\sigma-1)/2}e^{-\theta x}\,d\mu(\theta)\right](z)$$

$$= \frac{1}{\Gamma(\sigma)}\int_{\theta=0}^\infty \sum_{k\in M}(k+\theta)^{-\sigma}(1+z/(k+\theta))^{-\sigma}\,d\mu(\theta).$$

A routine calculation shows that $\phi_f(x) := \int_0^\infty x^{(\sigma-1)/2}e^{-\theta x}\,d\mu(\theta)$ is indeed in H_M which completes the proof. $\square$

Our interest in H_M, as previously noted, is that there is a sister operator $K = K_{M,\sigma}$ on H_M so that $S_{M,\sigma}K_{M,\sigma} = G_{M,\sigma}S_{M,\sigma}$. We now discuss K. The tools are Bessel functions and Laguerre polynomials.

The Bessel J functions of order $\alpha > -1$ are defined by

$$J_\alpha(u) := \sum_{n=0}^\infty \frac{(-1)^n(u/2)^{2n+\alpha}}{n!\Gamma(n+\alpha+1)}.$$

The Bessel I functions are defined by the same series as the Bessel J functions, except with no alternating signs; they will arise later in this section.

The J_α are linked to the generalized Laguerre polynomials

$$L_n^\alpha(u) := \sum_{j=0}^{n} \frac{(-1)^j u^j}{j!} \binom{n+\alpha}{j+\alpha}$$

by the identity

$$J_\alpha(2\sqrt{vw}) = (vw)^{\alpha/2} \sum_{k=0}^{\infty} L_k^\alpha(v) \frac{w^k e^{-w}}{\Gamma(k+\alpha+1)}.$$

Now we are ready to define K:

$$K_{M,\sigma}[\phi](w) := \int_{v=0}^{\infty} J_{\sigma-1}(2\sqrt{vw}) \phi(v) e_M(v)\, dv.$$

That is, K is an integral operator for which the kernel is the symmetric function $J_{\sigma-1}(2\sqrt{vw})$ and for which the measure with respect to which the integration is done is $d\mu_M(v) = \sum_{k \in M} e^{-vk}\, dv$. Ultimately we show that K is a compact symmetric operator on H_M with the concomitant nice spectral properties: an orthogonal system of eigenfunctions, eigenvalues real, and as in the case of finite dimensional operators, nothing more to the story. But first we must show that K is well defined.

Proposition 9.1 *If $\phi \in H_M$ and if $\zeta_M(\sigma) = \sum_{k \in M} k^{-\sigma} < \infty$, then for $x > 0$, $K[\phi](x)$ is defined. Moreover, the function $f = K[\phi]$ satisfies $\|f\|_B \ll \zeta_M(\sigma) \|\phi\|_M$.*

Proof. By the Cauchy-Schwarz inequality,

$$|K[\phi](x)| \le \|\phi\|_M \int_0^\infty \left(J_{\sigma-1}^2(2\sqrt{xy}) e_M(y) \right)^{1/2} dy.$$

Now

$$\int_0^\infty J_{\sigma-1}^2(2\sqrt{xy}) e^{-ky}\, dy = k^{-1} e^{-2x/k} I_{\sigma-1}(2x/k).$$

This is most easily checked using Mathematica [Ma]. A similar identity may be found in [Ask]; (formula 2.45W) it reads

$$\int_0^\infty J_\alpha(xz) J_\alpha(yz) e^{-t^2 z^2} z\, dz = \frac{1}{t^2} \exp(-(x^2+y^2)/(4t^2)) I_\alpha\left(\frac{xy}{2t^2}\right)$$

but this has a typographical error in that the first factor $1/t^2$ on the right hand side should read $1/(2t^2)$. With this repaired, the identity used here is equivalent after a change of variable to the case $x = y$ of 2.45. Thus,

$$|K[\phi](x)| \le \left(\sum_{k\in M} k^{-1} e^{-2x/k} I_{\sigma-1}(2x/k)\right)^{1/2} \|\phi\|_M.$$

Now $I_{\sigma-1}(2u) \le u^{\sigma-1} e^{2\sqrt{u}}$, so

$$|K[\phi](x)| \le \left(\sum_{k\in M} k^{-1} e^{-2x/k}(x/k)^{\sigma-1} e^{2\sqrt{x/k}}\right)^{1/2} \|\phi\|_M$$
$$\le e^{1/4} x^{(\sigma-1)/2} \zeta_M(\sigma)^{1/2} \|\phi\|_M.$$

Thus

$$\|K\phi\|^2 \le \|\phi\|^2 e^{1/2} \int_0^\infty x^{\sigma-1} \zeta_M(\sigma) e_M(x)\,dx$$
$$= \Gamma(\sigma) e^{1/2} \zeta_M(\sigma)^2 \|\phi\|_M^2 \Rightarrow$$
$$\Rightarrow \|K\phi\|_M \le e^{1/4} \sqrt{\Gamma(\sigma)} \zeta_M(\sigma) \|\phi\|_M.$$

Therefore, $K : H_M \to H_M$, and K is a bounded linear operator. $\square$

The claim that $SK = GS$ is verified by a straightforward evaluation of both sides of the equality, renaming v and x with x and v on the GS side, expanding $J_{\sigma-1}$ as a series, and integrating term by term.

The upshot is this: K is a compact, nuclear-of-order-zero operator, with an expansion in terms of Laguerre polynomials.

Theorem 9.3 *If $\sigma > 0$, $K = K_{M,\sigma}$, and $\phi \in H_M$, then*

$$K[\phi] = \sum_{k=0}^\infty \langle u_k, \phi\rangle e_k$$

where

$$u_k(w) = w^{(\sigma-1)/2} L_k^{s-1}(w) \text{ and } e_k(w) = \frac{w^{k+(\sigma-1)/2} e^{-w}}{\Gamma(k+\sigma)}.$$

Furthermore,

$$\|u_k\|^2 \ll \left(k^\sigma + \frac{k^2 \Gamma(\sigma)}{\sigma^2} \zeta_M(\sigma)\right)$$

and

$$\|e_k\|^2 \ll (2/3)^{2k}.$$

Functional-Analytic Techniques

Finally,

$$S^{-1}G[(z-1)^n] = \frac{x^{(\sigma-1)/2}n!}{\Gamma(n+\sigma)}L_n^{\sigma-1}(x).$$

Proof. First we establish the identity for $J_{\sigma-1}(2\sqrt{vw})$ claimed above.

$$J_{\sigma-1}(2\sqrt{vw}) = (vw)^{(\sigma-1)/2}\sum_{k=0}^{\infty}\frac{w^k e^{-w}}{\Gamma(k+\sigma)}L_k^{\sigma-1}(v)$$

because if $f(v,w) = (vw)^{-(\sigma-1)/2}J_{(\sigma-1)}(2\sqrt{vw})$ then

$$f(v,w) = \sum_{n=0}^{\infty}\frac{(-1)^n v^n w^n}{n!\Gamma(n+\sigma)}$$

$$= \sum_{n=0}^{\infty}\frac{(-1)^n v^n w^n}{n!\Gamma(n+\sigma)}\sum_{l=0}^{\infty}\frac{w^l e^{-w}}{l!}$$

$$= \sum_{n=0}^{\infty}\sum_{k=n}^{\infty}\frac{(-1)^n v^n w^k e^{-w}}{n!\Gamma(n+\sigma)(k-n)!}$$

$$= \sum_{k=0}^{\infty}\sum_{n=0}^{k}\frac{(-1)^n v^n}{n!}\binom{k+\sigma-1}{n+\sigma-1}\frac{w^k e^{-w}}{\Gamma(k+\sigma)} = \sum_{k=0}^{\infty}\frac{w^k e^{-w}}{\Gamma(k+\sigma)}L_k^{\sigma-1}(v).$$

Thus

$$K[\phi](w) = \int_0^{\infty}\phi(v)J_{\sigma-1}(2\sqrt{vw})\sum_{j\in M}e^{-jv}\,dv$$

$$= \int_0^{\infty}\phi(v)(vw)^{(\sigma-1)/2}\sum_{k=0}^{\infty}L_k^{(\sigma-1)}(v)\frac{w^k e^{-w}}{\Gamma(k+\sigma)}\sum_{j\in M}e^{-jv}\,dv$$

$$= \sum_{k=0}^{\infty}\frac{w^{k+(\sigma-1)/2}e^{-w}}{\Gamma(k+\sigma)}\langle v^{(\sigma-1)/2}L_k^{\sigma-1}(v),\phi\rangle = \sum_{k=0}^{\infty}\langle u_k,\phi\rangle e_k.$$

For the bound on $\|e_k\|$, we calculate

$$\|e_k\|_M^2 = \int_0^{\infty}\frac{x^{2k+\sigma-1}e^{-2x}}{\Gamma(k+\sigma)^2}\sum_{j\in M}e^{-jx}\,dx$$

$$= \frac{\Gamma(2k+\sigma)}{\Gamma(k+\sigma)^2}\sum_{j\in M}(j+2)^{-\sigma-2k}.$$

For the case $k=0$, this simplifies to $\|e_0\|_M^2 \le (\Gamma(\sigma))^{-1}\sum_{j\in M}j^{-\sigma}$. For $k\ge 1$, our bound simplifies to $\|e_k\|_M^2 \le \sum_{l=3}^{\infty}l^{-2-2k}\frac{\Gamma(\sigma+2k)}{\Gamma(\sigma+k)^2}$ since the sum here is $O(3^{-2k})$ while the ratio of Γ function values is $\ll (2e)^{\sigma-1}2^{2k}$ by Stirling's formula.

To bound $\|u_k\|$, we take $\alpha = \sigma - 1$ in an identity given in Askey [Ask] (p 15)

$$\int_0^\infty x^\alpha (L_k^\alpha)^2(x) e^{-x}\, dx = \Gamma(k+\alpha+1)/k!.$$

Next, we calculate that

$$\|u_k\|_M^2 = \sum_{j \in M} \int_0^\infty x^{\sigma-1}(L_k^{\sigma-1})^2(x) e^{-jx}\, dx.$$

The term corresponding to $j = 1$, which will be present if $1 \in M$, is thus $\Gamma(k+\sigma)/k!$. We now consider the terms corresponding to $k = 1$ and the contribution to those terms coming first from the range $0 \le x \le 1/k$ and second from $x > 1/k$.

The contribution coming from $x \le 1/k$ is bounded above by

$$\sum_{j \in M, j > 1} \int_0^{1/k} x^{\sigma-1} \left(\sum_{l=0}^k \frac{x^l}{l!} \binom{k+\sigma-1}{\sigma} \right)^2 e^{-jx}\, dx.$$

Now for $x \le 1/k$,

$$\sum_{l=0}^k \binom{k+\sigma-1}{\sigma} \ll \frac{k}{\sigma}.$$

Thus for $x \le 1/k$,

$$\sum_{j \in M, j > 1} \int_0^{1/k} x^{\sigma-1} \left(\sum_{l=0}^k \frac{x^l}{l!} \binom{k+\sigma-1}{\sigma} \right)^2 e^{-jx}\, dx$$
$$\ll \frac{k^2}{\sigma^2} \sum_{j \in M, j > 1} \int_0^{1/k} x^{\sigma-1} e^{-jx}\, dx \ll \frac{k^2 \Gamma(\sigma)}{\sigma^2} \sum_{j \in M} j^{-(\sigma-1)}.$$

For the contribution due to $x > 1/k$, and from $j > 1$, we calculate that

$$\int_{1/k}^{\infty} x^{\sigma-1} \left(L_k^{\sigma-1}(x)\right)^2 e^{-x} \sum_{j \in M, j > 1} e^{-(j-1)x} \, dx$$

$$\leq k \int_{1/k}^{\infty} x^{\sigma-1} \left(L_k^{\sigma-1}(x)\right)^2 e^{-x} \, dx$$

$$< k \int_0^{\infty} x^{\sigma-1} \left(L_k^{\sigma-1}(x)\right)^2 e^{-x} \, dx = \frac{\Gamma(k+\sigma)}{(k-1)!}$$

which is $\ll k^\sigma$. □

In a sense, the formula

$$K[\phi] = \sum_{k=0}^{\infty} \langle u_k, \phi \rangle e_k$$

gives us a matrix representation of K. Even though K acts on a Hilbert space rather than a finite dimensional space, and even though the 'matrix' is with respect to not one basis but two, we learn a lot. Grothendieck has worked all this out in greater generality, but in our Hilbert space setting matters are simpler. A compact symmetric operator has an orthogonal eigenvalue-eigenvector expansion just as in the finite dimensional case: $K\phi = \sum_{n=1}^{\infty} \lambda_n \gamma_n$, with $\|\gamma_n\| = 1$, $\langle \gamma_j, \gamma_k \rangle = 0$ for $j \neq k$, and with $\lambda_n \to 0$. Now observe that the image under K of the unit ball is a sort of ellipsoid, with the property that any subspace of dimension n meets the ellipsoid in an ellipsoid of minimal diameter $2|\lambda_n|$. On the other hand, the image under K of the span of $\{u_1, u_2, \ldots u_{n-1}\}$ is an $n-1$ dimensional subspace of H, and if we write $K = K_{n-1} + R_{n-1}$ with $K_{n-1}\phi = \sum_{j=1}^{n-1} \langle u_j, \phi \rangle e_j$, then $\|R_{n-1}\| \ll (2/3)^n$. Thus, if B is the unit ball in H, then the minimal diameter of $K_n B$ is $\ll (2/3)^n$. Therefore, $\lambda_n \ll (2/3)^n$. We now claim that the spectra of G as on operator on H_α and K on H_M are identical. In both cases, we are dealing with compact operators so the spectrum is pure point spectrum with the exception of an accumulation point at zero. In neither case is zero itself an eigenvalue. Now if $K\phi = \lambda\phi$, let $f = S\phi \in H_2$. Then

$$Gf = GS\phi = SK\phi = S\lambda\phi = \lambda S\phi = \lambda f$$

so that λ is an eigenvalue of G as well. If $Gf = \lambda f$, let $\phi = S^{-1}Gf \in H$. Then

$$K\phi = KS^{-1}Gf = S^{-1}G^2 f = S^{-1}\lambda Gf = \lambda S^{-1}Gf = \lambda\phi$$

so that λ is an eigenvalue of K as well. This completes the proof that G and K have the same spectrum. We leave to the reader the verification that there is no instance of $(G-\zeta I)f_1 = f_2, (G-\zeta I)f_2 = f_3, \ldots (G-\zeta I)f_r = 0$. (Since K is symmetric, nothing of the sort can happen with K.) From this it follows that all eigenvalues of $G_{M,\sigma}$ are real.

There is some numerical evidence in support of the conjecture that the eigenvalues decline exponentially, (we already know that $\lambda_n \ll (2/3)^n$), that they alternate in sign, and that the geometric multiplicity of each eigenvalue is one. (This multiplicity is the dimension of the space of all f so that $(G-\lambda I)^k f = 0$ for some $k > 0$, but under the circumstances, there can be no nilpotent action in G since there is none in K, so that if $(G-\lambda I)^k f = 0$ then $(G - \lambda I)f = 0$.) Thus, the geometric multiplicity of an eigenvalue is simply the dimension of the eigenspace corresponding to the eigenvalue in question. In particular, recent work of Flajolet and Vallée [FV2] reports strong numerical evidence that for the successive eigenvalues of $G_{\mathbb{Z}^+,2}$, the ratio λ_{n+1}/λ_n tends to $-((\sqrt{5}-1)/2)^2)$. Bolstering the numerical evidence is the fact that G is the sum of operators $G_k : f \to (k+z)^{-\sigma} f(1/(k+z))$ each of which has a geometric sequence spectrum [May2]. In the case at hand, the leading eigenvalue is $(1 + \theta/k)^{-\sigma}$ where $\theta = (-k + \sqrt{k^2 + 4})/2$ is the positive solution to $\theta = 1/(k+\theta)$, and the ratio of consecutive eigenvalues is $-\theta^2$, from which it would seem that this one component of G controls the behavior of the deep eigenvalues.

9.6 The Uniform Spectral Gap

In this section we show that there is a positive, continuous function $\rho(\sigma)$ so that for all M with $|M| \geq 2$ for which $\sum_{k \in M} k^{-\sigma} < \infty$, the spectral radius of G, as an operator on H_A or H_B, as well as the radius of K on H_M, when restricted to the complement of the eigenspace of multiples of $g_{M,\sigma}$ or its counterpart in H_M, is less than $(1 - \rho(\sigma))\lambda_M(\sigma)$.

That is, the operator G has one big positive eigenvalue, of multiplicity one; $\lambda_M(\sigma)$. The rest of the spectrum sits inside a circle of radius $(1 - \rho(\sigma))\lambda_M(\sigma)$. This has the important consequence that there is a complex analytic extension of $\lambda_M(\sigma)$ into a neighborhood of the real interval on which $\lambda_M(\sigma)$ is convergent. The consequence is important because the derivatives of $\lambda_M(\sigma)$ at selected values of σ determine the dynamics of the continued fraction algorithm and related processes, but the existence of these derivatives must first be established, and that is done by showing that $\lambda_M(\sigma)$ is complex analytic, through perturbation theory, which, if it is to apply, requires this spectral gap.

Recall that we have shown that for $\sigma > 0$ and M satisfy the standard conditions, there exists a function $g = g_{M,\sigma} : [0,1] \to \mathbb{R}^+$ of the form

$$g(t) = g_{M,\sigma}(t) = \int_0^1 (1+\theta t)^{-\sigma} d\mu_{M,\sigma}(\theta)$$

so that

$$G_{M,\sigma} g_{M,\sigma}(t) = \lambda_M(\sigma) g_{M,\sigma}(t).$$

From its form, it is clear that $g \in H_A$. Now the idea is that the action of $G^2_{M,\sigma}$ on H_A amounts to a scaling by $\lambda_M^2(\sigma)$ combined with a compression toward multiples of $g_{M,\sigma}$. The pace of this compression is the issue. We begin by defining another cone. Positive functions with bounded log derivative are in a sense preserved by G. Let

$$\mathcal{A}_{M,\alpha} := \{f \in H_A : f > 0 \text{ on } [0,1] \text{ and } |f'(t)| \le \alpha f(t) \text{ for } 0 \le t \le 1\}.$$

Then $\mathcal{A}_{M,\alpha}$ is closed under addition and multiplication by a positive scalar.

Lemma 9.12

$$G^2_{M,\sigma} : \mathcal{A}_{M,2\sigma} \to \mathcal{A}_{M,(3/2)\sigma}.$$

Proof. Clearly if f is positive on $(0,1)$ then so is $G^2 f$. Taking derivatives and simplifying, we see that the claim holds provided that for all $(j,k) \in M$,

$$\frac{\sigma k}{(jk+1+tk)^{\sigma+1}} + \frac{2\sigma}{(jk+1+tk)^{\sigma+2}} < \frac{(3/2)\sigma}{(jk+1+kt)^\sigma}$$

and clearing terms and simplifying reduces this inequality to one which is true by inspection for $0 \le t \le 1$ and for any positive integers j,k. $\square$

Recall that $P_{M,\sigma}$ is the projection operator taking f to the component of g in f, or more exactly,

$$P_{M,\sigma}[f] = \lim_{r \to \infty} \lambda_{M,\sigma}^{-r} G_{M,\sigma}^r[f]$$

and that

$$P_{M,\sigma}[f] = \frac{\int_0^1 f \, d\mu_{M,\sigma}}{\int_0^1 g_{M,\sigma} \, d\mu_{M,\sigma}}.$$

It will now be convenient to introduce also the related linear functional $L = L_{M,\sigma}$ given by $L_{M,\sigma}[f] = \int_0^1 f \, d\mu_{M,\sigma}$.

192   *Continued Fractions*

Theorem 9.4 *Uniformly over σ and M satisfying the standard conditions, and uniformly over $f \in H_A$,*

$$\lambda_{M,\sigma}^{-r} G_{M,\sigma}^r[f] = \frac{\int_0^1 f \, d\mu_{M,\sigma}}{\int_0^1 g_{M,\sigma} \, d\mu_{M,\sigma}} g_{M,\sigma}$$
$$+ O\left((1+\sigma)(1-e^{-3\sigma}/6)^{r/2}\|f\|_A\right).$$

Proof. We begin with some lemmas.

Lemma 9.13 *Let $C_\sigma := 3e^\sigma(1 + 1/\sigma)$. If $h = h[f] = f + C_\sigma \|f\|_A g_{M,\sigma}$ then $h \in \mathcal{A}_{2\sigma}$.*

Proof. We have $|f(t)| \leq \|f\|_A$ and $|f'(t)| \leq \|f\|_A$ for $0 \leq t \leq 1$. Now $g_{M,\sigma}(t) = \int_0^1 (1+\theta t)^{-\sigma} \, d\mu(\theta)$ so

$$0 > \frac{g'(t)}{g(t)} > -\frac{\sigma}{1+t}.$$

Thus $g \in \mathcal{A}_\sigma$ and $1 \geq g(t) \geq (1+t)^{-\sigma}$. A simple calculation shows that

$$h[f](t) \geq -\|f\|_A + C_\sigma \|f\|_A g(t) \geq \frac{2}{3} C_\sigma g(t),$$

while on the other hand

$$|h[f]'(t)| \leq (1 + \sigma C_\sigma g(t))\|f\|_A \leq \frac{4}{3}\sigma C_\sigma g(t)\|f\|_A.$$

From the two inequalities above, $|h'| \leq 2\sigma h$ so $h \in \mathcal{A}_{2\sigma}$. □

Lemma 9.14 $G_{M,\sigma}^2 : \mathcal{A}_{2\sigma} \to \mathcal{A}_{3\sigma/2}$.

Proof. All that is needed is to expand $G^2 f$ and check the details:

$$G^2 f[t] = \sum_{j,k \in M} (jk + 1 + jt)^{-\sigma} f(1/(j + 1/(k+t)))$$

and

$$(G^2 f)' = \sum_{j,k \in M} (jk + 1 + jt)^{-\sigma-2}.$$
$$(f'(1/(j+1/(k+t))) - \sigma j(jk+1+jt) \cdot$$
$$f(1/(j+1/(k+t))).$$
  □

Our next lemma is the heart of the matter: peeling off a little multiple of g from a function in $\mathcal{A}_{3\sigma/2}$ leaves a function in $\mathcal{A}_{2\sigma}$.

Lemma 9.15 *If $u \in \mathcal{A}_{3\sigma/2}$ then*

$$u - (1/6)e^{-3\sigma}L_{M,\sigma}[u]g_{M,\sigma} \in \mathcal{A}_{2\sigma}.$$

Proof. It suffices to show two things: first, that $u(0) \geq e^{-3\sigma/2}L[u]$, and second, that if $\zeta \leq (1/6)e^{-3\sigma/2}u(0)$, then $u - \zeta g \in \mathcal{A}_{2\sigma}$.

The first is simple: since $u \in \mathcal{A}_{3\sigma/2}$, $u(t) \leq e^{3\sigma/2}u(0)$ so that

$$\int_0^1 u \, d\mu_{M,\sigma} \leq u(0)e^{3\sigma/2}.$$

For the second claim, suppose $\zeta \leq (1/6)e^{-3\sigma/2}u(0)$. Then

$$|(u - \zeta g)'(t)| \leq |u'(t)| + \frac{1}{6}e^{-3\sigma/2}u(0)|g'(t)|$$
$$\leq \frac{3\sigma}{2}u(t) + \frac{1}{6}e^{-3\sigma/2}u(0)g(t)$$
$$\leq 2\sigma(u(t) - \frac{1}{6}e^{-3\sigma/2}u(0)g(t)).$$

Hence,

$$|(u - \zeta g)'(t)| \leq 2\sigma(u(t) - \zeta g(t)).$$

$\square$

Now consider the sequence determined by $h_0 = h[f] = f + C_\sigma g$ and recursively by

$$h_{r+1} = \lambda_M^{-2}(\sigma)G_{M,\sigma}^2[h_r] - e_r g_{M,\sigma}$$

where $e_r = (1/6)e^{-3\sigma}L_{M,\sigma}[h_r]$.

Lemma 9.16 *For all $r \geq 1$,*

$$h_r \in \mathcal{A}_{2\sigma}$$

and

$$L_{M,\sigma}[h_{r+1}] = (1 - (1/6)e^{-3\sigma}L_{M,\sigma}g_{M,\sigma})L_{M,\sigma}[h_r].$$

Proof. By 9.13, $h_0 \in \mathcal{A}_{2\sigma}$. By 9.14, if $h_r \in \mathcal{A}_{2\sigma}$ then $\lambda^{-2}G^2[h_r] \in \mathcal{A}_{3\sigma/2}$. To apply 9.15 and conclude that $h_{r+1} \in \mathcal{A}_{2\sigma}$, we need to know that $e_r \leq (1/6)e^{-3\sigma/2}h_r[0]$. But $L[h_r] = \int_0^1 h_r \, d\mu \leq \int_0^1 h(0)e^{3\sigma t/2} \, d\mu \leq$

$h(0)e^{3\sigma/2}$ so this lemma applies. Thus inductively $h_r \in \mathcal{A}_{2\sigma}$ for all r. For the second claim, we calculate

$$L[h_{r+1}] = \int_0^1 \lambda^{-2} G^2 h_r - (1/6)e^{-3\sigma} L[h_r] g \, d\mu$$

$$= \int_0^1 h_r \, d\mu - (1/6)e^{-3\sigma} L[h_r] L[g] = L[h_r](1 - (1/6)e^{-3\sigma} L[g]). \quad \square$$

We now proceed to the main proof. We have

$$L[h_0] = \int_0^1 (f + C_\sigma \|f\|_A g) \, d\mu \le \|f\|_A (1 + 3(1 + 1/\sigma)e^\sigma \int_0^1 g \, d\mu)$$
$$\le (1 + 3(1 + 1/\sigma)e^\sigma) \|f\|_A.$$

Thus

$$L[h_r] \le (1 - (1/6)e^{-3\sigma})^r (4 + 3/\sigma)e^\sigma \|f\|_A.$$

Now

$$\lambda^{-2r} G^{2r}[h_0] = (\sum_{j=0}^{r-1} e_j) g + h_r.$$

Because $e_j = (1/6)e^{-3\sigma} L[h_j]$ declines exponentially, $\sum_{j=0}^\infty e_j$ is a convergent geometric series with sum $\lambda[h_0]/\lambda[g]$ and this fits in with the requirement that $\lambda^{-2r} G^{2r} h_0 \to (L[h_0]/L[g])g$. Consequently,

$$\lambda^{-2r} G^{2r}[h_0] = \frac{L[h_0]}{L[g]} g + h_r - (1 - (1/6)e^{-3\sigma})^r \frac{L[h_0]}{L[g]} g.$$

For any $h \in \mathcal{A}_{2\sigma}$, $\|h\|_A \le L[h] e^{2\sigma}(1 + 2\sigma)$.

Thus since $L[h_r] = (1 - (1/6)e^{-3\sigma} L[g])^r L[h_0]$,

$$\|h_r\|_A \le L[h] e^{2\sigma}(1 + 2\sigma),$$

and so

$$\lambda^{-2r} G^{2r}[f] = \frac{L[f]}{L[g]} g + p_r,$$

where

$$\|p_r\|_A \le (1 - (1/6)e^{-3\sigma} L[g])^r (L[f] + C_\sigma \|f\|_A L[g]) e^{2\sigma}(1 + 2\sigma) + $$
$$(1 - (1/6)e^{-3\sigma} L[g])^r (1 + \sigma) \frac{L[h_0]}{L[g]}.$$

Now $L[f] \leq \|f\|_A$, and $L[h_0] = L[f] + 3(1 + 1/\sigma)e^\sigma \|f\| L[g]$, so

$$\frac{L[h_0]}{L[g]} = \frac{L[f]}{L[g]} + 3(1 + 1/\sigma)e^\sigma \|f\|_A.$$

Putting this all together gives

$$\|p_r\|_A \leq 8(1 + 2\sigma)e^{3\sigma}(1 - (1/6)e^{-3\sigma}L[g])^r \|f\|_A$$

and this completes the proof of Theorem 9.4. □

Stripped to its essentials, this says that $\lambda_{M,\sigma}^{-1} G_{M,\sigma}$ has 1 as an eigenvalue of multiplicity 1, and the rest of the spectrum sits inside a circle of radius $\rho(\sigma) \leq (1 - (1/6)e^{-3\sigma}L[g])^{1/2}$.

What of the spectral gap in H_B? It is at least as large, because we know that $G_{M,\sigma}$ acting on H_B has pure point spectrum apart from an accumulation point at zero. If there were any eigenfunction $g_2 \in H_B$ for which the corresponding eigenvalue λ_2 satisfied $\rho(\sigma) < |\lambda_2|/\lambda_M(\sigma) < 1$, this g_2 would also be an element of H_A and an eigenfunction of $G_{M,\sigma}$ acting on A, contrary to the conclusion of theorem 9.4. Since the spectrum of $K_{M,\sigma}$ acting on H_M is identical to that of $G_{M,\sigma}$ acting on H_B, this operator too has the same spectral gap. By standard perturbation results [CR], it therefore follows that $\lambda_M(\sigma)$ extends to an analytic function in a neighborhood of σ, for any pair M, σ satisfying the standard conditions.

From this it now follows that

Theorem 9.5 *For all M and σ satisfying the standard conditions, the operator $G_{M,\sigma}$ has spectral radius $\lambda_M(\sigma)$, and as $r \to \infty$,*

$$\lambda_M^{-r}(\sigma) G_{M,\sigma}[f](t) = C_f g_{M,\sigma}(t) + O\left(\left(1 - \frac{1}{12}e^{-7\sigma/2}\right)^r \|f\|_A\right)$$

uniformly over $f \in H_A$ and over σ and M.

9.7 Log Convexity of λ_M

There is a natural interval $I_M = (\alpha_M, \infty)$ on which λ_M is defined: $\sum_{n \in M} n^{-s} < \infty$ for $s > \alpha_M$ and $\sum_{n \in M} n^{-s} = \infty$ for $s < \alpha_M$. The sum is convergent on all of $\mathbb{R}$ if M is finite, and $\alpha_M = 1$ if $M = \mathbb{Z}^+$. It can converge, or diverge, at $s = \alpha_M$; more commonly, it diverges.

Log convexity of λ_M means that for $s \in \mathbb{R}$ and $h > 0$ such that $s - h > \alpha_M$,

$$\lambda_M(s-h)\lambda_M(s+h) \geq \lambda_M^2(s).$$

We say that a positive function is *strongly log concave* on an interval if its log is twice differentiable and positive on that interval. We then have

Theorem 9.6 *For any $M \subset \mathbb{Z}^+$ with at least two elements λ_M is strongly log convex on I_M.*

Proof. First we show that λ_M is at least log concave in the weaker sense. Recall that

$$\lambda_M(s) = \lim_{r \to \infty} \left(\sum_{\mathbf{v} \in M^r} |\mathbf{v}|^{-s} \right)^{1/r},$$

so it is sufficient to show that for all r, M, and relevant s and h, we have

$$\sum_{\mathbf{u} \in M^r} \sum_{\mathbf{v} \in M^r} |\mathbf{u}|^{-s+h}|\mathbf{v}|^{-s-h} \geq \sum_{\mathbf{u} \in M^r} \sum_{\mathbf{v} \in M^r} |\mathbf{u}|^{-s}|\mathbf{v}|^{-s}.$$

But this follows immediately from the fact that

$$|\mathbf{u}|^{-s+h}|\mathbf{v}|^{-s-h} - 2|\mathbf{u}|^{-s}|\mathbf{v}|^{-s} + |\mathbf{u}|^{-s-h}|\mathbf{v}|^{-s+h} \geq 0.$$

Now on I_M, λ_M is analytic, positive, and not constant, so $\log \lambda_M$ can at worst only have isolated zeros of its second derivative. The rest of the proof is somewhat lengthy but the idea is simple: the digits (X_i) of a continued fraction decomposition are correlated, but the correlation between X_i and X_j decays exponentially in $|i - j|$. This long-range near-independence permits us to establish that for some $0 < \theta < 1$, and uniformly over $E \in \mathbb{R}$,

$$\sum_{\mathbf{u} \in M^n} |\mathbf{u}|^{-s}(\log |\mathbf{u}| - E)^2 \gg n^\theta \lambda_M^n(s).$$

From this, we can show that

$$\sum_{\mathbf{u,v} \in M^n} |\mathbf{u}|^{-s}|\mathbf{v}|^{-s}(\log |\mathbf{u}| - \log |\mathbf{v}|)^2 \gg n^\theta \lambda_M^{2n}(s).$$

On the other hand, differentiating the identity

$$\lambda^n(s)g(0) = \sum_{\mathbf{u} \in M^n} |\mathbf{u}|^{-s} g([\mathbf{u}])$$

once, then again, and combining terms leads to an identity of the form

$$n\lambda_M^{2n}(s)(\lambda_M'' \lambda_M - {\lambda_M'}^2)(s)$$
$$= \tfrac{1}{2} \sum_{\mathbf{u},\mathbf{v}\in M^n} |\mathbf{u}|^{-s}|\mathbf{v}|^{-s} g_s([\mathbf{u}]) g_s([\mathbf{v}])(\log|\mathbf{u}| - \log|\mathbf{v}|)^2 + \text{error terms}.$$

With adequate control of the error terms, the possibility that $(\lambda_M'' \lambda_M - {\lambda_M'}^2)(s) = 0$ at some particular value of s is excluded.

We now present the details. Let $v(f)$ denote the total variation of f on $[0,1]$. Consider the space $\mathcal{V}$ of functions of bounded variation on $[0,1]$, with norm $\|f\| = v(f) + \sup_{[0,1]} |f|$. The operator L_s maps $\mathcal{V}$ into itself. It is analytic, and in particular, infinitely differentiable. Thus, for any $s_0 \in (\alpha_M, \infty)$, there is a complex neighborhood of s_0 in which $L_{M,s}$ has the form $L + (s-s_0)L' + \tfrac{1}{2}(s-s_0)^2 L'' + L^*$, with L, L', and L'' bounded operators on $\mathcal{V}$ that depend on s_0 but not s, and with $\|L^*\| \ll |s-s_0|^3$. Furthermore, $L = L_{M,s_0}$ has a decomposition of the form $L = \lambda_M(s_0) P_{M,s_0} + Q_{M,s_0}$, where $P_{M,s}$ is a projection of $\mathcal{V}$ onto the one dimensional subspace of $\mathcal{V}$ spanned by $g_{M,s}$, and $Q_{M,s}$ is an operator which takes g_s to zero and has spectral radius strictly less than $\lambda_M(s)$.

From this it follows by standard perturbation theory that the eigenfunction g_s (here normalized so that $g_s(0) = 1$ for all s) and corresponding eigenvalue $\lambda_M(s)$ are analytic in s. In the case of g_s, this means that for each $s_0 > \alpha_M$, there exist $f_1 = \partial g_s / \partial s$ and $f_2 = \partial^2 g_s / \partial s^2 \in \mathcal{V}$ such that in some complex neighborhood of s_0,

$$g_s = g_{s_0} + (s-s_0) f_1 + \tfrac{1}{2}(s-s_0)^2 f_2 + O((s-s_0)^2).$$

We recall that for real $s > \alpha_M$, g_s has the form $g_s(t) = \int_0^1 (1+\theta t)^{-s} d\mu(\theta)$, where μ is a probability measure on $[0,1]$. Thus $g_s(t)$ is a strictly decreasing function of t, and $g_s(1)/g_s(0) \geq 2^{-s}$.

The measure μ, which depends on M and s, is given by

$$\mu([0,x]) = \lim_{n\to\infty} \lambda_M^{-n}(s) \sum_{\mathbf{v} \in M^n, [\mathbf{v}] \leq x} |\mathbf{v}|^{-s} g(\{\mathbf{v}\}).$$

Equivalently, let $[[\mathbf{v}]] = \{[\mathbf{v}+t] : 0 \leq t \leq 1\}$. Then $\mu = \mu_{M,s}$ is characterized by the condition that for $\mathbf{v} \in M^n$, $\mu([[\mathbf{v}]]) = \lambda_M^{-n}(s)|\mathbf{v}|^{-s} g_{M,s}(\{\mathbf{v}\})$. The projection operator $P_{M,s}$ then has two characterizations. If $C = C_{M,s} = \int_0^1 g_{M,s}(t)\, d\mu_{M,s}(t)$, then

$$P_{M,s} f = \lim_{n\to\infty} \lambda_M^{-n}(s) L_{M,s}^n f = C_{M,s}^{-1} \left(\int_0^1 f\, d\mu_{M,s} \right) g_{M,s}.$$

Let $\mathcal{N}_{M,s}$ be the kernel of $P_{M,s}$. Because of the spectral gap, the restriction of $(I - \lambda_M(s)^{-1} L_{M,s})$ to $\mathcal{N}$ is invertible and bounded, with the inverse given by $\sum_{k=0}^{\infty}(\lambda^{-1}L)^k$.

With this background in place, we are now in a position to take derivatives and draw conclusions.

Fix M and s and let $L = L_{M,s}$. Let $g = g_{M,s}$, and let g' and g'' denote the first and second derivatives of g with respect to s. Let $\lambda(s) = \lambda_M(s)$. On setting $t = 0$ in the formula $L^n g(t) = \lambda^n(s) g(t)$ and taking derivatives with respect to s, we have

$$\lambda^n(s) = (L^n g)(0) = \sum_{\mathbf{v} \in M^n} |\mathbf{v}|^{-s} g([\mathbf{v}]),$$

$$n\lambda^{n-1}(s)\lambda'(s) = \sum_{\mathbf{v} \in M^n} |\mathbf{v}|^{-s}(-g([\mathbf{v}])\log|\mathbf{v}| + g'([\mathbf{v}])), \text{ and}$$

$$n(n-1)\lambda^{n-2}(s)(\lambda'(s))^2 + n\lambda^{n-1}(s)\lambda''(s)$$
$$= \sum_{\mathbf{v} \in M^n} |\mathbf{v}|^{-s}(g([\mathbf{v}])\log^2|\mathbf{v}| - 2g'([\mathbf{v}])\log|\mathbf{v}| + g''([\mathbf{v}])).$$

Thus, on multiplying the first line above by the last and subtracting the square of the middle line, we have

$$n\lambda^{2n-2}(s)(\lambda(s)\lambda''(s) - \lambda'^2(s))$$
$$= \sum_{\mathbf{u},\mathbf{v} \in M^n} |\mathbf{u}|^{-s}|\mathbf{v}|^{-s}\{g([\mathbf{u}])g([\mathbf{v}])\log^2|\mathbf{v}| - 2g([\mathbf{u}])g'([\mathbf{v}])\log|\mathbf{v}| + g([\mathbf{u}])g''([\mathbf{v}])$$
$$- \log|\mathbf{u}|g([\mathbf{u}])\log|\mathbf{v}|g([\mathbf{v}])$$
$$+ g'([\mathbf{u}])g'([\mathbf{v}]) - \log|\mathbf{u}|g([\mathbf{u}])g'([\mathbf{v}]) - \log|\mathbf{v}|g([\mathbf{v}])g'([\mathbf{u}])\}.$$

This simplifies to $n\lambda^{2n}(s)(\log \lambda(s))'' = A + B + C$, say, where, on taking advantage of the symmetry between $\mathbf{u}$ and $\mathbf{v}$ in summation,

$$A = \tfrac{1}{2} \sum_{\mathbf{u},\mathbf{v} \in M^n} |\mathbf{u}|^{-s}|\mathbf{v}|^{-s} g([\mathbf{u}])g([\mathbf{v}])(\log|\mathbf{u}| - \log|\mathbf{v}|)^2,$$

$$B = \sum_{\mathbf{u},\mathbf{v} \in M^n} |\mathbf{u}|^{-s}|\mathbf{v}|^{-s}(\log|\mathbf{u}| - \log|\mathbf{v}|)(g([\mathbf{u}])g'([\mathbf{v}]) - g'([\mathbf{u}])g([\mathbf{v}]),$$

and where C, the sum of the terms not involving $\log|\mathbf{u}|$ or $\log|\mathbf{v}|$, is $O(\lambda^{2n}(s))$ because g' and g'' are bounded uniformly in t.

Now by the Cauchy-Schwarz inequality together with the fact that g and its derivatives are bounded,

$$B \ll \sum_{\mathbf{u},\mathbf{v}\in M^n} |\mathbf{u}|^{-s}|\mathbf{v}|^{-s}|(\log|\mathbf{u}|-\log|\mathbf{v}|)|$$

$$\ll \left(\sum_{\mathbf{u},\mathbf{v}\in M^n} |\mathbf{u}|^{-s}|\mathbf{v}|^{-s}\right)^{1/2} A^{1/2} \ll \lambda^n(s)\sqrt{A}.$$

It now remains to give a lower bound on A that is strong enough to exclude the possibility that $\log \lambda''(s) = 0$. Anything that swamps $O(\lambda^{2n}(s))$ will serve, in view of the bounds on B and C above. And then, once it is established that $\log \lambda''(s) > 0$, we can conclude, roundabout, that A was comparable to $n\lambda^{2n}(s)$. But this is getting ahead of the story. For now, we begin by reducing A to a one-dimensional sum. Let $E_{M,n}(s) = \lambda^{-n}(s)\sum_{\mathbf{u}\in M^n} |\mathbf{u}|^{-s}g([\mathbf{u}])\log|\mathbf{u}|$ be the "expected value of $\log|\mathbf{u}|$." Then

$$A = \tfrac{1}{2}\lambda^n(s) \sum_{\mathbf{u}\in M^n} |\mathbf{u}|^{-s}g([\mathbf{u}])(\log|\mathbf{u}| - E_{M,n}(s))^2.$$

If only $\log|\mathbf{u}|$ were the sum of n independent random variables, each with variance comparable to 1! Our wish is not granted in full, as there is this pesky short-range correlation in the digits of $\mathbf{u}$. However, there is enough truth to this wish to serve our needs. This stems essentially from the fact that for any $r > 1$, any $\mathbf{w} \in M^r$, and any $q < r$, the value of a convergent depends only weakly on the deep digits: If $\gamma = 2/(1+\sqrt{5})$, then $[\mathbf{w}] = (1+O(\gamma^q))[w_1 w_2 \ldots w_q]$ and $\{\mathbf{w}\} = (1+O(\gamma^q))\{w_{r-q+1} w_{r-q+2} \ldots w_r\}$. We next recall that $\mathbf{u}$ is a string of n positive integers, say, $\mathbf{u} = u_1 u_2 \ldots u_n$, and that

$$\log|\mathbf{u}| = \sum_{1}^{n} \log(u_j + \{u_1 \ldots u_{j-1}\}). \tag{9.1}$$

Although these terms are not independent, one from the next, if we skip through the list, breaking it into lists of length $\sqrt{n}$, then those at the end of each batch are effectively independent of each other. Moving from the metaphorical to the computational, we take $k = \lfloor\sqrt{n}\rfloor$ and write n as the sum of k terms, each k or $k+1$ in case $k^2 + k \geq n$, or each $k+1$ or $k+2$, in case $k^2 + k < n$, and with the smaller numbers first. We then write $\mathbf{u}$ as the concatenation of k strings of the form $\mathbf{v}_j X_j$. (For instance, if $n = 13$ and $u = (1234123512362)$, then $k = 3$ and $\mathbf{u}$ is decomposed as $\mathbf{v}_1 X_1 \mathbf{v}_2 X_2 \mathbf{v}_3 X_3$

with $\mathbf{v}_1 = 123$, $X_1 = 4$, $\mathbf{v}_2 = 123$, $X_2 = 5$, $\mathbf{v}_3 = 1236$, and $X_3 = 2$.) We let $\mathcal{V}(M,n)$ denote the set of such lists of k lists of integers in M, of lengths $k-1$ and k, or k and $k+1$ with the shorter lists first, and with the sum of the list lengths equal to $n-k$.

Now by (9.1), we have

$$|\mathbf{u}| = (1+O(n\gamma^k))\prod_1^k |\mathbf{v}_j| \prod_1^k (X_j + \{\mathbf{v}_j\} + [\mathbf{v}_{j+1}]),$$

where $[\mathbf{v}_{k+1}] = 0$ by convention.

Now let $\theta_j = \{\mathbf{v}_j\} + [\mathbf{v}_{j+1}]$. For a list $\mathbf{v}$ of lists of $k-2$ or $k-1$ elements of M, such that the sum of the lengths of the $\mathbf{v}_j$ is $n-k$, and for a list X of k elements of M, let $\mathbf{u}[\mathbf{v}, X]$ denote the list $\mathbf{v}_1 X_1 \ldots \mathbf{v}_k X_k \in M^n$. Let $\|[\mathbf{v}, X]\| = \prod_1^k |\mathbf{v}_j|(X_j + \theta_j)$, so that $\|[\mathbf{v}, X]\| = |\mathbf{u}[\mathbf{v}, x]|(1+O(n\gamma^k))$. Let

$$B_{M,n,\mathbf{v}}(s) = \frac{\sum_{X\in M^k} \|[\mathbf{v}, X]\|^{-s} \log\|[\mathbf{v}, X]\|}{\sum_{X\in M^k} \|[\mathbf{v}, X]\|^{-s}}.$$

That is, $B_{M,n,\mathbf{v}}(s)$ is the expected value of $\log|\mathbf{u}[\mathbf{v}, X]|$ when the probability of X is given by $\|\mathbf{u}[\mathbf{v}, X]\|^{-s}/\sum_{Y\in M^k}\|\mathbf{u}[\mathbf{v}, Y]\|^{-s}$.

If we cut a probability space into segments, and sum the weighted square departure from the segment mean instead of the overall mean, and then take a weighted average of these variances, we shall arrive at a number that is in general less than the overall variance, agreeing if and only if the segment means are all equal. Thus

$$\lambda_M^{-n}(s) \sum_{\mathbf{u}\in M^n} |\mathbf{u}|^{-s} g([\mathbf{u}])(\log|\mathbf{u}| - B_{M,n})^2$$

$$= (1+O(n\gamma^k))\lambda_M^{-n}(s) \sum_{\mathbf{v}\in\mathcal{V}(M,n)} \sum_{X\in M^k} \|[\mathbf{v}, X]\|^{-s} g([\mathbf{v}_1])(\log\|[\mathbf{v}, X]\| - B_{M,n})^2$$

$$\geq \lambda_M^{-n}(s) \sum_{\mathbf{v}\in\mathcal{V}(M,n)} \sum_{X\in M^n} \|[\mathbf{v}, X]\|^{-s} g([\mathbf{v}_1])(\log\|[\mathbf{v}, X]\| - B_{M,n,\mathbf{v}})^2.$$

But this last is

$$\lambda_M^{-n}(s) \sum_{\mathbf{v}\in\mathcal{V}(M,n)} g([\mathbf{v}_1]) \prod_1^k |\mathbf{v}_j|^{-s}$$

$$\cdot \sum_{X\in M^k} \prod_1^k (X_j + \theta_j(\mathbf{v}))^{-s} \left(\sum_1^k (\log(X_j + \theta_j) - C_\mathbf{v}\right)^2,$$

where $C_{\mathbf{v}} = B_{M,n,\mathbf{v}} - \sum_1^k \log|\mathbf{v}_j|$. Now a little calculation shows that also

$$C_{\mathbf{v}} = \left(\frac{\sum_{X \in M^k} \prod_1^k (X_j + \theta_j(\mathbf{v}))^{-s} \sum_1^k \log(X_j + \theta_j(\mathbf{v}))}{\sum_{X \in M^k} \prod_1^k (X_j + \theta_j)^{-s}} \right).$$

Next, we observe that

$$\lambda_M^{-n}(s) \sum_{\mathbf{v},X} g([\mathbf{v}_1]) \|[\mathbf{v}, X]\|^{-s} \left(\sum_1^k \log(X_j + \theta_j) - C_{\mathbf{v}} \right)^2$$

$$= \lambda_M^{-n}(s) \sum_{\mathbf{v},X} g([\mathbf{v}_1]) \prod_1^k |\mathbf{v}_j|^{-s}$$

$$\cdot \frac{\sum_{X \in M^k} \prod_1^k (X_j + \theta_j)^{-s} \left(\sum_1^k \log(X_j + \theta_j) - C_{\mathbf{v}} \right)^2}{\sum_{X \in M^k} \prod_1^k (X_j + \theta_j)^{-s}}.$$

Now this is a weighted sum of the variances of random variable sums of the form $\sum_1^k \log(X_j + \theta_j)$, taken over the various $\mathbf{v} \in \mathcal{V}(n)$, where for each $\mathbf{v} \in \mathcal{V}(n)$, the random variables $X_1, \ldots, X_k$ (which implicitly depend also on $\mathbf{v}$, M, and s) are functions on the discrete probability space $\mathcal{U}_{\mathbf{v}}$ with event space M^k, ($k = \lfloor \sqrt{n} \rfloor$), and assigning to $x \in M^k$ the probability $p_{\mathbf{v},M,s}(x) = \alpha_{\mathbf{v},M,s}(x)/\sum_{\mathbf{y} \in M^k} \alpha_{\mathbf{v},M,s}(\mathbf{y})$, where $\alpha_{\mathbf{v},M,s}(x) = \prod_1^k (X_j + \theta_j(\mathbf{v}))^{-s}$. The weights total $1 + O(n\gamma^k)$. For each $\mathbf{v}$, the X_j are independent, and the variance of each X_j is bounded below, independent of $\mathbf{v}$ and j, though not of s or M. Thus, for each $\mathbf{v}$, the variance of the sum is comparable to k, and so also is the weighted average. Therefore,

$$\lambda_M^{-n}(s) \sum_{\mathbf{u} \in M^n} |\mathbf{u}|^{-s} (\log |\mathbf{u}| - B_{M,n})^2 \gg n^{1/2}.$$

From this, together with our earlier bounds for the error terms linking this sum to the second derivative of $\log \lambda_M(s)$, it follows that this second derivative is strictly positive on the entire interval (α_M, ∞), as claimed. □

Note that we have, in a somewhat roundabout way, proved that $\sum_{\mathbf{u},\mathbf{v} \in M^n} |\mathbf{u}|^{-s} (\log |\mathbf{u}| - \log |\mathbf{v}|)^2 \gg n\lambda_M^n(s)$, just as would have been the case had the 'digits' u_k been entirely uncorrelated in the sums in question.

9.7.1 Applications of strict log convexity

Take a random number X in $[0,1]$, using the usual measure $m[0,x] := x$. Let $P_r[X]$ and $Q_r[X]$ denote the numerator and denominator of the continued fraction expansion of X taken to r terms. How is $\log Q_r$ distributed? What if we restrict attention to those X for which the first r digits belong to a fixed set M. What is the conditional distribution of $\log Q_r$?

These questions amount to asking for the asymptotics of the measures

$$\mu_{M,r} := \frac{\sum_{\mathbf{v}\in M^r}|\mathbf{v}|^{-2}(1+\{\mathbf{v}\})^{-1}\delta(\log|\mathbf{v}|)}{\sum_{\mathbf{v}\in M^r}|\mathbf{v}|^{-2}(1+\{\mathbf{v}\})^{-1}},$$

where $\delta(z)$ denotes a unit point mass at z and where

$$\{\mathbf{v}\} = \{(v_1, v_2, \ldots v_r)\} := [v_r, v_{r-1}, \ldots v_1].$$

The moment generating function, or characteristic function, of this probability measure is

$$\frac{\sum_{\mathbf{v}\in M^r}|\mathbf{v}|^{-2+s}(1+\{\mathbf{v}\})^{-1}}{\sum_{\mathbf{v}\in M^r}|\mathbf{v}|^{-2}(1+\{\mathbf{v}\})^{-1}}.$$

There is a theorem due to Hwang (cited in [Va1]):

Theorem 9.7 *If Ω_r is a sequence of random variables for which the moment generating functions*

$$F_r(s) := E[s^{\Omega_r}] = \exp[V_r(s) \cdot (1 + O(1/w_r))],$$

with the implicit constant uniform in a complex disk about the origin, if $v_r(s) = \phi_r U(s) + V(s)$ with $U(s)$ and $V(s)$ analytic in that disk, if $U(0) = V(0) = 0$, if $U''(0) \neq 0$, and if ϕ_r and w_r tend to infinity as $r \to \infty$, then with the notation $\mu_r := \phi_r U'(0)$ and $\sigma_r^2 := \phi_r U''(0)$,

$$E[\Omega_r] = \mu_r + V'(0) + O(1/w_r)$$

and

$$\mathrm{Var}[\Omega_r] = \sigma_r^2 + V''(0) + O(1/w_r).$$

Furthermore, the distribution of Ω_r is asymptotically Gaussian, with distribution

$$P_r = \mathrm{prob}\left[\frac{\log Q_r - A_r}{\sqrt{B_r}} < x\right] \to \Phi(X) + O(1/S_r)$$

where $A_r = E[\log Q_r]$, $B_r = \mathrm{Var}[\log Q_r]$, and $S_r = \min[\sqrt{\phi_r}, W_r]$.

We shall now need some of the details of the perturbation theory by which we establish the existence of analytic continuations of $G_M(s,t)$ in s, in neighborhoods of real $s = \sigma$. Our application involves the particular case $\sigma = 2$ but the ideas would work more generally.

Let $P_{s,M}$ be the (analytic) projection operator given by

$$P_s[f](t) := \frac{1}{2\pi i} \oint [((1+\xi)I - G_{M,s})^{-1}[f]](t)\, d\xi.$$

We set $Q_{M,s} := I - P_{M,s}$ and observe that $P_{M,s}$, $Q_{M,s}$ and $G_{M,s}$ all commute. We also set $N_{M,s} := Q_{M,s} H_B$.

This gives us the following:

(i) $G_{M,s}[g_M(s,t)] = \lambda_M(s) g_M(s,t)$
(ii) $\lambda_M(s) I - G_{M,s} : H_B \to N_{M,s}$
(iii) $((\lambda_M(s)I - G_{M,s})_{N_{M,s}})^{-1}$ exists and is bounded.
(iv) $\lambda'_M(s) = (P_{M,s} G'_{M,s}[g_M(s,t)](0)/g_M(s,0)$
(v) $\|Q^r_{M,s}\| \ll (1-\epsilon)^r \lambda^r_M(s)$

this last, uniformly over s in some neighborhood of $s = \sigma$, with some fixed ϵ depending only on M and σ.

We must adapt a slightly different definition of $g_{M,s}$ than our usual normalization which was to set $g(0) = 1$. This time, we choose the normalization

$$g_M(s,t) := P_{s,M} g_M(2,t).$$

(Indeed, we had no definition of $g_M(s,t)$ for complex s.) With this terminology, we have

$$G_{M,s}[1/(1+t)] = \lambda_M(s) P_{M,s}[1/(1+t)] g_M(s,t) + G_{M,s}(t) Q_{M,s}[1/(1+t)].$$

From this it follows that the conditional moment generating function for the random variable $\Omega_r[X] = \log[Q_r[X]]$, where X has uniform distribution on $[0,1]$, satisfies

$$MGF[s] = \frac{\sum_{\mathbf{v} \in M^r} |\mathbf{v}|^{-2+s}(1+\{\mathbf{v}\})^{-1}}{\sum_{\mathbf{v} \in M^r} |\mathbf{v}|^{-2}(1+\{\mathbf{v}\})^{-1}}$$
$$= \frac{\lambda^r_M(2-s) P_{2-s,M}[1/(1+t)](0) g_M(2-s,0)}{\lambda^r_M(2) P_{2,M}[1/1(+t)](0) g_M(2,0)} + O((1-\epsilon)^r),$$

an estimate which fits the requirements of Hwang's theorem, with $\phi_r = r$, $U'(0) = (-\log \lambda_M)'(2))$, and $U''(0) = (\log \lambda_M)''(2)$.

In Hwang's theorem, then, we have

(i) $U(s) = \log[\lambda_M(2-s)/\lambda_M(2)]$

(ii) $V(s) := \log P_{M,2-s}[1/(1+t)](0) - \log P_{M,2}[1/(1+t)](0) + \log[g_M(2-s,0)/g_M(2,0)]$.

Thus, the conditional distribution of $\log Q_r[X]$ given that the first r digits of the continued fraction of X are in M, is asymptotically Gaussian, with mean asymptotic to $-r\lambda'_M(2)/\lambda_M(2)$ and variance asymptotic to $r \log \lambda''_M(2)$. In particular, if $M = \mathbb{N}$, this says that the mean of $\log Q_r$ is asymptotic to $(\pi^2/(12 \log 2))r \approx 1.18656911041562545$ and the variance to approximately 0.004826620315037090. (There is no known connection between the value of the second derivative of $\log \lambda_{\mathbb{Z}^+}(s)$ at $s = 2$ and familiar constants. The importance of this quantity to questions of continued fractions has prompted interest in it, and Loïck Lhote has studied the related quantity and given a polynomial time algorithm for computing it. This related quantity gives the scaling coefficient in the variance in the distribution of the number of steps in the Euclidean algorithm taken over pairs (u, v) with $0 \leq u < v \leq x$. This variance is comparable to $\log x$.)

The French school, we may call it, has taken great strides in the past decade, and has developed a distinctive approach integrating functional analysis, complex analysis, and the perspective of dynamical systems, into the study of continued fractions and related topics. This is the topic of our next chapter.

Chapter 10

The Generating Function Method

Several continued-fraction style algorithms have been studied in recent years with great success by a loose team of investigators, for the most part French, and in particular by B. Vallée of the University of Caen, using the perspective of dynamical systems, generating functions, and complex and functional analysis. Here, we concentrate on their results and methods as they apply to the standard continued fraction algorithm. Everything in this chapter comes from their work.

The various continued fraction algorithms all fall within the scope of *dynamical systems*, that is, systems in which one has a compact metric space termed the *state space* X, and an endomorphism T of that state space, termed the *shift mapping*. The *trajectory* of a point $x \in X$ is the sequence of iterates $(x, Tx, T^2x \ldots)$. Here, we take $X = [0,1]$. The textbook example of a dynamical system is $T_2 : x \to 2x - \lfloor 2x \rfloor = \{2x\}$. The basic dynamical system associated with continued fractions involves the mapping $T : x \to 1/x - \lfloor 1/x \rfloor = \{1/x\}$. Both are of course piecewise continuous.

In the case of T_2, there are two inverse branches, $h_1(x) = x/2$ and $h_2(x) = (x+1)/2$, and there are 2^r inverse branches of depth r, corresponding to the 2^r elements of $\{h_1, h_2\}^r$. The images of $[0,1]$ under these mappings constitute the dyadic partition of $[0,1]$ into the 2^r subintervals $[2^{-r}(k-1), 2^{-r}k]$, $1 \le k \le 2^r$.

In the case of T, there are countably many inverse branches $h_k(x) = 1/(k+x)$, and the inverse branches of depth r have the form $h = h_1 \circ \ldots \circ h_r$, and are linear fractional mappings of $[0,1]$ into intervals $u_h = \{h(x) : 0 \le x \le 1\}$ which again form a partition of $[0,1]$.

The rth iterate under T of x has the form $T^r x = h_r \circ \cdots \circ h_1 x$, and one studies the sequence $(h_1, h_2, \ldots)$ of h's that unfold from x, or what is essentially the same, the sequence of integers by which these h were indexed.

This sequence in effect encodes x.

In the case of T_2, we get the binary decimal expansion of x, and it terminates if and only if x is a dyadic rational. In the case of the continued fraction mapping T, if x is rational, we get a finite sequence, while if x is irrational, it is infinite; the sequence of indexing integers is just the sequence of continued fraction digits.

10.1 Entropy

The *entropy* of a partition of $[0,1]$ into a set U of intervals is defined to be $-\sum_{u \in U} |u| \log |u|$; thus, the entropy of the rth order dyadic partition of $[0,1]$ is $r \log 2$. The entropy of the *dynamical system* D with partitions of depth r indexed by the inverse rth order branch set $\mathcal{H}_D^r$ is defined to be

$$h_D = -\lim_{r \to \infty} \frac{1}{r} \sum_{h \in \mathcal{H}_D^r} |u_h| \log |u_h|.$$

According to a classical formula of Rohlin [V3], h_D is also given by

$$h_D = \int_0^1 \log |T_D'(x)| \psi_D(x)\, dx$$

where ψ_D is the probability density function for the measure with respect to which T_D is ergodic. (Or, equivalently, the density function which is an eigenfunction for the *transfer operator* $f \to \sum_{h \in \mathcal{H}_D} |h'(x)| f \circ h(x)$.)

The entropy of our first example is thus $\log 2$, both directly and from Rohlin's formula. For the continued fraction dynamical system, it is

$$h = -2 \int_0^1 \log(x) \psi(x)\, dx = \frac{-2}{\log 2} \int_0^1 \frac{\log x}{1+x}\, dx = \frac{\pi^2}{6 \log 2}.$$

Remark 10.1 *For the centered continued fraction algorithm, incidentally, the invariant density is $\psi_c(x) = \frac{1}{\log \phi} \left(\frac{1}{\phi+x} + \frac{1}{1+\phi-x} \right)$, where $\phi = (1+\sqrt{5})/2$, and it works out that the entropy is $\pi^2/(6 \log \phi)$. The speed of the algorithm, in terms of the number of steps in typically needs, is directly proportional to the entropy, so the centered algorithm is faster by a factor of $(\log 2)/(\log \phi)$ than the classical algorithm.*

10.2 Notation

For rational inputs $x = u/v$, let $l(u/v)$ and $l(u,v)$ both denote the number of steps of the form $x \to T(x)$ it takes to carry x to 0. Let

$$\Omega = \{(u,v) : 0 \le u \le v \text{ and } \gcd(u,v) = 1\},$$
$$\Omega(x) = \{(u,v) : (u,v) \in \Omega \text{ and } v \le x\}.$$

Let $\Omega^{[r]} = \{(u,v) : (u,v) \in \Omega \text{ and } l(u,v) = r\}$, and define $\Omega^{[r]}(x)$ in the obvious way.

The *expected value* of $l(u,v)$, taken over all pairs $(u,v) \in \Omega(x)$, is

$$E_x = \frac{1}{|\Omega(x)|} \sum_{r \ge 1} r \left| \Omega^{[r]}(x) \right|,$$

and the higher moments are

$$E^{[k]}(x) = \frac{1}{|\Omega(x)|} \sum_{r \ge 1} r^k \left| \Omega^{[r]}(x) \right|.$$

Let $S(s,w) = \sum_{r \ge 1} w^r \sum_{(u,v) \in \Omega^{[r]}} v^{-s}$. (This is the 'generating function' of the title of this section.) Let

$$G^{(k)}(s) = \sum_{n \ge 1} n^{-s} \sum_{(u,n) \in \Omega} (l(u,n))^k.$$

Then

$$G^{(k)}(s) = \frac{\partial^k}{\partial z^k} S(s, e^z)|_{z=0}.$$

Let $A_\infty(\mathcal{V})$ denote the space of functions from $\mathcal{V} = \{z : |z - 1| \le 3/2\}$ into $\mathbb{C}$ that are continuous on $\mathcal{V}$ and holomorphic on the interior. We equip $A_\infty(\mathcal{V})$ with the sup norm. Note that the image under any linear fractional mapping $h \in \mathcal{H}$ of $\mathcal{V}$ is a complex disk strictly inside $\mathcal{V}$, and carries the real interval $[0,1]$ into its own (real) interior. $A_\infty(\mathcal{V})$ is a Banach space.

Let H_s and F_s denote linear operators given by

$$H_s[f](x) = \sum_{n=1}^\infty (n+x)^{-s} f(1/(n+x)),$$

$$F_s[f])(x) = \sum_{n=2}^\infty (n+x)^{-s} f(1/(n+x)).$$

For $\Re s > 1$, H_s maps $A_\infty(\mathcal{V})$ into $A_\infty(\mathcal{V})$. Furthermore, H_s is compact [Shp] and *nuclear of order zero* [V3]. In consequence, they have discrete spectrum apart from a (possible) accumulation point at 0.

Furthermore, for real $s > 1$, H_s is a *positive operator*, that is, if f is positive on $[0, 1]$ then so is $H_s f$. It enjoys a stronger condition along the same lines, [V3], and in consequence, for $s > 1$, there is a unique dominant eigenvalue $\lambda(s)$, and a corresponding (positive) eigenfunction ψ_s, also unique up to scaling. We scale so that $\int_0^1 \psi_s(x)\,dx = 1$, and we note that ψ_2, call it just ψ, is our old friend the Gauss density $\psi(x) = 1/((1+x)\log 2)$, while $\lambda(2) = 1$. It turns out that $\lambda(s)$ is also given as $\lim_{r\to\infty} \sum_{h\in\mathcal{H}^r} |h|^{-s}$, where $|h|$ is the denominator of $h(0)$, or equivalently, $|h| = \sqrt{h'(x)}|_{x=0}$.

Finally, let $(I - wH_s)^{-1}[f](x)$ be the mapping that takes $f \in \mathcal{X}$ to that $F \in \mathcal{X}$, if it exists, such that $(I - wH_s)F = f$. For $|w|$ sufficiently small, $(I - wH_s)^{-1}[f](x) = \sum_{k=0}^{\infty} w^k (H_s^k[f])(x)$. The existence of an inverse in other circumstances is not obvious, but perturbation theory is relevant. At $(s, w) = (2, 1)$, $(I - wH_s)$ is singular, with a null space consisting of the multiples of $1/(1 + x)$, and a complementary invariant subspace consisting of all functions f so that $\int_0^1 f(x)\,dx = 0$. This fact constitutes a kind of pivot for the functional analysis and perturbation theory that figures prominently in this approach.

10.3 A Sampling of Results

Out of all this, Vallée distills the following result, here specialized to the classical continued fraction algorithm.

Theorem 10.1 *Let p be a positive integer, and let $f \in A_\infty(\mathcal{V})$, with $f > 0$ on $\mathcal{V} \cap \mathbb{R}$. Then $I - H_s$ is an invertible operator in $A_\infty(\mathcal{V})$, and for $x \in \mathcal{V} \cap \mathbb{R}$, for s in a neighborhood of 2,*

$$(I - H_s)^{-p}[f](x) \approx \frac{1}{(s-2)^p}\left(\frac{2}{h}\right)^p \psi(x) \int_0^1 f(t)\,dt.$$

Furthermore, the function $s \to \lambda(s)$ is strictly decreasing on the real interval $(1, \infty)$, while for $s = \sigma + it$ with $\sigma > 1$ and $t \in \mathbb{R}$, the spectral radius of H_s is strictly less than $\lambda(\sigma)$ except when $t = 0$.

There is a Tauberian theorem due to Delange which now applies. This theorem reads as follows:

Theorem 10.2 Let $\sigma > 0$. Let $F(s) = \sum_0^\infty a_n s^n$ be a Dirichlet series with nonnegative coefficients (a_n) that converges for $\Re(s) > \sigma$. Assume further that $F(s)$ extends to an analytic function on a domain that includes all of the vertical line $\Re(s) = \sigma$ with the exception of $s = \sigma$, and finally, that for some $\gamma > 0$, $F(s)$ has the form $F(s) = A(s)(s-\sigma)^{-\gamma-1} + C(s)$ where A and C are analytic at σ with $A(\sigma) \neq 0$. Then as $x \to \infty$,

$$\sum_{n \leq x} a_n = \frac{A(\sigma)}{\sigma \Gamma(\sigma+1)} x^\sigma \log^\gamma x (1 + o(1)).$$

With $w = e^z$, we have

$$S(s, e^z) = \sum_{(u,v) \in \Omega} v^{-s} e^{z\, l(u,v)}.$$

Thus $G^{(k)}(s) = \frac{\partial^k}{\partial z^k} S(s, e^z)|_{z=0}$. But

$$S(s, e^z) = e^z F_s[1](0) + e^{2z}(F_s \circ (I - e^z H_s)^{-1} \circ H_s)[1](0).$$

Differentiating this k times with respect to z gives an expression in which the main term is

$$(F_s \circ (I - H_s)^{-k-1} \circ H_s^{k+1})[1](0).$$

It now follows from the Tauberian theorem that the kth moment of the number of steps in the Euclidean algorithm is asymptotic to $(2/h)^k \log^k x = (12 \log 2/\pi^2)^k \log^k x$. Similar results are obtained for the centered algorithm and other variants; in particular, the average number of steps taken by the centered Euclidean algorithm, on input pairs (u, v) with $u < v \leq x$, is asymptotic to $(12 \log \phi/\pi^2)$.

From these estimates, we are led to expect that the distribution of the number of steps $l(u,v)$ on Ω_x is bunched near $(12 \log 2/\pi^2) \log x$. It turns out that not only is this the case, but that the distribution is in fact asymptotically Gaussian, or normal, with a variance also comparable to $\log x$. The proportionality constant for the variance is related to the *second* derivative of $\log \lambda(s)$. This result was first established, but with a weak error term and only for the classical Euclidean algorithm, in [He4]. More recently, Baladi and Vallée have established that a wide range of performance statistics for a wide range of Euclidean algorithms exhibit Gaussian distributions, and their results also provide essentially best-possible error estimates. We now sketch their results and give some indication of their approach, again limiting ourselves to the case of the number of steps taken by the classical

Euclidean algorithm. The following two results are the specialization to this case of theorems in [BV].

Let $\Lambda(s) = \log \lambda(2s)$, so that $\Lambda(1) = 0$. Let $\mu = 2/|\Lambda'(1)| = 12 \log 2/\pi^2$, and let $\delta^2 = 2\Lambda''(1)/|\Lambda'(1)|^3$.

(The constant δ^2 is also referred to in the literature as γ_H or "Hensley's constant".) It has been shown to be computable in polynomial time as a function of the number of digits required, and computed rigorously to some 20 places, by Loïcke Lhote, who obtained the value

$$\gamma_H = 0.51606240889999180681.$$

Theorem 10.3 *(Baladi and Vallée) Uniformly for $y \in \mathbb{R}$, as $x \to \infty$,*

$$\frac{1}{|\Omega_x|} \# \left\{ (u,v) \in \Omega_x : \frac{l(u,v) - \mu \log x}{\delta \sqrt{\log x}} \leq y \right\}$$
$$= \frac{1}{\sqrt{2\pi}} \int_{-\infty}^{y} e^{-t^2/2} \, dt + O\left(\frac{1}{\sqrt{\log x}}\right).$$

Their second main result gives an estimate for how many pairs $(u,v) \in \Omega_x$ there are for which $l(u,v)$ takes a given value.

Theorem 10.4 *Uniformly over $y \in \mathbb{R}$, as $x \to \infty$,*

$$\frac{1}{|\Omega_x|} \# \left\{ (u,v) : \mu \log x + \delta y \sqrt{\log x} - 1/2 < l(u,v) \right.$$
$$\left. \leq \mu \log x + \delta y \sqrt{\log x} + 1/2 \right\} = \frac{1}{\delta \sqrt{2\pi \log x}} e^{-y^2/2} + O\left(\frac{1}{\log x}\right).$$

The proof depends heavily on a key innovation: the complex analysis is extended from treating just a neighborhood of the critical point ($s = 2$ or $s = 1$, depending on how we define λ or Λ), to treating at least a punctured half plane.

One defines again a linear operator $H_{s,w}$ acting on a space of functions, but this time, a family of topologically equivalent norms is used: $\|f\|_{1,t} = \sup |f| + \frac{1}{|t|} \sup |f'|$. Take

$$H_{s,w}[f](x) = e^w \sum_{n \geq 1} (n+x)^{-2s} f\left(\frac{1}{n+x}\right).$$

Their crucial technical result is this:

Theorem 10.5 *For any ξ with $0 < \xi < 1/5$, there is a (real) neighborhood O_ξ of $(1,0)$, and an $M > 0$, such that for all (complex) (s,w) satisfying $(\Re s, \Re w) \in O_\xi$ and $|t| \geq ((1+\sqrt{5})/2)^4$, $\|(I - H_{s,w})^{-1}\|_{1,t} \leq M|t|^\xi$.*

By means of this result, it becomes possible to carry through a calculation using residues and contours that deform from a vertical line.

Chapter 11

Conformal Iterated Function Systems

There is a body of work, due in large measure to Daniel Mauldin and Mariusz Urbański, that extends and generalizes continued fractions in a direction different from what we have seen so far. Our continued fraction Cantor sets E_M can be seen as instances of sets associated with *conformal iterated function systems*. These were introduced by Mauldin and Urbański in [MU1] and developed further in [MU2]. A *conformal iterated function system* involves a countable index set I with at least two elements, a compact set X of a Euclidean space, and a set S_I of injective contractions ϕ_j on X satisfying certain technical conditions. The restriction of these conditions to the case in which $X = [0,1]$ is that

(1) Uniform contraction: there must be a constant $c < 1$ such that for all $j \in I$ and all $x, y \in X$, $|\phi_j(x) - \phi_j(y)| \le c|x-y|$.
(2) Analytic extension: there must be an open interval V containing X such that all ϕ_j extend to monotone continuously differentiable injections of V into V.
(3) Bounded distortion: The bounded distortion property concerns not only the ϕ_j but their compositions. There must be a constant K such that, for all functions ϕ formed by composing a finite number of elements of S_I, (in any order and allowing repetition), and for all $x, y \in V$, $|\phi'(y)| \le K|\phi'(x)|$.

These properties are essentially satisfied by the system in which $I = M \subseteq \mathbb{Z}^+$ and ϕ_j is the function $x \to 1/(j+x)$; the snag is that if $1 \in M$, the uniform contraction condition applies only to compositions of two or more ϕ_j; this however turns out to be sufficient for all purposes.

The connection between this system and E_M is straightforward. Let M^* denote $\cup_{n=1}^{\infty} M^n$. Given a word $w \in M^{\infty}$, let $(w \mid n) \in M^*$ denote the truncation $(w_1 w_2 \ldots w_n)$ of w to its first n entries, and let $\phi_{w|n}$ denote the composition $\phi_1 \circ \cdots \circ \phi_n$. Then

$$E_M = \bigcup_{w \in M^\infty} \bigcap_{n=1}^{\infty} \phi_{w|n}(X).$$

(Note that $E_M = \cup_{j \in M} \phi_j(E_M)$. This condition, together with the stipulation that E_M be a nonempty subset of X, seems to characterize E_M.)

The fine structure of fractal sets can be looked at with tools other than *dimensions* of one sort or another. What about density? The story here is simple; Urbański proves that if $M \neq \mathbb{Z}^+$, then E_M is nowhere dense. But there are other ways to distinguish between sets that have little holes only, and sets that have relatively big holes at all scales.

A set $E \subseteq X$ is *porous* if there exists $c > 0$ such that for all $\epsilon > 0$ and all $x \in X$ there exists $y \in X$ such that the open ball about y of radius $c\epsilon$ is disjoint from E yet contained in the open ball about x of radius ϵ. Urbański proves [U] that

Theorem 11.1 *E_M is porous if and only if there exists $0 < \theta < 1$ and $x \geq 0$ such that for every $j \in M$ and every $x \leq p \leq j$, either the integer interval $[j - p, j]$ or the integer interval $[j, j + p]$ contains a run of at least $\lfloor \theta p \rfloor$ consecutive non-elements of M.*

When j and θ are small, $\theta p < 1$ so nothing is required; thus, a corollary is that for all finite M, E_M is porous.

Our next result from the work of Mauldin and Urbański has to do with measures associated in some cases with E_M. To describe the cases in which these measures exist, one needs the *pressure function* P_M of a system.

The general definition of P_I is that it is $\lim_{n \to \infty} \log \left(\sum_{w \in I^n} |\phi'_w|^s \right)$. In the particular case of a continued fraction conformal iterated function system, $P_M(s) = \log \lambda_M(2s)$. They prove that this P_M is strictly decreasing on the interval where it is finite, that for all M, $\dim E_M$ is the infimum of the set of all s such that $P_M(s) \leq 0$, and that if there exists an s such that $P_M(s) = 0$, or equivalently, $\lambda_M(2s) = 1$, then necessarily that s is the Hausdorff dimension.

When M is finite, such an s does exists. When M is infinite, there is the possibility that $\sum_{n \in M} n^{-2s}$ diverges for all $s < \alpha_M$, and converges for all $s \geq \alpha_M$, and $P_M(\alpha_M) < 0$, and then no such s exists. The Hausdorff dimension of E_M is then α_M, but we do not get the next nice feature: a natural measure associated with E_M and h_M.

When there exist s such that $P_M(s) > 0$ but finite, we do get such a measure. (We have already encountered this measure, which we termed $\mu_{M,\sigma}$, in connection with an integral representation of $g_{M,\sigma}(t)$.) Mauldin and Urbański prove that if h_M is the Hausdorff dimension of E_M, and if $P_M(h_M) = 1$, then there is a probability measure μ on $[0,1]$ such that if $j \neq k \in M$, then $\mu(1/(X+j) \cap 1/(X+k)) = 0$, and for all Borel sets A of $[0,1]$, and all $j \in M$,

$$\mu(1/(A+j)) = \int_A (j+t)^{-2h_M} d\mu.$$

To this we add that if g is the eigenfunction associated with the leading eigenvalue, (normalized so that $g(0) = 1$), of the operator $L_{M,2h_M}$, if $\mathbf{v} \in M^*$, and if $[[\mathbf{v}]]$ denotes the interval with endpoints $[\mathbf{v} + \min E_M]$, $[\mathbf{v} + \max E_M]$, then

$$\mu([[\mathbf{v}]]) = |\mathbf{v}|^{-2h_M} g(\{\mathbf{v}\}),$$

which also characterizes μ. Finally, there is a connection to the structure of L_{M,h_M} and $\mathcal{V}$. Let $\mathcal{N}$ be the set of all $f \in \mathcal{V}$ such that $L_{M,h_M}^n f \to 0$ as $n \to \infty$. Then $\mathcal{V} = \mathbb{R}g + \mathcal{N}$, and both $\mathbb{R}g$ and $\mathcal{N}$ are invariant under L_{M,h_M}. There is a projection π_M linear operator from $\mathcal{V}$ onto $\mathbb{R}g$ that satisfies $\pi_M \circ L_{M,h_M} = L_{M,h_M} \circ \pi_M = \pi_M$, and $\mathcal{N}$ is its kernel. If $c_M = \int g\, d\mu$, then this projection has the integral represent

$$\pi_M f = \left(c_M^{-1} \int f\, d\mu\right) g.$$

What sort of measure is this? It is supported on E_M. If we expand an interval about some x not in E_M, the measure of that interval may jump from zero to some positive value. If, on the other hand, we consider the measure of an interval $(x - \epsilon, x + \epsilon)$ about $x \in E_M$, then as a result of one of our characterizations of μ, it has positive measure because there is some $\mathbf{v} \in M^*$ such that $[[\mathbf{v}]]$ is contained in our interval. What happens if we expand the interval?

A measure ν on X is said to have the *doubling property* if there exists a constant $c > 0$ such that for almost all $x \in X$ (with respect to ν), and for all $0 < \epsilon < 1$,

$$\nu([x - 2\epsilon, x + 2\epsilon]) \leq c\nu([x - \epsilon, x + \epsilon]).$$

It is not terribly difficult to see that if M is finite then μ_M has this doubling property, so we sketch the proof before describing the results of Mauldin and Urbański concerning infinite sets M.

Fix a finite set M, and consider $x \in E_M$ and $\epsilon > 0$. There is a longest $\mathbf{v} \in M^*$ such that $x \in [[\mathbf{v}]]$ and $[[\mathbf{v}]] \subseteq (x - \epsilon, x + \epsilon)$. For this $\mathbf{v}$, $|\mathbf{v}|$ will be comparable to $\epsilon^{-1/2}$, and $\mu([[\mathbf{v}]]) \gg \epsilon^{h_M}$. On the other hand, if $y \notin [[\mathbf{v}]]$, but $y \in E_M$ and $|x-y| < 2\epsilon$, then there exists $\mathbf{u} \in M^*$ such that $x = [\mathbf{u}j + \tau$ and $y = [\mathbf{u}k + t]$, with $t, \tau \in E_M$ and $j \neq k \in M$. Since M is finite, it follows that

$$|x - y| \gg |\mathbf{u}|^{-2}(1 + \min E_M - \max E_M)$$

and taking $\mathbf{w}$ to be the shortest of these $\mathbf{u}$'s, over all $y \in E_M \cap (x-2\epsilon, x+2\epsilon)$, we conclude that $|\mathbf{w}|^{-2} \ll \epsilon$ and $(x - 2\epsilon, x + 2\epsilon) \cap E_M \subseteq [[\mathbf{w}]]$. Thus $\mu((x - 2\epsilon, 2 + \epsilon)) = \mu((x - 2\epsilon, x + 2\epsilon) \cap E_M) \leq \mu([[\mathbf{w}]]) \ll \epsilon^{h_M}$.

In [MU2], it is shown that if M satisfies any one of a list of technical conditions, then μ_M exists and satisfies the doubling property. The simplest of these conditions is that $1 \notin M$. Examples are also given of M for which μ_M exists but does not satisfy the doubling property.

Chapter 12

Convergence of Continued Fractions

Simple continued fractions $\beta = [b_1, b_2, \ldots] = 1/(b_1 + 1/(b_2 + \ldots))$ in which the partial quotients $b_1, b_2 \ldots$ are positive integers, always converge, and there are good, simple, explicit bounds for $|[b_1, b_2, \ldots, b_r] - \beta|$. If, however, we allow the b_j to depart from these conditions, widening the domain of possible values to, say, the positive real numbers, then convergence is not assured. There are also continued fractions of the form

$$\alpha = \cfrac{a_1}{b_1 + \cfrac{a_2}{b_2 + \cfrac{a_3}{b_3 + \ldots}}} = [(a_1, b_1), (a_2, b_2), (a_3, b_3), \ldots],$$

in which the a_j need not all be one, and again, convergence is problematic. There are interesting and important continued fractions of this type; for instance, [JT]

$$\tanh(z) = \cfrac{z}{1 + \cfrac{z^2}{3 + \cfrac{z^2}{5 + \cfrac{z^2}{7 + \ldots}}}}$$

is a continued fraction which reduces to a simple continued fraction with positive integer partial quotients when $z = 1$. But to understand the restricted identity, one must understand the more general identity, and for that, we must have some control of issues of convergence. The classic references for this topic are Perron's *Kettenbrüchen* [Pe], and more recently, Jones and Thron's *Continued Fractions: Analytic Theory and Applications*, [JT].

12.1 Some General Results and Techniques

Let

$$K[(a_1, b_1), \ldots, (a_r, b_r)] = K_r = [(a_1, b_1), \ldots, (a_r, b_r)] = \cfrac{a_1}{b_1 + \cfrac{a_2}{b_2 \cdot \cdot \cdot + \cfrac{a_r}{b_r}}}$$

denote the finite continued fraction 'convergent' at depth r, associated with a sequence of a's and b's. Let A_r and B_r be the formal numerator and denominator of the rth convergent, so that $A_0 = 0$, $A_1 = a_1$, $B_0 = 1$, $B_1 = b_1$, and $A_n = b_n A_{n-1} + a_n A_{n-2}$, $B_n = b_n B_{n-1} + a_n B_{n-2}$. Let $w_n = B_n/B_{n-1}$.

The exact form of a continued fraction is to some extent a matter of presentation. A continued fraction can be rewritten in such a way that the convergents are unchanged, for instance,

$$\cfrac{a_1}{b_1 + \cfrac{a_2}{b_2 + \cfrac{a_3}{b_3 + \ldots}}} = \cfrac{1}{(b_1/a_1) + \cfrac{1}{(a_1 b_2/a_2) + \cfrac{1}{(a_2 b_3/a_1 a_3) + \ldots}}}.$$

We begin with an old result of Stern and Stolz.

Theorem 12.1 *If $\sum_1^\infty |b_j| < \infty$, then the sequence (K_r), given by*

$$K_r = \cfrac{1}{b_1 + \cfrac{1}{b_2 \cdot \cdot \cdot + \cfrac{1}{b_r}}}$$

fails to converge.

Proof. We give what may be a new proof, which yields en passant an explicit lower bound for the difference $|K_{2r} - K_{2r+1}|$:

$$|K_{2r} - K_{2r+1}| \geq \exp\left(-\sum_1^{2r+1} |b_j|\right).$$

We use the spherical representation of $\mathbb{C}$, and associate to any $z \in \mathbb{C}$ a point $p(z) \in S \subseteq \mathbb{R}^3$, where S is the sphere $x_1^2 + x_2^2 + (x_3 - 1/2)^2 = 1/4$ about $(0,0,1/2)$ of radius $1/2$: For finite $z = x + iy$ with x and y real,

$$p(x + iy) = \frac{(x, y, x^2 + y^2)}{1 + x^2 + y^2}$$

is the point in S, other than $(0,0,1)$, at which the line joining $(x,y,0)$ to $(0,0,1)$ meets S. The natural completion of this representation is to associate the complex point at infinity with the unused point $(0,0,1)$ of S, and we do this. Next, let $\rho[z_1, z_2] = \|p(z_1) - p(z_2)\|$ denote the Euclidean norm in $\mathbb{R}^3$ of $p(z_1) - p(z_2)$. Then ρ is a metric on $\mathbb{C}$, and $|z_1 - z_2| \geq \rho[z_1, z_2]$, with equality if and only if $z_1 = z_2$.

The key step in the proof is the following

Lemma 12.1 *If u, v, and w are complex numbers, then*

$$|\log \rho[u + z, v + z] - \log \rho[u, v]| \leq |z|.$$

Proof. We use an identity which is, while simple enough to state, rather a chore to verify. For real t,

$$\frac{d}{dt} \log \rho[u + tz, v + tz] = -\Re\left[\frac{\bar{z}(u + tz)}{1 + |u + tz|^2} + \frac{\bar{z}(v + tz)}{1 + |v + tz|^2}\right].$$

It is sufficient for this claim, that if u_1, u_2, v_1, v_2, and z_1, z_2 are the real and complex parts of u, v, and z respectively, then

$$\frac{d}{dt}\left[\frac{\rho^2[u + tz, v + tz]}{\rho^2[u, v]}\right]\bigg|_{t=0} = -2\left(\frac{u_1 z_1 + u_2 z_2}{1 + u_1^2 + u_2^2} + \frac{v_1 z_1 + v_2 z_2}{1 + v_1^2 + v_2^2}\right).$$

Verifying this last statement is a matter of multiplying everything out and cancelling, which is a fit task for a computer algebra system. We choose *Mathematica*. In actual use, intermediate steps for the code below may be viewed by suppressing the semicolons, and instead *executing* the lines, one by one, by entering on the keyboard a *Return* command in place of each semicolon.

```
p[{x_,y_}]:={x,y,x^2+y^2}/(1+x^2+y^2);
rhosqrd[u_,v_]:=(p[u]-p[v]).(p[u]-p[v]);
expn=rhosqrd[t{z1,z2}+{u1,u2},t{z1,z2}+{v1,v2}];
expnprime=(D[expn,t])/.t->0; expn0=expn/.t->0;
logprime=expnprime/expn0
```

and then

```
claimed=-2({z1,z2}.{u1,u2}/(1+u1^2+u2^2)+
{z1,z2}.{v1,v2}/(1+v1^2+v2^2));
```

and finally,

```
Simplify[logprime-claimed]
```

and when zero is returned, we have verified the identity. The utility of this identity comes from the fact that in general, for real z_1, z_2, w_1, and w_2, we have $(z1, z2) \cdot (w_1, w_2) \leq \sqrt{z_1^2 + z_2^2}\sqrt{w_1^2 + w_2^2}$. Thus in the right hand side of the identity we just verified, the expression to be multiplied by -2 is no greater than $\frac{1}{2}|z|$, so that the expression is no less than $-|z|$. As this holds for all t in $[0, 1]$, it follows by integration from zero to one that $|\log \rho[u + z, v + z] - \log \rho[u, v]| \leq |z|$ as claimed. □

Armed with this lemma, we now consider two finite lists, both depending on n and on the initial segment $(b_1, b_2, \ldots, b_{2n+1})$ of our sequence (b). We set $z_0[n] = 0$, $w_0[n] = \infty$, and then use the recurrence relations

$$z_k[n] = \frac{1}{b_{2n+2-k} + z_{k-1}[n]}, \quad w_k[n] = \frac{1}{b_{2n+2-k} + w_{k-1}[n]}$$

for $1 \leq k \leq 2n + 1$ with the result that $z_{2n+1}[n] = K_{2n+1}$ is the $2n + 1$st convergent of our continued fraction, while $w_{2n+1}[n] = K(2n)$ is the immediately preceding $2n$th convergent. Initially, $\rho[z_0[n], w_0[n]] = 1$. As k increases, the pair (z_{k-1}, w_{k-1}) is replaced first by $(z_{k-1} + b_{2n+2-k}, w_{k-1} + b_{2n+2-k})$, which has the effect of multiplying $\rho[z, w]$ by a factor of at least $\exp[-|b_{2n+2-k}|]$, followed by inversion of both z and w, which has no effect, as $\rho[z, w] = \rho[1/z, 1/w]$. Thus,

$$\rho[z_k, w_k] \geq \exp\left[-\sum_{j=1}^{k} b_{2n+2-j}\right]$$

and in particular, $\rho[K_{2n+1}, K_{2n}] \geq \exp\left[-\sum_1^{2n+1} |b_j|\right]$. □

See also Theorem 7.4 of [Be] for a recently discovered conceptual proof. There is a partial converse to the theorem of Stern and Stolz, due to Van Vleck. [VV].

Theorem 12.2 *Let* $(b) = (b_1, b_2, \ldots) \in \mathbb{C}^*$, $\epsilon > 0$, *and assume that for all* $n \geq 1$, $-\pi/2 + \epsilon < \arg[b_n] < \pi/2 - \epsilon$. *Let* $K_n = [b_1, b_2, \ldots, b_n]$ *be the convergent of depth* n *in the continued fraction* $1/(b_1+1/(b_2+1/(b_3+\ldots)))$. *Then*

(1) Both (K_{2n}) *and* (K_{2n+1}) *converge to finite complex values.*

(2) Each K_n is finite and nonzero, and $-\pi/2 + \epsilon < \arg[K_n] < \pi/2 - \epsilon$.
(3) There is a complex number K so that
$$\lim_{n\to\infty} K_{2n} = \lim_{n\to\infty} K_{2n+1} = K$$
if and only if $\sum |b_n| = \infty$.

Here, we give Jerry Lange's constructive proof of his recent strengthened version of this result [LJ].

Theorem 12.3 *Suppose (b_n) is a sequence of nonzero complex numbers within the wedge $V[\epsilon] = \{z \in \mathbb{C} \mid z \neq 0 \text{ and } |\arg(z)| \leq \pi/2 - \epsilon\}$. Let $f_n = [b_1, \ldots, b_n]$. Then*

(1) *(f_{2n}) and (f_{2n-1}) both converge to finite values, f_E and f_O say, in $V[\epsilon]$.*
(2) *All f_n are finite complex numbers in $V[\epsilon]$.*
(3) *(f_n) converges, that is, $f_E = f_O = f$, if and only if $\prod_{k=2}^\infty (1 + |b_k w_{k-1}| \sin \epsilon)$ diverges.*
(4) *$\prod_{k=2}^\infty (1 + |b_k w_{k-1}| \sin \epsilon)$ converges if and only if $\sum_1^\infty |b_k|$ diverges.*
(5) *If this sum diverges, then*
$$|f - f_{n-1}| \leq \frac{1}{\Re[B_n \overline{B_{n-1}}]} \leq \frac{1}{\Re[b_1] \prod_{k=2}^n (1 + |b_k w_{k-1}| \sin \epsilon)}.$$

Proof. We first prove the last item. Recall that if $M_0 = \begin{pmatrix} 1 & 0 \\ 0 & 1 \end{pmatrix}$ and if $M_n = M_{n-1} \begin{pmatrix} 0 & 1 \\ 1 & b_n \end{pmatrix}$ then $M_n = \begin{pmatrix} A_{n-1} & A_n \\ B_{n-1} & B_n \end{pmatrix}$ where

$$\frac{A_n}{B_n} = \cfrac{1}{b_1 + \cfrac{1}{b_2 + \cdots}} = [b_1, \ldots, b_n].$$

We now use an idea which has played a significant role in both the study of the more general analytic continued fractions currently under discussion, and in the dynamical analysis of algorithms related to continued fractions.

The linear fractional transformations on $\mathbb{C}$, given by $h_b : z \to 1/(b+z)$, are intimately connected to continued fractions. If we have a sequence $(b_1, b_2, \ldots)$ in mind, we let $t_n = h_{b_n}$. In general, a linear fractional transformation on $\mathbb{C}$ is a mapping $\phi : z \to (az+b)/(cz+d)$, where $\begin{vmatrix} a & b \\ c & d \end{vmatrix} \neq 0$.

The set of linear fractional transformations on $\mathbb{C}$, with the operation of function composition, is a group. The identity element is the identity function. Linear fractional transformations are meromorphic on $\mathbb{C}$ and map the set of lines and circles in $\mathbb{C}$ onto the set of lines and circles. Their specific relevance to continued fractions comes from the fact that if

$$T_{(b,n)} = T_{b_1,b_2,\ldots b_n} = h_{b_1} \circ h_{b_2} \circ \cdots \circ h_{b_n} = t_1 \circ \cdots \circ t_n$$

then

$$T_{(b,n)}(z) = \frac{A_n + zA_{n-1}}{B_n + zB_{n-1}}.$$

Let us fix (b), and abbreviate $T_{(b,n)}$ to T_n. Lange introduces the notation $f_n = \frac{A_n}{B_n} = T_n(0) = T_{n+1}(\infty)$. Now

$$\begin{vmatrix} A_{n-1} & A_n \\ B_{n-1} & B_n \end{vmatrix} = |M_n| = (-1)^n \text{ and } \begin{vmatrix} A_{n-2} & A_n \\ B_{n-2} & B_n \end{vmatrix} = (-1)^{n-1} b_n.$$

Lange sets $h_n = -T_n^{-1}(\infty) = \frac{B_n}{B_{n-1}} = 1/[b_n, \ldots, b_1]$. Observe that $H_0 = \infty$, $h_1 = b_1$, and for $n \geq 2$, $h_n = b_n + 1/h_{n-1}$, because $[b_n, \ldots, b_1] = 1/(b_n + [b_{n-1}, \ldots, b_1])$. Furthermore,

$$T_n(h_n) = \frac{1}{2}\left(\frac{A_n}{B_n} + \frac{A_{n-1}}{B_{n-1}}\right) = \frac{1}{2}(f_n + f_{n-1}).$$

Next, we define a finite list of subsets of $\mathbb{C}$. Let $0 < \theta < \pi/2$ and let V_0 be the closed right half plane of $\mathbb{C}$, that is,

$$V_0 = \{z \in \mathbb{C} \mid z = x + iy \text{ with } x \geq 0, y \in \mathbb{R}\}.$$

Then let

$$V_1 = t_1 V_0 = 1/(b_1 + V_0) = \{z \in \mathbb{C} \mid \exists u \in \mathbb{C}, z = t_1(u)\},$$
$$V_2 = t_1 t_2(V_0) = T_2 V_0, \ldots, V_n = [b_1, b_2, \ldots, b_n + V_0].$$

Then $V_0 \supseteq V_1 \supseteq V_2 \ldots$, and $f_m \in V_n$ for $m \geq n$. Thus

$$m \geq n \Rightarrow |f_m - f_n| \leq \text{diam} V_n.$$

This brings us to the crux of the matter: Why is V_n small?

Well known properties of linear fractional transformations on $\mathbb{C}$ give us the following start: V_n is a closed disk in $\mathbb{C}$ containing, on its boundary, the points $A_n/B_n = T_n(0)$, $(A_n + iA_{n-1})/(B_n + iB_{n-1}) = T_n(i)$, and $T_n(-i)$. A somewhat painful calculation, which we relegate to another *Mathematica* session, now establishes that the diameter of the circle through these three points is $1/\Re(B_n\overline{B_{n-1}})$.

A lower bound for $D_n = \Re(B_n\overline{B_{n-1}})$ will serve to bound $\operatorname{diam} V_n$, so we calculate $D_0 = 0$, $D_1 = \Re[b_1]$, and then

$$\begin{aligned} D_n &= \Re[B_n\overline{B_{n-1}}] = \Re[(b_nB_{n-1} + B_{n-2})\overline{B_{n-1}}] \\ &= D_{n-1} + \Re[b_n]|B_{n-1}|^2] = D_{n-1} + \Re[b_n]|h_{n-1}||B_{n-1}\overline{B_{n-2}}| \\ &\geq D_{n-1}(1 + \Re[b_n]|w_{n-1}|). \end{aligned}$$

Thus, for $n \geq 1$,

$$D_n \geq D_{n-1}(1 + \Re[b_n]|w_{n-1}|)$$

so that

$$\begin{aligned} D_n &\geq D_1 \prod_{k=2}^n (1 + \Re[b_k]|w_{k-1}|) \\ &= \Re[b_1] \prod_{k=2}^n (1 + \Re[b_k]|w_{k-1}|). \end{aligned}$$

This completes the proof of the last item in Lange's Van Vleck theorem. Now if $\sum |b_n|$ converges, then by the Stern-Stolz theorem, (f_n) diverges, and indeed, as we have seen, $\operatorname{diam} V_n \geq \exp[-\sum_1^\infty |b_k|]$, so that

$$\prod_{k=2}^\infty (1 + |b_kw_{k-1}|\sin\epsilon) \leq \frac{1}{\Re[b_1]} \exp\left(\sum_1^\infty |b_k|\right).$$

This proves one direction of the third item. In the other direction, if the product at issue converges, we set $d_k = b_kw_{k-1}$, and calculate that $(1+d_k = w_kw_{k-1})$, so that

$$b_{2n} = \frac{d_{2n}}{b_1}\frac{\prod_2^n(1 + d_{2k-2})}{\prod_2^n(1 + d_{2k-1})} \quad b_{2n+1} = b_1d_{2n+1}\frac{\prod_2^n(1 + d_{2k-1})}{\prod_1^n(1 + d_{2k})}.$$

Now $\prod_2^n(1 + |b_kw_{k-1}|\sin\epsilon)$ is presumed to converge, so $\sum |d_k|$ converges, so the products of the twin identities above for b_{2n} and b_{2n+1} converge, so that b_n is comparable to d_n in absolute value. Thus, $\sum |b_n|$ converges. This proves the other direction of the third claim in the theorem.

The second item is an immediate consequence of the fourth. This leaves the first item, or strictly speaking, that subcase in which $\sum |b_k|$ converges.

Now, let D_n denote the reciprocal of the diameter of V_n. Here, we calculate that

$$|f_n - f_{n-2}|\sin\epsilon = \left|\frac{b_n}{B_n B_{n-2}}\right|\sin\epsilon$$
$$\leq \frac{\Re[b_n]|b_{n-1}|^2}{|B_n B_{n-1}||B_{n-1} B_{n-2}|}$$
$$\leq \frac{D_n - D_{n-1}}{D_n D_{n-1}} = \left(\frac{1}{D_{n-1}} - \frac{1}{D_{n-2}}\right) \leq \left(\frac{1}{D_{n-2}} - \frac{1}{D_n}\right).$$

Since (D_n) is a monotone increasing sequence, the sequence of values of $(1/D_{n-2} - 1/D_n)$ converges. □

There are numerous other results on convergence of continued fractions. We give one of these, but without proof. The new classic in this realm is Jones and Thron's Continued Fractions: Analytic Theory and Applications. [JT].

Theorem 12.4 *The continued fraction $a_1/(1 + a_2/(1 + a_3/(1 + \ldots)))$ converges if and only if at least one of the series*

$$\sum_1^\infty \left|\frac{a_2 a_4 \ldots a_{2n}}{a_3 a_5 \ldots a_{2n+1}}\right|, \quad \sum_1^\infty \left|\frac{a_3 a_5 \ldots a_{2n+1}}{a_4 a_6 \ldots a_{2n+2}}\right|$$

diverges.

Finally, we mention again the work of Beardon, [Be], who has linked continued fractions to the theory of Möbius maps of $\mathbb{C} \cup \infty$ and their extensions to maps of hyperbolic 3-space $\mathbb{H}^3$, where they become isometries. This allows him to provide new proofs of classical theorems and to extend some of them to multiple-dimensional analogues of continued fractions.

12.2 Special Analytic Continued Fractions

One begins the study of analytic continued fractions with the observation that it is perfectly legitimate to consider fractions of the form $a_0 + b_1/(a_1 + b_2/(a_2 + \ldots))$ where the b_i need not be 1. The calculation of the corresponding p_r and q_r proceeds almost as in the case of simple continued fractions, with the difference that the recurrence relation becomes $p_r = a_r p_{r-1} + b_r p_{r-2}$, $q_r = a_r q_{r-1} + b_r q_{r-2}$. There is a matrix version of this: Let $M_{-1} := \begin{pmatrix} 0 & 1 \\ 1 & 0 \end{pmatrix}$ and for $r \geq 0$ let

$$M_r = M_{r-1} \begin{pmatrix} 0 & b_r \\ 1 & a_r \end{pmatrix}$$

where b_0 is by convention 1. Particular interest attaches to cases in which b_r involves a complex variable z. For a modern survey of the subject, see [LL].

Space does not permit a complete development of this topic, but it is possible to give the flavor, and to then explain some classical and famous continued fraction identities such as the one for e.

We begin with a discussion of analytic continued fractions in general. An analytic continued fraction is a continued fraction

$$\cfrac{b_0(z)}{a_1(z) + \cfrac{b_1(z)}{a_2(z) + \cfrac{b_2(z)}{a_3(z) + b_3(z)/(a_4(z) + \ldots)}}}$$

where the functions $a_j(z)$ and $b_j(z)$ are complex analytic functions. All the applications made here concern the case in which all $a_j(z)$ and $b_j(z)$ are polynomials, so that the convergents are rational functions of z. This makes them relatively tractable objects computationally, and one may hope that for certain complex analytic functions, a continued fraction approximation furnishes faster convergence, over a wider domain, than do partial sums of the Laurent expansion of a function. There are now some general theorems in this vein. There is the basic notion of 'correspondence': a continued fraction corresponds to a meromorphic function $f = \sum_{n=N}^{\infty} c_n z^n$ if the Laurent series of the convergents agree, coefficient by coefficient, with that for f, to a depth which tends to infinity as the depth of the continued fraction expansion increases.

One of the earliest discovered remarkable identities in continued fractions is that

$$\tanh(1) = 1 + 1/(3 + 1/(5 + 1/(7 + \ldots))).$$

There are numerous identities along these lines, and most of them are now understood to be special cases of a few more general theorems in analytic continued fractions. The first order generalization of the above identity is

$$\tanh(z) = \cfrac{z}{1 + \cfrac{z^2}{3 + \cfrac{z^2}{5 + \cfrac{z^2}{7 + \cdots}}}}$$

valid for all z other than poles of $\tanh(z)$. Although some continued fraction formulas are slow to converge, this one has excellent and rapid convergence. Plots of tanh over $\mathbb{C}$ would require four-dimensions so we show the log of the absolute value as a contour plot, both for the five-layered continued fraction and for the function $\log|\tanh(y + ix)|$ (switching x and y for convenience of display):

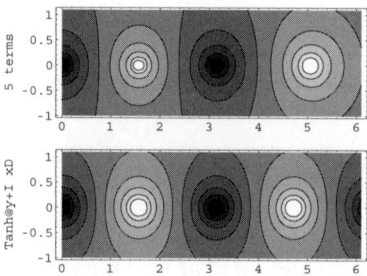

Fig. 12.1 tanh and its continued fraction approximation.

But this identity is still just the tip of the iceberg. The yet more general result (in Volume 2, page 104 of [Pe]) is that if

$$\Psi_1(\gamma; x) := (1/\Gamma(\gamma))\left[1 + \sum_{k=1}^{\infty} \frac{x^k}{k!} \prod_{j=0}^{k-1} \frac{1}{\gamma + j}\right]$$

then with the exception of those x for which $\Psi_1(\gamma + 1; x) = 0$,

$$\frac{\Psi_1(\gamma; x)}{\Psi_1(\gamma + 1; x)} = \gamma + \cfrac{x}{\gamma + 1 + \cfrac{x}{\gamma + 2 + \cfrac{x}{\gamma + 3 + \cdots}}}.$$

Taking $\gamma = 1/2$ and setting $x = z^2/4$ in the formula directly above gives a result equivalent to the one for $\tanh(z)$.

The formal correctness of this more general formula is an immediate consequence of the recursive identities

$$\Psi_1(\gamma; x) = \gamma \Psi_1(\gamma + 1; x) + x \Psi_1(\gamma + 2; x)$$

and the consequent recursive system

$$\Psi_1(\gamma + n; x) = (\gamma + n)\Psi_1(\gamma + n + 1; x) + x\Psi_1(\gamma + n + 2; x).$$

Perron shows that one has equality, rather than the weaker formal power series agreement, giving this case as an example of Theorem 2.46, criterion B p.97.

Still more general power series identities involve the hypergeometric functions.

Let $F(\alpha, \beta, \gamma; x) := 1 + \sum_{k=1}^{\infty} \frac{x^k}{k!} \prod_{j=0}^{k-1} \frac{(\alpha+j)(\beta+j)}{\gamma+j}$. These are the *hypergeometric* functions, and they will come up again later in this chapter. For now, their importance is that

$$\frac{F(\alpha, \beta, \gamma; x)}{F(\alpha, \beta + 1, \gamma + 1; x)} \sim 1 + \cfrac{a_1 x}{1 + \cfrac{a_2 x}{1 + \cfrac{a_3 x}{1 + \cdots}}}$$

where $a_{2k} = -(\beta + k)(\gamma - \alpha + k)/(\gamma + 2k - 1)(\gamma + 2k)$ and $a_{2k+1} = -(\alpha + k)(\gamma - \beta + k)/(\gamma + 2k)(\gamma + 2k + 1)$. Here, "$U \sim V$" means that a given power series and a given rational function agree formally, coefficient by coefficient.

The proof of this formal identity is inductive, based on two simple power series identities involving hypergeometric functions:

$$F(\alpha, \beta, \gamma; x) = F(\alpha, \beta + 1, \gamma + 1; x) - \frac{\alpha(\gamma - \beta)}{\gamma(\gamma + 1)} x F(\alpha + 1, \beta + 1, \gamma + 2; x)$$

and

$$F(\alpha, \beta, \gamma; x) = F(\alpha + 1, \beta, \gamma + 1; x) - \frac{\beta(\gamma - \alpha)}{\gamma(\gamma + 1)} x F(\alpha + 1, \beta + 1, \gamma + 2; x),$$

which are readily checked by referring to the definition of F. Now with

$$P_{2n}(x) := \frac{F(\alpha + n, \beta + n, \gamma + 2n; x)}{F(\alpha + n, \beta + n + 1, \gamma + 2n + 1; x)}$$

and

$$P_{2n+1}(x) := \frac{F(\alpha + n, \beta + n + 1, \gamma + 2n + 1; x)}{F(\alpha + n + 1, \beta + n + 1, \gamma + 2n + 2; x)}$$

an application of the identities above using $(\alpha+n, \beta+n, \gamma+2n)$ or $(\alpha+n, \beta+n+1, \gamma+2n+1)$ in place of (α, β, γ) gives

$$P_{2n}(x) = 1 + \frac{a_{2n+1}x}{P_{2n+1}(x)} \text{ and } P_{2n+1}(x) = 1 + \frac{a_{2n+2}x}{P_{2n+2}(x)},$$

where $a_{2n+1} = -\frac{(\alpha+n)(\gamma-\beta+n)}{(\gamma+2n)(\gamma+2n+1)}$ and $a_{2n+2} = -\frac{(\beta+n+1)(\gamma-\alpha+n+1)}{(\gamma+2n+1)(\gamma+2n+2)}$. From this it follows that

$$\frac{F(\alpha, \beta, \gamma; x)}{F(\alpha, \beta+1, \gamma+1; x)} \sim 1 + \frac{a_1 x}{1 + \frac{a_2 x}{1 + \frac{a_3 x}{1+\dots}}}$$

as claimed.

The continued fraction expansion of e^z is

$$e^z = 1 + \frac{2z}{2 - z + \frac{z^2}{2 + \frac{z^2}{6 + \frac{z^2}{10 + \frac{z^2}{14+\dots}}}}}.$$

A detailed treatment of this identity with attention to convergence may be found in Perron, in the section on Padé approximations. It is implicit in the proof that convergence is uniform on compact subsets of $\mathbb{C}$, with a small error inside $|z| \leq \log n$ for the nth convergent. It turns out that the error remains small for $|z| \leq n/2$. Since the nth rational approximation has poles within $|z| \leq 3n/2$ for large n, this is essentially best possible.

To prove this, first let $p_n(z)$ and $q_n(z)$ be the numerator and denominator in the expansion of the continued fraction above to depth n, scaled so that the constant term in each polynomial is 1. Equivalently, $p_n(z)$ is the polynomial given by $p_0 = 1, p_1 = 1 + z/2$, and for $n \geq 2$, by

$$p_n(z) = p_{n-1}(z) + \frac{z^2}{4(2n-1)(2n-3)} p_{n-2}(z)$$

and $q_n(z) = p_n(-z)$. These polynomials are known explicitly and have been studied extensively; they are also given by $p_n(z) = \sum_{k=0}^{n} \frac{(2n-k)! \, n! \, z^k}{(2n)! \, k! \, (n-k)!}$.

Theorem 12.5 *For all $n \geq 1$ and all z with $|z| \leq n/2$,*

$$\left| \frac{p_n(z)}{q_n(z)} e^{-z} - 1 \right| \leq 2^{-n}$$

Proof. We use the classical identity

$$p_n(z) - q_n(z)e^z = \frac{(-1)^n z^{2n+1}}{(2n)!} \int_0^1 e^{tz}(1-t)^n t^n \, dt$$

and we use a result of Saff and Varga [SV] concerning the zeros of $q_n(z)$: all zeros of $q_n(z)$ lie in the right half plane, and outside the parabola $y^2 \leq 4(n+1)(n+1-x)$. From this it follows immediately that $|\zeta_{n,k}| \geq n+1$ for each of the zeros $\zeta_{n,k}, 1 \leq k \leq n$ of q_n. It is sufficient to prove the result with an error of 2^{-n-1} in the right half plane because if $u \approx 1$ then $1/u \approx 1$, and taking $u = p_n(z)e^{-z}/q_n(z)$ with $\Re[z] > 0$ we have $|p_n(-z)e^z/q_n(-z) - 1| = |q_n(z)/(p_n(z)e^{-z}) - 1| < 2^{-n}$ because $|p_n(z)e^{-z}/q_n(z) - 1| < 2^{-n-1}$. Turning now to the proof of the bound for $\Re(z) \geq 0$ and $|z| \leq n/2$, we first observe that from the integral identity for $p_n - q_n e^z$ we have

$$|p_n(z)e^{-z} - q_n(z)| \leq \frac{(n/2)^{2n+1}}{(2n)!} \int_0^1 \left|(1-t)^n t^n e^{t(z-1)}\right| dt \leq \frac{(n/2)^{2n+1}(n!)^2}{(2n)!(2n+1)!}.$$

Now $q_n(z) = \frac{n!}{(2n)!}(-1)^n \prod_{k=1}^n (z - \zeta_{k,n})$ so that for $|z| \leq n/2$, $|1/q_n(z)| \leq \frac{(2n)!}{n!}(1+n/2)^{-n}$. Thus

$$\left|\frac{p_n(z)}{q_n(z)}e^{-z} - 1\right| \leq \frac{n^{2n+1}(n!)2^{-2n-1}}{(2n+1)!}\frac{2^n}{n^n} \leq \frac{n^{n+1}2^{-n-1}}{2n+1}\prod_{j=1}^n (n+j)^{-1} < 2^{-n-2}.$$

$\square$

Saff and Varga have studied the zeros of p_n. They obtain remarkably complete information [SV] for the asymptotic behavior of the zeros of p_n. (The zeros of q_n, which are the poles of the Padé rational approximations to e^z, are the negatives of these zeros). The zeros of p_n lie close to the curve $2n\Gamma$ where $\Gamma = \{z : \Re(z) > 0$ and $\left|\frac{ze^{\sqrt{1+z^2}}}{1+\sqrt{1+z^2}}\right| = 1\}$. The relative density of zeros along this curve is proportional to the derivative of the argument of this modulus-one quantity with respect to arc length as one moves along Γ. The nearest point to zero along this curve lies within distance $2/3$ of the origin so that poles of p_n/q_n will be found within distance $4n/3$ of the origin for large n. The polynomials $p_n(2x)$ and $p_n(-2x)$ are themselves serviceable approximations to e^x and e^{-x} respectively; $f_n(x) := p_n(2x)$ satisfies the differential equation $f'_n(x) = f_n(x) - \frac{x}{2n-1} f_{n-1}(x)$, from which it follows that for $0 < x < n/2$, $f_n(x) = e^{x-\theta x^2/(2n-1)}$ where $0 < \theta < 1$.

It seems likely that similar results could be established for a wide class of classical analytic continued fraction expansions.

We now give a small list of some other continued fraction identities involving well-known functions. (16) p 103 Perron: (Laplace, Jacobi, Seidel)

$$2\xi e^{\xi^2} \int_\xi^\infty e^{-t^2}\, dt = \cfrac{1}{1+\cfrac{1/2\xi^2}{1+\cfrac{2/2\xi^2}{1+\cfrac{3/2\xi^2}{1+\ldots}}}}$$

$$e^x = 1 + x/(1 - (x/2)/(1 + (x/6)/(1 - (x/6)/(1 + (x/10)/(1 - \ldots)))))$$

$$\log((1+x)/(1-x)) =$$
$$2x/(1 - x^2/(3 - 4x^2/(5 - 9x^2/(7 - 16x^2/(9 - \ldots)))))$$

(for $|x| < 1$)

$$e^z = 1/(1 - z/(1 + z/(2 - z/(3 + 2z/(4 - 2z/(5 + 3z/(6 - 3z \ldots)))))))$$

[JT] page 207.

$$\log(1+z) = \cfrac{z}{1+\cfrac{z}{2+\cfrac{z}{3+\cfrac{4z}{4+\cfrac{4z}{5+9z/(7+9z/\ldots)}}}}}.$$

Equivalently, $\log(1+z) = z/(1 + a_1/(z + a_2/(z + \ldots)))$, where $a_{2n} = n/(2(2n+1))$ and $a_{2n+1} = (n+1)/(2(2n+1))$, and for z not in $(-\infty, -1]$.

Other special functions for which continued fraction expansions are known include, but are not limited to, $\arctan z$, the Bessel function ratio $J_{\nu+1}(z)/J_\nu(z)$, $(1+z)^\alpha$ (for non-integer α), and $\gamma(a, z) = \int_0^z t^{a-1} e^{-t}\, dt$.

Remark 12.1 *Francis Clarke, Continued Fraction Expansions and the Legendre Polynomials, Bull. London Math. Soc.* **18** *(1986) 255-260, gives a clean proof of the (classical) identities*

$$e^{2/k} = 1 + \cfrac{2}{k - 1 + \cfrac{1}{3k + \cfrac{1}{5k + \cfrac{1}{7k + \ldots}}}}$$

(valid for integer $k \geq 1$;) these are based on specializing the expansion for e^z. The classical continued fraction expansion of e, (case $k = 2$ above) follows from this by an induction showing that the convergents of the above continued fraction constitute every third convergent to this classical expansion.

Bibliography

Anderson, P. G., Brown, T. C., and Shiue, P. J.-S. (1995). A Simple proof of a remarkable continued fraction identity, *Proc. AMS* **123**, 7, pp. 2005-2009.

Askey, R. (1975). Orthogonal Polynomials and Special Functions, *Regional Conference Series in Applied Mathematics* SIAM.

Astels, S. (1999). Cantor sets and numbers with restricted partial quotients, *Transactions AMS* **352**, 1, pp. 133-170 .

Babenko, K. I. (1978). On a problem of Gauss, *Soviet Mathematical Doklady* **19**, 1, pp. 136-140.

Baladi, V. and Vallée, B. (2003). Systèmes dynamiques et algorithmique, *Algorithms Seminar 2001-2002*, INRIA, pp. 121-150.

Baladi, V. and Vallée, B. (2005). Euclidean algorithms are Gaussian, *J. Number Theory* **110**, 2, pp. 331-386.

Bandtlow, O. and Jenkinson, O. Invariant measures for real analytic expanding maps (to appear).

Beardon, A. F. (2001). Continued fractions, discrete groups and complex dynamics, *Computational Methods and Function Theory* **1**, No. 2, 535-594.

Besicovitch, A. S. and Taylor, S. J. (1954). On the complementary intervals of a linear closed set of zero Lebesgue measure, *J. London Math Soc.* **29**, pp. 449-459.

Blachman, N. M. (1984). The continued fraction as an information source, *IEEE Transactions on Information Theory* **IT-30**, 4, pp. 671-674.

Bombieri, E. and van der Poorten, A. J. (1993). Continued fractions of algebraic numbers, *Number Theory Research Reports*, ceNTRe, Macquarie University, NSW, Australia, pp. 93-138.

Böhmer, P.E. (1926). Über die Transzendenz gewisser dyadischer Brüche, *Math. Ann.* **96**, pp. 367-377.

Böhmer, P.E. (1926). Erratum, *Math. Ann.* **96**, pp. 735.

Borosh, I. (1975). More numerical evidence on the uniqueness of Markov numbers, *BIT* **15**, pp. 351-357.

Bosma, W., Jager, H., and Wiedijk, F. (1983). Some metrical observations on the approximation by continued fractions, *Indag. Math.*, **45**, no. 3, pp. 281-299.

Brent, R. P., van der Poorten, A. J., and te Riele, Herman J. J. (1996). A

comparative study of algorithms for computing continued fractions of algebraic numbers, Lecture Notes in Comput. Sci., no. 1122, Springer, Berlin pp. 35-47.
Brown, T. C. and Shiue, P. J. (1995). Sums of fractional parts of integer multiples of an irrational, *J. of Number Theory* **50**, 2, pp. 181-192.
Bumby, R. (1982). Hausdorff dimensions of Cantor sets, *J. für die reine und angewandte Mathematik* **331**, pp. 192-206.
Burger, E. (2000). Diophantine Olympics and World Champions: Polynomials and Primes Down Under, *Am. Math. Monthly* **107**, no. 9, pp. 822-829.
Cassels, J. (1955). Simultaneous diophantine approximation II, *Proc. London Math. Soc.*, 3 5, pp. 435-448
Clarke, F. (1986). Continued fraction expansions and the Legendre Polynomials, *Bull. London Math. Soc.* **18**, no. 3, pp. 255-260
Clarke, F.W., Everitt, W.N., Littlejohn, L.L., Vorster, S.J.R. (1999). H. J. S. Smith and the Fermat two squares theorem, *Amer. Math. Monthly* **106**, no. 7, pp. 652-665.
Cohen, H. (1993). A course in computational algebraic number theory, GTM 138, Springer Verlag, Berlin.
Cohn, H. (1996). Symmetry and specializability in continued fractions, *Acta Arith.* **LXXV.**, 4, pp. 297-320.
Crandall, M. G. and Rabinowitz, P. H. (1973). Bifurcation, perturbation of simple eigenvalues, and linearized stability, *Arch. Rational Mech. Anal.* **52**, 2, pp. 161-180.
Cusick, T. W. (1977). Continuants with bounded digits, *Mathematika* **24**, 2, pp. 166-172.
Cusick, T. W. (1978). Continuants with bounded digits II, *Mathematika* **25**, 1, pp. 107-109.
Cusick, T. W. (1985). Continuants with bounded digits III, *Monatsh. Math.* **99**, 2, pp. 105-109.
Cusick, T. W. and Flahive, M. (1989). The Markoff and Lagrange Spectra, Mathematical Surveys and Monographs, 30, *AMS,* Providence, RI.
Dajani, K. and Kraaikamp, C. (1997). A Note on the approximation by continued fractions under an extra condition, *Proc. of the New York Journal of Mathematics Conf.*, pp. 69-80.
Dajani, K. and Kraaikamp,C. (2002). Ergodic theory of numbers, *Carus Mathematical Monographs no. 29,* MAA.
Daudé, H., Flajolet, P., and Vallée, B. (1997). An average-case analysis of the Gaussian algorithm for lattice reduction, *Combinatorics, Probability and Computing* **6**, 4, pp. 397-433.
Diamond, H. G. and Vaaler, J. D. (1986). Estimates for partial sums of continued fraction partial quotients, *Pacific J. of Math* **122**, pp. 73-82.
Diviš, B. (1973). On the sums of continued fractions, *Acta Arith.* **22**, pp. 157-173.
Dixon, J. D. (1970). The number of steps in the Euclidean algorithm. *J. of Number Theory* **2**, pp. 414-422.
Drmota, M. and Tichy, R. F. (1997). *Sequences, Discrepancies and Applications*, Lecture Notes in Math., no. 1651, Springer Verlag, Berlin.

Efrat, I. (1993). Dynamics of the continued fraction map and the spectral theory of $SL(2, \mathbb{Z})$, *Inventiones Mathematicae* **114**, pp. 207-218.

Entacher, K. (2005). On the Beauty of Uniform Distribution mod One, *The Mathematica Journal* **9**, pp. 583-597.

Flajolet, Philippe and Vallée, Brigitte (1996). Continued fraction algorithms, functional operators, and structure constants, INRIA Technical Report RR2931.

Flajolet, P. and Vallée, B. (1998). Continued fraction algorithms, functional operators, and structure constants, *Theoretical Computer Science* **194**, 1-2, pp. 1-34.

Flahive, M. and Hensley, D. (1984). Approximation to complex ξ from within a wedge, *J. Number Theory* **19**, 1, pp. 81-84.

Freiman, G. A. (1973). On the beginning of Hall's ray, Teorija cisel (number theory), Kalininskii Gosudarstvennii Universitet, Moscow, pp. 87-113.

Friesen, C. and Hensley, D. (1996). The statistics of continued fractions for polynomials over a finite field, *Proc. AMS* **124**, 9, pp. 2661-2673.

Gauss, K. E. F. (1917). Collected Works Teubner, Leipzig, Vol. X, p. 372.

Gohberg, I.C. and Krein, M. G. (1969). Introduction to the theory of linear nonselfadjoint operators, Vol 18 Translations of Mathematical Monographs, AMS, Providence.

Grabiner, D. (1992). Farey nets and multidimensional continued fractions, *Monatsh. Math.* **114**, pp. 35-60.

Guivarc'h, Y. and Le Jan, Y. (1993). Asymptotic winding of the geodesic flow on modular surfaces and continuous fractions, *Ann. Scient. Éc. Norm. Sup.*, *4th série* **26**, pp. 23-50.

Hall Jr., M. (1947). On the sum and product of continued fractions, *Annals of Math.* **48**, 4, pp. 966-993.

Halmos, P. R. (1956). Lectures on ergodic theory, Chelsea, New York.

Harman, G. and Wong, K. C. (2000). A note on the metrical theory of continued fractions, *Am. Math. Monthly* **107**, pp. 834-837.

Heilbronn, H. (1969). On the average length of a class of finite continued fractions. Number Theory and Analysis (Papers in Honor of Edmund Landau) Plenum, New York pp. 87-96

Hensley, D. (1977). Simple continued fractions and special relativity, *Proc. AMS* **67**, 2, pp. 219-220.

Hensley, D. (1994). The Hausdorff Dimensions of some continued fraction Cantor sets. *J. of Number Theory* **49**, 2, pp. 182-198.

Hensley, D. (1994). The distribution mod n of fractions with bounded partial quotients, *Pacific J. of Math* **166**, 1, pp. 43-54.

Hensley, D. (1994). The number of steps in the Euclidean algorithm. *J. of Number Theory* **49**, 2, pp. 142-182.

Hensley, D. (1991). The largest digit in the continued fraction expansion of a rational number, *Pacific J. of Math* **151**, 2, pp. 237-255.

Hensley, D. (1990). The distribution of badly approximable rationals and continuants with bounded digits II, *J. of Number Theory* **34**, 3, pp. 293-334.

Hensley, D. (1996). A polynomial time algorithm for the Hausdorff dimension of continued fraction Cantor sets. *J. of Number Theory* **58**, 1, pp. 9-45.

Hensley, D. (1998). Metric diophantine approximation and probability, *New York J. of Math.* **4**, pp. 249-258 (an online journal).

Hensley, D. (2002) On the Spectrum of the Transfer Operator for Continued Fractions with Restricted Partial Quotients, Number Theory for the Millennium, Vol II, pp. 175-194, edited by M.A. Bennett, B.C. Berndt, N. Boston, H.G. Diamond, A.J. Hildebrand, and W. Philipp, A.K. Peters, Natick, MA.

Hensley, D. and Su, F. E. (2004). Random walks with badly approximable numbers, Unusual applications of number theory, DIMACS Ser. Discrete Math. Theoret. Comput. Sci. **64**, pp. 95-101.

Hurwitz, A. (1888). Über die Entwicklung Complexer Grössen in Kettenbrüche, *Acta Math.* **11**, pp. 187-200.

Hurwitz, J. (1895) Über eine besondere Art der Kettenbruch-entwicklung complexer Grössen, Thesis, Universität Halle-Wittenberg, Halle, Germany.

Hwang, H.-K. (1994). Théorèmes limites pour les structures combinatoires et les fonctions arithmetiques, PhD Thesis, École Polytechnique (in French).

Jarnik, V. (1928). Zur metrischen theorie der diophantische approximationen, *Proc. Mat. Fiz.* **36**, pp. 91-106.

Jenkinson, O.and Pollicott, M. (2001). Computing the dimension of dynamically defined sets: E_2 and bounded continued fractions, *Ergod. Th. & Dynam. Sys* **21**, pp. 1429-1445.

Jenkinson, O. (2004). On the density of Hausdorff Dimensions of bounded type continued fraction sets; the Texan Conjecture, *Stochastics and Dynamics* **4**, pp. 63-76.

Jones, W. B. and Thron, W. J. (1980). Continued fractions. Analytic theory and applications. Encyclopedia of Mathematics and its Applications, **11**. Addison-Wesley Publishing Co., Reading, Mass.

Kessenböhmer, M. and Zhu, S., Dimension sets for infinite IFS's: the Texan conjecture. (to appear)

Khinchin, A. Ya. (1964). Continued Fractions, Univ. of Chicago Press.

Krasnoselskii, M. (1964). Positive solutions of operator equations. P. Noordhoff, Groningen.

Kmošek, M. (1979). Continued fraction expansion of some irrational numbers, M. Thesis, Uniwersytet Warszawski, Warszawa. (in Polish).

Knopfmacher, A. (1991). The length of continued fraction expansions for a class of rational functions in $\mathcal{F}_q(X)$, *Proc. Edinburgh Math. Soc.* **34**, pp. 7-17.

Knopfmacher, A. and Knopfmacher, J. (1988). The exact length of the euclidean algorithm in $\mathbb{F}_q[X]$, *Mathematika* **35**, pp. 297-304.

Kuipers, L. and Niederreiter, H. (1974). Uniform distribution of sequences, John Wiley & Sons, New York.

Kuz'min, R.O. (1928). A problem of Gauss, *Dokl. Akad. Nauk.*, Ser. A ,pp. 375-380.

Lagarias, J. (1994). Geodesic multidimensional continued fractions, *Proc. London Math. Soc.* **3**, 69, pp. 464-488.

Lagrange, J. L. Sur la Manière d'Approcher de la Valeur Numérique des Racines des Équations par les Fractions Continues, Oeuvres VIII, pp. 73-131.

Lange, L. J. (1999). A generalization of Van Vleck's theorem and more on complex continued fractions, *Contemporary Mathematics* **236**, (AMS series), pp. 179-192.

Lapidus, M. L. and van Frankenhuysen, M. (2000). Fractal geometry and number theory, complex dimensions of fractal strings and zeros of zeta functions, Birkhäuser, Boston.

Lenstra, Jr. H. W. and Shallit, J. O. (1997). Continued fractions and linear recurrences, *Journal de Theorie des Nombres de Bordeaux* **9**, pp. 267-279.

Lévy, P. (1952). Fractions continues aléatoires, *Rend. Circ. Mat. Palermo* **2**, 1, pp. 170-208.

Lhote, L. (2002). Modélisation et approximation des sources complexes, Mémoire de DEA, Université de Caen.

Lorentzen, L. (2000). Convergence of corresponding continued fractions, Number theory for the millennium, II (Urbana, IL, 355-373, A K Peters, Natick, MA, 2002.

Mathematica, Wolfram Research, Champaign Il.

Mauldin, R. D. (1991). Infinite Iterated Function Systems: Theory and Applications, *Progress in Probability* **37**, pp. 91-110.

Mauldin, R. D. and Urbański, M. (1996). Dimensions and measures in infinite iterated function systems, *Proc. London Math. Soc.* **73**, 3, pp. 105-154.

Mauldin, R. D. and Urbański, M. (1999). Conformal iterated function systems with applications to the geometry of continued fractions, *Trans. AMS* **351**, pp. 4995-5025.

Mayer, D. (1991). Continued fractions and related transformations, Chapter 7 in Ergodic Theory, Symbolic Dynamics and Hyperbolic Spaces, edited by T. Bedford, M. Keane and C. Series, Oxford.

Mayer, D. (1980). On composition Operators on Banach spaces of Holomorphic Functions, *J. Functional Analysis* **35**, pp. 191-206.

Mayer, D. (1990). On the thermodynamic formalism for the Gauss map, *Commun. Math. Phys.* **130**, pp. 311-333.

Mayer, D. and Roepstorff, G. (1987). On the relaxation time of Gauss' continued fraction map. I. The Hilbert space approach. *J. Stat. Phys.* **47**, 1/2, pp. 149-171.

Mayer, D. and Roepstorff, G. (1988). On the relaxation time of Gauss' continued fraction map II. The Banach space approach (transfer operator approach), *J. Stat. Phys.* **50**, 1/2, pp. 331-344.

Nair, R. (1997). On metric diophantine approximation and subsequence ergodic theory, New York J. Math 3A (1997) Proceedings of the New York Journal of Mathematics Conference June 9-13, pp. 117-124.

Narasimhan, R. (1971). Several Complex Variables, University of Chicago Press, Chicago.

Niederreiter, H. (1986). Dyadic fractions with small partial quotients, *Monatsh. Math.* **101**, pp. 309-315.

Niederreiter, H. (1987). Rational functions with partial quotients of small degree in their continued fraction expansion, *Monatsh. Math.* **103**, pp. 269-288.

Niederreiter, H. Continued fractions with small partial quotients, Proc. 1986 Nagasaki Number Theory conference, (Y. Morita, ed.) pp. 1-11, Tōhoku Univ., Sendai, Japan 1987.

Niederreiter, H. (1977). Pseudo-random numbers and optimal coefficients, *Advances in Math.* **26** pp. 99-181.

Niederreiter, H. (1978). Quasi-Monte Carlo methods and pseudorandom numbers, *Bull. Am. Math. Soc.* **84**, pp. 957-1041.

Niederreiter, H. (1987). Point sets and sequences with small discrepancy, *Monatsh. Math.* **104**, pp. 273-337.

Perron, O. (1954). Die Lehre von der Kettenbrüchen, vol. 1, Teubner.

Philipp, W. (1988). Limit theorems for sums of partial quotients of continued fractions, *Monatsh. Math.* **105**, pp. 195-206.

Philipp, W. (1976). A conjecture of Erős on continued fractions, *Acta Arith.* **XXVII**, pp. 379-386.

Rudin, W. (1987). Real and complex analysis, McGraw Hill, New York.

Ruelle, D. (1978). Thermodynamic formalism, Addison-Wesley, Reading, MA.

Rockett, A. and Szüsz, P. (1992). Continued fractions, World Scientific, Singapore.

Saff, E. B. and Varga, R. S. (1978). On the zeros and poles of Padé approximants to e^z. III, *Numerische Math.* **30**, pp. 241-266

Sander, J. W. (1987). On a conjecture of Zaremba, *Monatsh. Math.* **104**, 2, pp. 133-137.

Schmidt, A. (1975). Diophantine approximation of complex numbers. *Acta Math.* **134**, pp. 1-85

Schmidt, A. (1982). Ergodic theory for complex continued fractions. *Monatsh. Math.*, **93**, pp. 39-62

Schmidt, W. (1980). Diophantine approximation, Lecture Notes in Math. **785**, Springer Verlag.

Schmidt, W. (1966). On badly approximable numbers and certain games, *Trans. AMS* **123**, pp. 178-199.

Schoißengeier, J. (1984). On the discrepancy of $(n\alpha)$, *Acta Arith.* **44**, pp. 241-279.

Shallit, J. (1979), Simple continued fraction expansions for some irrational numbers, *J. of Number Theory* **11**, pp. 209-217.

Shallit, J. (1992). Real numbers with bounded partial quotients: A survey, *Enseign. Math.* **38**, pp. 151-187.

Shapiro, J. (1997). Compact composition operators on spaces of boundary regular holomorphic functions, *Proc. AMS* **100**, pp. 49-57.

Smith, H. J. S. (1876). Note on continued fractions, *Messenger Math.* **6**, pp. 1-14.

Stilwell, J. (2001). Modular miracles, *Am. Math. Monthly*, **108**, p. 70.

Su, Francis Edward (1998). Convergence of random walks on the circle generated by an irrational rotation, *Trans. AMS* **350**, pp. 3717-3741.

Tausworthe, R. C. (1965). Random numbers generated by linear recurrences modulo two, *Math. Comp.* **19**, pp. 201-209.

Urbański, M. (2001). Porosity in conformal infinite iterated function systems, *J. Number Theory* **88**, 2, pp. 283-312.

Vallée, B. (2000). Digits and continuants in Euclidean algorithms, Ergodic versus Tauberian theorems, *J. de Théorie des Nombres de Bordeaux* **12**, pp. 531-570.

Vallée, B. (1995). Opérateurs de Ruelle-Mayer généralisés et analyse des algorithmes d'Euclide et de Gauss. Rapport de Recherche de l'Université de Caen, Les Cahiers du GREYC # 4.

Vallée, B. (1998). The complete analysis of the binary Euclidean algorithm, *Algorithmic Number Theory* (Portland, OR 1998), pp. 77-94. Lecture Notes in Comput. Sci. 1423, Springer, Berlin.

Vallée, B. (1998) Fractions continues á contraintes périodiques, *J. Number Theory* **72**, pp. 183-235.

Vallée, B. (2003). Dynamical analysis of a class of Euclidean algorithms, *Theoretical Computer Science* **297**, pp. 447-486.

van der Poorten, A. J. (1993). Continued fractions of formal power series, *Proc. of the Third Conf. of the Canadian Number Theory Association*, ANT, pp. 453-466.

van der Poorten, A. J. (1988). Solution de la conjecture de Pisot sur le quotient de Hadamard de deux fractions rationelles. *C. R. Acad. Sci. Paris* **306**, pp. 97-102.

van der Poorten, A. J. (1984). p-adic methods in the study of Taylor coefficients of rational functions, *Bull. Austral. Math Soc.* **29**, pp. 109-117.

van der Poorten, A. J. and Shallit, J. (1993). A specialized continued fraction. *Canadian J. of Mathematics* **45**, 5, pp. 1067-1079.

Van Vleck, E. B. (1901). On the convergence of continued fractions with complex elements, *Trans. AMS* **2**, pp. 215-233.

Vardi, I. (1998). Archimedes' cattle problem, *American Math. Monthly* **105**, pp. 305-319.

Wirsing, E. (1974). On the theorem of Gauss-Kusmin-Lévy and a Frobenius-type theorem for function spaces. *Acta Arith.* **24**, pp. 507-528.

Zaremba, S. K. (1966). Good lattice points, discrepancy, and numerical integration, *Ann. Mat. Pure Appl.* **73**, pp. 293-317.

Index

absolutely continuous, 79, 86
algebraic extension, 125, 126
algebraic integer, 26, 28, 29, 126, 127, 129, 137, 141–143
algebraic integer-quartic, 82
algebraic number, 24, 100, 125, 127, 136
algorithm, 1, 2, 7, 11, 31, 34, 36, 41, 44, 45, 150
algorithm-CF, 61
algorithm-Euclidean, 3
algorithm-Lagarias, 128
algorithm: Fincke-Pohst, 115, 116
algorithm: Gauss, 102
algorithm: Jacobi-Perron, 127
algorithm: LLL, 29, 103, 116
analytic continued fraction, 224
approximation error, 8, 70, 71, 75, 127
approximation error-best possible, 14, 104
approximation error-normalized, 49
approximation-simultaneous, 101
Askey, 185, 188
Astels, 145, 151

Böhmer, 19
Babenko, 181
badly approximable, 143
badly approximable vector, 18, 100, 125, 127, 128
Baladi, 209

Banach space, 54, 87, 162, 167, 207
Bandtlow, 87
basis: lattice, 29, 102
basis: reduced, 102
Bessel function, 184, 230
binary shift, 3
Borel set, 215
Borosh, 154
Bosma, 49
bounded distortion property, 213
bounded linear operator, 90, 186
bounded partial quotients, 33, 34, 37, 39, 45, 46, 161
bridge, 145, 150
Burger, 21

Cantor set, 145, 213
Cassels, 100
Cauchy's inequality, 89
Cauchy-Schwarz inequality, 182, 185, 199
centered continued fraction, 40, 41, 72, 206
CF algorithm, 61
Clarke, 230
Cohen, 28
Cohn, H., 19
compact operator, 87, 90, 91, 169, 170, 189
compact operator: s-numbers of, 170
compact set, 177
conformal, 93, 94

conjugates-of an algebraic number, 134
continuant, 6, 76
continued fraction algorithm-Hurwitz, 71
continued fraction Cantor set, 36, 145, 161, 163
continued fraction expansion, 225, 229–231
continued fraction expansion-tanh(z), 225
continued fraction expansion-e^i, 71
continued fraction expansion-for $\log(1+z)$, 230
continued fraction expansion-of e^z, 228
continued fraction expansion-unusual, 19, 20, 82
continued fraction expansions-miscellaneous, 230
continued fraction-analytic, 225
continued fraction-convergent, 10–12, 14, 17, 21, 24
continued fraction-nonconvergence of, 218
continued fraction-periodic, 8, 9
continued fraction-purely periodic, 9, 10
continued fraction-rewriting, 218
continued fraction-terminating, 8
convergent, 14, 72, 75, 218, 225, 231
convex polytope, 125
convex set, 117, 119
convex-polyhedron, 119
correspondence-of a continued fraction to a function, 225
Cusick, 153, 164
cylinder, 134

Daudé, 31
degree-of a field extension, 133
degree-of a polynomial, 24, 26
degree-of an algebraic number, 127, 129, 131, 133–138, 140, 141
Delange, 208
determinant, 6, 101, 113, 118, 119

determinant-of a lattice, 21, 104
dimension-of a lattice, 102
Diophantine approximation, 13, 100, 102, 111
Dirichlet pigeonhole principle, 127
Dirichlet series, 209
discrepancy, 13, 18, 39, 45, 46
discrete spectrum, 208
Diviš, 151
dominant eigenvalue, 92, 156, 166
doubling property, 216
doubling property-of a measure, 215
Drmota, 100, 127
dynamical system, 50–52, 205
dynamical system-extension, 58

eigenfunction, 53, 206, 215
eigenfunction-dominant, 32
eigenfunction-positive, 208
eigenvalue-dominant, 53, 215
ellipse, 128
Encyclopedia of Integer Sequences, 44
entropy, 206
ergodic, 49–51, 57, 63, 155, 206
Euclid, 1
Euclidean algorithm, 1, 3, 4, 7, 30, 103
Euclidean algorithm-binary, 3
Euclidean algorithm-centered, 3
Euclidean algorithm-extended, 4
Euclidean domain, 23, 24
Euler, 5

field-algebraic, 28
field-extension, 29
Fincke-Pohst algorithm, 31, 115
finite field, 5, 7, 24, 25, 33
Flahive, 153
Flajolet, 31, 190
folding lemma, 20
fractal, 46, 81, 163
fractal set, 214
fractal string, 166
Frankenhuysen, 166
Freiman, 152
Friesen, 5, 26

fundamental set of units, 132
fundamental unit-of an algebraic number field, 129

Gauss, 102, 108, 155
Gauss density, 79, 158
Gauss measure, 51, 56
Gauss-Kuz'min theorem, 73, 79, 156, 157
Gaussian (normal) distribution, 2, 209
Gaussian integer, 23, 26, 27, 30, 67, 74, 77, 80, 94
gcd, 1, 2, 4, 7, 13, 26, 28, 113, 118, 119
gcd-of ideals, 29
generating function, 207
geodesic, 106
geometric zeta function, 166
golden mean, 19
good denominator, 127, 129, 132, 135–137, 139, 143
good unit, 133–137
Gram-Schmidt algorithm, 102, 103
Grothendieck, 189
group, 4, 29, 222
group-generator, 4

Hadamard, 12
Hall, 145
Halton sequence, 46
Hammersley sequence, 46
Harman, 21
Hausdorff dimension, 36, 37, 100, 163–166, 174, 214, 215
Hensley's constant, 210
Hermite, 32, 105, 107
Hermite approximation, 107
Hermite denominator, 32, 107, 108, 111–115, 125, 128
Hilbert space, 168, 181, 189
holomorphic, 207
holomorphic function, 90
Hurwitz algorithm, 71, 72, 79, 80
Hurwitz, A., 8, 72
Hurwitz, J., 74

hyperbola, 129
hyperbolic tangent, 225
hypergeometric function, 227

ideal, 28, 29
ideal-finitely generated, 29
individual ergodic theorem, 51, 52
integral basis, 29
integral basis-of an ideal, 29
invariant density, 79, 86, 155, 206
invariant measure, 56
invariant subspace, 208
inverse branch, 205
iterated function system, 213, 214

Jacobi, 101, 229
Jacobi symbol, 4
Jacobi-Perron algorithm, 101, 127
Jager, 49
Jenkinson, 87, 175
Jones, 217, 224

Kessenböhmer, 176
Kmošek, 19
Knopfmacher, 61
Krasnoselskii, 87, 167
Krasnoselskii generalization, 176, 177, 179
Kronecker algorithm, 4

Lévy metric, 177
Lagarias, 119
Lagarias geodesic algorithm, 67, 128
Lagrange, 13
Lagrange spectrum, 145, 153
Laguerre polynomial, 182, 184, 186
Laguerre polynomial-generalized, 185
Lange, 221–223
Lapidus, 166
Laplace, 155, 229
lattice, 21, 30, 34, 102, 132
lattice basis, 29, 102, 103, 108
lattice determinant, 104
lattice reduction, 21, 28, 29, 31, 32, 102–104, 108
lattice reduction-Gaussian, 31

Laurent series, 225
Lebesgue measure, 79, 86, 100, 161, 166
Legendre, 4
Legendre polynomial, 230
Lenstra, A., 103
Lenstra, H., 12, 49, 50, 103
Lhote, 204, 210
linear fractional, 94, 95, 110, 205, 207, 221
linear functional, 54
linear operator, 53, 156, 162, 207, 210
LLL algorithm, 29, 31, 103
LLL reduction, 102, 103
LLL-reduced, 103, 104
log convexity, 195
log convexity-strict, 202
Lovasz, 103

Markoff spectrum, 145, 153
Markoff triple, 154
Mathematica, 1, 2, 185, 219, 222
Mauldin, 175, 213–216
Mayer, 170
meromorphic, 222
Minkowski dimension, 163, 166
Minkowski reduced, 108, 112–114, 116, 118
Minkowski reduction, 28, 102
Minkowski's theorem, 104
Minkowski-reduced, lexicographic, 114
Minkowski-reduced: forward-reduced, 114–116, 118
mixing, 51, 57, 59–61, 87
modular inverse, 5
modular tiling, 105
Montel's theorem, 91

Nair, 50, 51
Niederreiter, 25, 33, 34, 36, 46
Niederreiter net, 46
NP-hard, 31, 115, 125
nuclear of order zero, 186, 208
null space, 179, 180, 208

ordered derivation, 149
orthogonal set-of eigenvectors, 189
orthogonal vectors, 30
orthonormal set, 64, 170

Padé approximation, 228, 229
parabola theorem, 224
PARI, 158
partial quotient, 7, 11, 17, 111, 217
partition-of $\mathbb{C}$, 67
Pell's equation, 10, 11
Perron, 101, 217, 227–229
Perron-Frobenius theorem, 87, 167, 176
perturbation theory, 208
pigeonhole principle, 127
Pisot-Vijayaraghavan number, 26, 131
point spectrum, 189
polynomial-primitive, 25
porous-fractal set, 214
positive cone, 92
positive operator, 87, 156, 176, 208
positive: u_0-positive, 87, 176
pressure function, 214
prime-sum of two squares, 20
principal ideal domain, 30
probability density, 79, 87, 155, 156, 158, 206
projection, 87, 180, 215
proper cone, 176
pseudo-random numbers, 25, 34
pseudorandom, 82

quadratic irrational, 8, 9, 12, 18
quadratic nonresidue, 20, 31
quadratic residue, 4

random number, 155
random walk, 18
reduced basis, 28, 102
regular chain, 67, 68
relatively prime, 5
reproducing cone, 176
restricted partial quotients, 50
Riemann sum, 56
Riesz representation theorem, 179

Rohlin, 206
Rohlin's formula, 206

Saff, 229
Sander, 36
scaled error, 82, 83, 85, 127, 130, 139, 142
scaled error-typical, 49
Schmidt, A., 27, 67, 72
Schmidt, W., 100
Schoißengeier, 46
Seidel, 229
Shallit, 12, 19
Sierpinski gasket, 100
signature-of a field extension, 132, 133
signature-of an algebraic number, 138, 141
simultaneous diophantine approximation, 32, 101
Sloane, 44
Smith, 19
spectral gap, 167, 198
spectral gap-uniform, 190
spectral radius, 32, 87, 163, 165, 208
spectrum, 182, 189, 190
spectrum-discrete, 208
spherical representation of $\mathbb{C}$, 219
statistics, 25
statistics-of an algorithm , 209
Stern, 218
Stilwell, 106
Stirling's formula, 188
Stolz, 218
Su, 17
sum of two squares-prime, 20
sum-of sets, 145
surface, 138, 139, 141, 142
surface-for scaled errors, 138

Tauberian theorem, 208, 209
Tessler, 45
Thron, 217, 224
Tichy, 100, 127
trajectory, 205
transfer operator, 63, 87, 91, 155
tree, 12, 13, 93, 97, 145, 150, 154, 164
tree-binary, 145, 146
tree-bridge, 145
tree-formal, 145, 146
tree-gap, 145, 146
tree-interval, 146

unimodular matrix, 113
unit-of an algebraic number field, 129
Urbański, 175, 213–216

Vallée, 4, 31, 181, 190, 205, 208, 209
van der Poorten, 12
van der Poorten-Shallit constant, 20
Van Vleck, 220, 223
Vandermonde matrix, 131
Vardi, 11
Varga, 229
variance, 209
Vitali convergence theorem, 91

weak mixing, 51, 52
Weyl, H., 172
Wiedijk, 49
Wirsing, 156–158
Wirsing constant, 157
Wong, 21
word, 40, 41, 213

Zaremba, 34, 162
Zaremba's conjecture, 34, 38
Zhu, 176